welding projects

A Design Approach

welding projects

A Design Approach

by **JAMES A. PENDER**
Technical Director
Tabor Park Secondary School
Scarborough, Ontario

and

FRANK N. MASSON
Welding Instructor
H. B. Beal Secondary School
London, Ontario.

McGRAW-HILL RYERSON LIMITED
Toronto, Montreal, New York, London, Sydney, Johannesburg, Mexico, Auckland, Panama, Düsseldorf, Singapore, São Paulo, Kuala Lumpur, New Delhi

WELDING PROJECTS: A DESIGN APPROACH

ISBN 0-07-077330-0

345678910 BP 543210987

DENIS TROWER: TECHNICAL ILLUSTRATOR

Canadian Cataloguing in Publication Data

Pender, James A., 1924-
Welding projects

For use in schools.
ISBN 0-07-077330-0 pa.

1. Welding — Laboratory manuals. I. Masson, Frank N., 1927- II. Title

TT211.P45 671.5'2'04 C76-017124-6

Pages 35, 36 (H.1.b. and H.1.c.), 44, 76 (L.3.), 79 (L.12.), 80, 82, 84 (L.18.), 87 (L.23. and L.25.) : illustrations courtesy of Di-Acro Division of Houdaille Industries, Inc.

Pages 36 (H.2.a.), 37, 38 (H.6.), 39, 45, 46 (H.19.b.), 47 (H.21.), 76 (L.4.), 81 (L.14.), 83, 87 (L.24) : illustrations courtesy of Hossfeld Manufacturing Co.

Pages 38 (H.3.), 46 (H.20.a.), 47 (H.20.b.), 77 (L.6.), 78 (L.10.), 88 (M.1.), 89 (M.6.) : illustrations courtesy of Creative Metalcraft, Creative Educational Services Canada Ltd.

Printed and bound in Canada

TABLE OF CONTENTS

INTRODUCTION

This book is intended to develop skills and knowledge by augmenting classroom lessons with related projects, an excellent source of motivational material. It will develop welding and associated skills, such as interpreting drawings and blueprints. Step by step it will help the student understand and apply concepts in planning, designing, and laying out projects.

Many welding projects start from basic shapes which with a few simple changes can be used in a variety of more complex articles. The basic designs and their incorporation will be stressed here.

Work must be carefully planned. Because many alternatives will be presented, the student must decide which is the best approach for the project in mind.

It is assumed that each student has been instructed in the basic fundamentals of welding and is aware of the safety practices that must be observed.

PREFACE
METRICATION IN WELDING

Almost all the countries of the world presently use some version of the metric system for measurement purposes. Here in North America where "inch-pound" systems of measurement are still in use, increasing utilization of the metric system known as SI (short for "Le Système International d'Unités") will probably result in our traditional measurement units becoming far less popular by early in the next decade.

The ministers of education of all provinces have agreed to a policy which encourages boards of education to implement SI units in all subjects. Thus, SI will be the normal measurement system at all education levels by the late 70's.

SI will also prevail in everyday life. Weather reports, highway signs, product packaging, shoe sizes, pharmaceuticals and a multitude of other common objects and services will appear in SI units. Today's students, therefore, must be able to "think SI."

At the same time the student cannot entirely forget the "inch-pound" system. Because of the complex position certain products occupy in our technological world they cannot be converted as readily as others. Such basic domestic objects as screws, nuts, bolts, lumber, and metal sheets, bars, and tubes presently manufactured to "inch-pound" standards may not be available in SI sizes for some time.

In general, industries engaged in manufacturing welding products have not yet been able to convert to the SI system. New standard sizes for all manufactured products related to welding must first be decided upon. Equally important, a market for the new sizes must be found. Only when both these factors are met will welding products meeting SI standards and tolerances be made widely available.

The welding student must therefore be able to think and work in SI units while using the metal products built to meet old "inch-pound" standards. This duality is purposely reflected in this text. The reader will notice that SI units are used whenever possible. "Inch-pound" measurements are used when they refer to products presently manufactured, bought, and sold only in "inch-pound" standards.

By using "inch-pound" measurements only in reference to "inch-pound" products and by employing SI in all other situations, the student will clearly see the differences between the two systems. The problems that this duality poses for the individual and the community, as well as many of the potential advantages of SI, will also be apparent. Section F helps prepare the student for work in SI and further explains the reasons for use of both SI and "inch-pound" systems in the text.

BASIC SHAPES IN METAL

Very early in recorded history, people discovered metal and realized how useful it could be. Weapons made of copper dating back to 4000 B.C. have been found in Egypt. Bronze pots and pans dating back to 2000 B.C. have also been found. Iron seems to have become popular around 1500 B.C. and at one time was considered more valuable than gold.

Metals are divided into two categories: ferrous metals — which contain iron — and non-ferrous metals, which contain no iron. The following table shows some of the metals in each category.

FERROUS METALS.	NON-FERROUS METALS.	
Alloy steel.	Aluminum.	Nickel.
Cast iron.	Brass.	Pewter.
Galvanized iron.	Bronze.	Platinum.
Iron.	Copper.	Silver.
Steel.	Gold.	Tin.
Tin plate.	Lead.	Zinc.

The most commonly used of all the above in metalworking is low carbon steel (mild steel). Carbon steels are processed as hot rolled steel (H.R.S.) or cold rolled steel (C.R.S.). The metal used for almost all of the projects in this book will be low carbon steel, hot rolled. If the student wishes to substitute any other metal (ferrous or non-ferrous), he or she should check for the manufacturer's specific welding recommendations.

One of the most important properties of any metal is its ability to be bent, rolled, squeezed, or hammered into almost any shape. Fig. A.1. (page 2) shows some of the more common basic shapes that are manufactured today.

MATERIAL SELECTION

It is important that the student be able to identify the material required for a specific project. The following is a description of the more commonly used metal shapes and the methods used to specify them. As all these shapes are presently manufactured only in "inch-pound" standards, their sizes are given in that measurement system, both here and throughout this text. Please also note that these "inch-pound" products are currently priced, bought, and sold by the pound. Whether steel will be priced by its mass under new SI standards is not yet certain.

Angle Iron or angles are specified by the length of the leg and the thickness of the material. Where the angle iron has unequal leg lengths, the largest leg is usually specified first, e.g., equal leg — 2 in. x 2 in. x ¼ in., unequal leg — 3 in. x 2 in. x ¼ in.

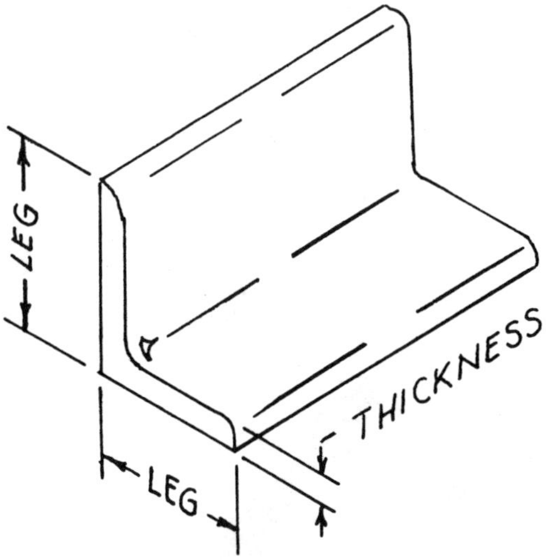

Fig. A.2. Angle iron

BEAMS

Beams are specified by size and mass. They are also identified as "I", "H", or "W.F." (wide

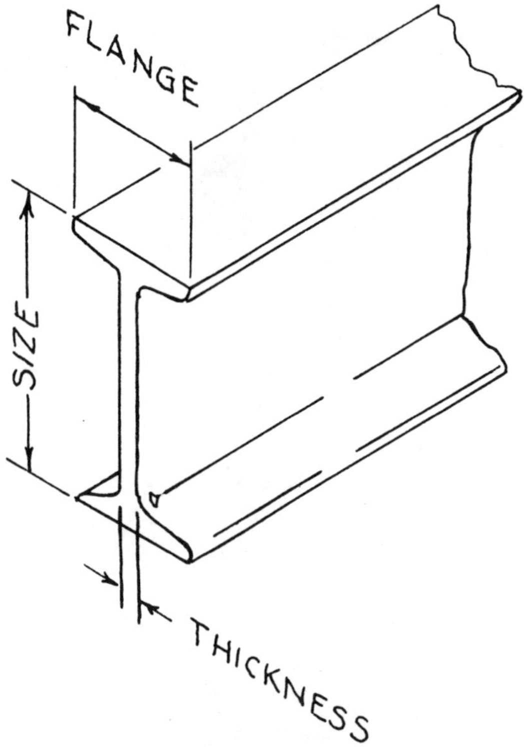

Fig. A.3. "I" beam

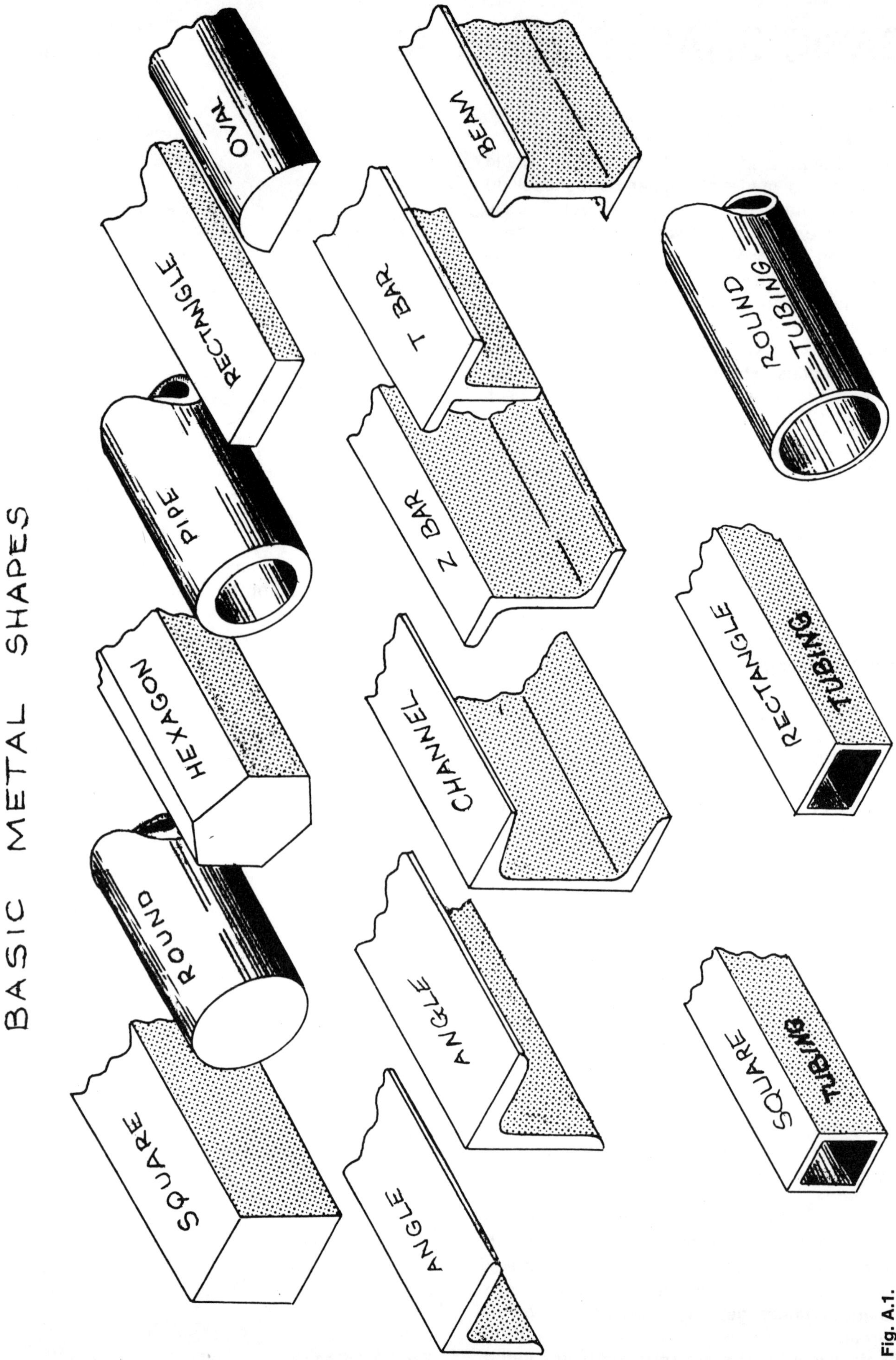
BASIC METAL SHAPES
SQUARE
ROUND
HEXAGON
PIPE
RECTANGLE
OVAL
ANGLE
ANGLE
CHANNEL
Z BAR
T BAR
BEAM
SQUARE TUBING
RECTANGLE TUBING
ROUND TUBING

Fig. A.1.

"H" & W.F. BEAMS

Section and size in in.	Mass in lb./ft.	Depth of section in in.	Flange width in in.	Flange thickness in in.	Web thickness in in.
8 x 8	31.0	8.00	8.00	0.433	0.288
	34.3	8.00	8.00	0.459	0.375
	35.0	8.12	8.027	0.493	0.315
	40.0	8.25	8.077	0.558	0.365
	48.0	8.50	8.117	0.683	0.405
	58.0	8.75	8.222	0.808	0.510
	67.0	9.00	8.287	0.933	0.575
24 x 12	100.00	24.00	12.00	0.775	0.468

"I" BEAMS

Size in in.	Mass in lb./ft.	Thickness of web in in.	Width of flange in in.
8	18.4	0.270	4.0
	23.0	0.441	4.171

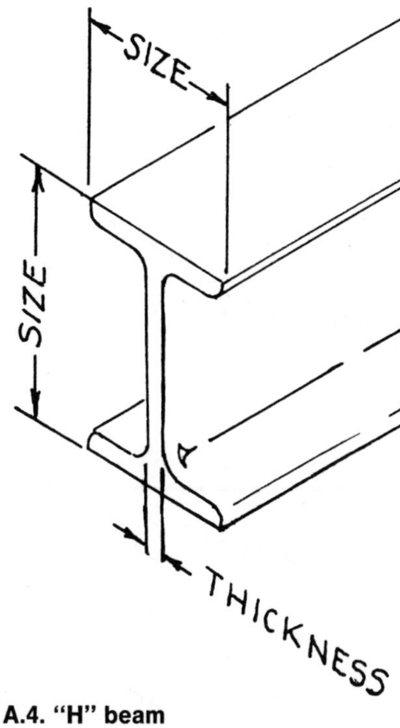

Fig. A.4. "H" beam

flange), and as the name implies, the end section of the beam can be identified by its similarity to the letters "I" and "H". These beams (as with all standard manufactured steel shapes) are manufactured to *close* but not *exact* tolerances; also within each size there can be variations in depth of section, flange widths, and web thicknesses. It is therefore, important that students obtain and study manufacturers' steel catalogues before ordering any beams or materials. As a rule of thumb however, "I" beams have a narrower flange than "H" beams of the same size. "H" beams will have approximately the same width of flange as height of section, e.g., 8 in. x 8 in. The section and size identification become important when there is a wide variation in dimensions of "H" beams, e.g., 24 in. x 12 in. In this case the section (24 in.) is usually placed first.

Channels are specified by depth of section and mass although it may also be necessary to specify web and flange thickness for the same size channel.

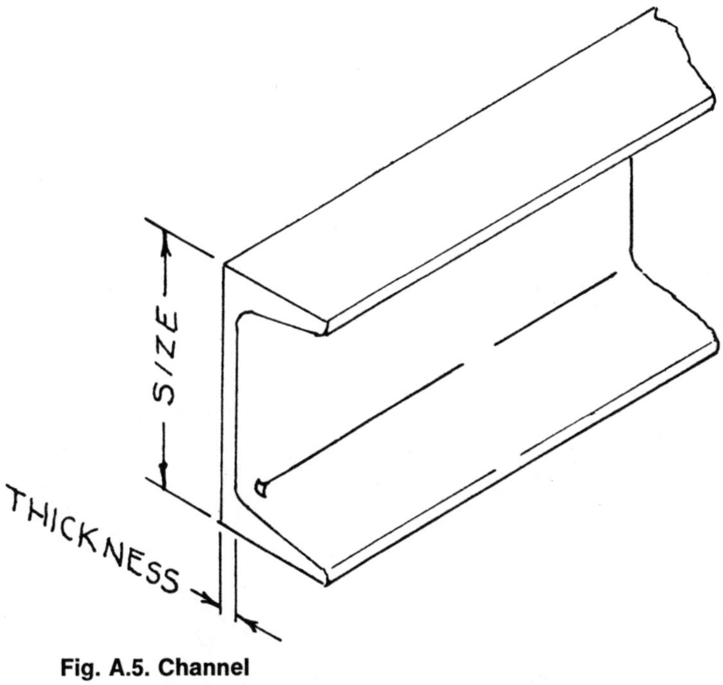

Fig. A.5. Channel

Flat bars or "flats" are specified by the thickness first and the width secondly, for example: ¼ in. x ¾ in.

Ovals or "half ovals" are specified by width and height, in that order, for example: 2 in. x ½ in.

Plate and sheet steel are specified sequentially by thickness, width, and length, for example: 10 (ga) x 72 in. x 96 in.

Pipe is specified by a size wherein the outside diameter (O.D.) will always remain the same but the wall thickness of internal diameter (I.D.) will change as designated by a standard such as the American Standards Association's terms like schedule 40, Extra Heavy (E.H.), Double

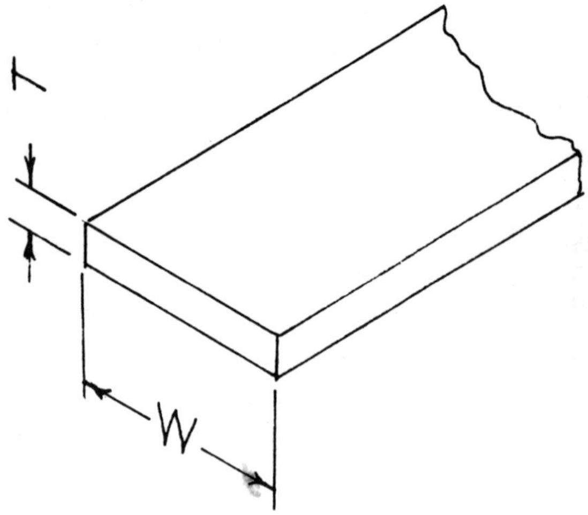

Fig. A.6. Flat bar

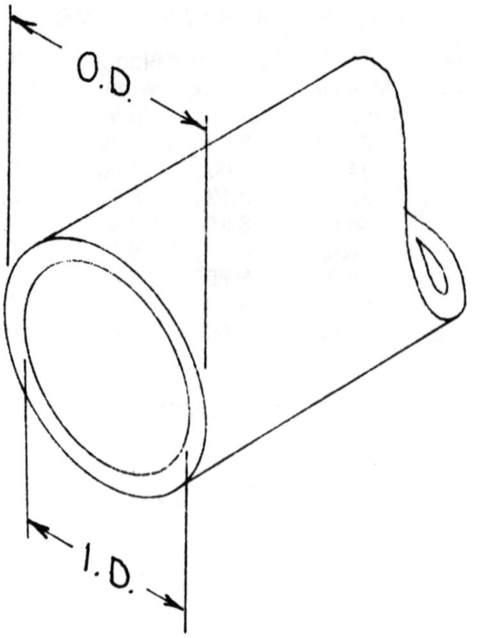

Fig. A.9. Pipe

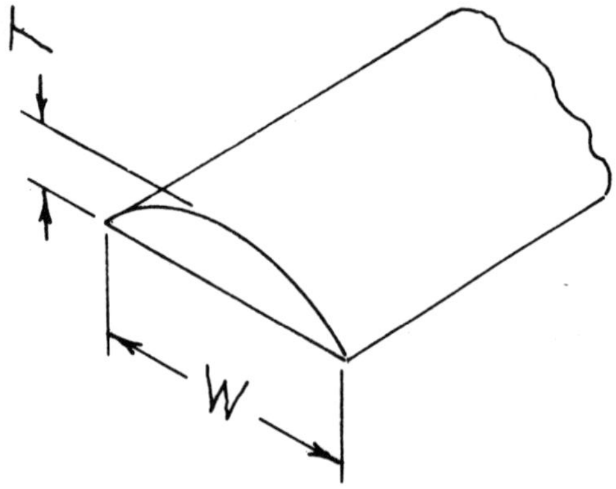

Fig. A.7. Oval

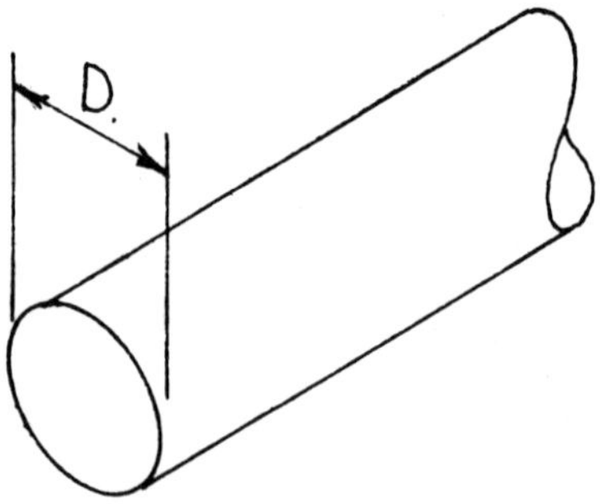

Fig. A.10. Round bar

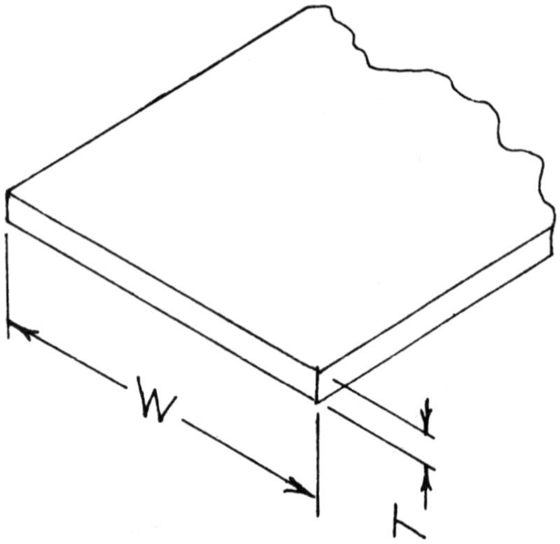

Fig. A.8. Plate and sheet steel

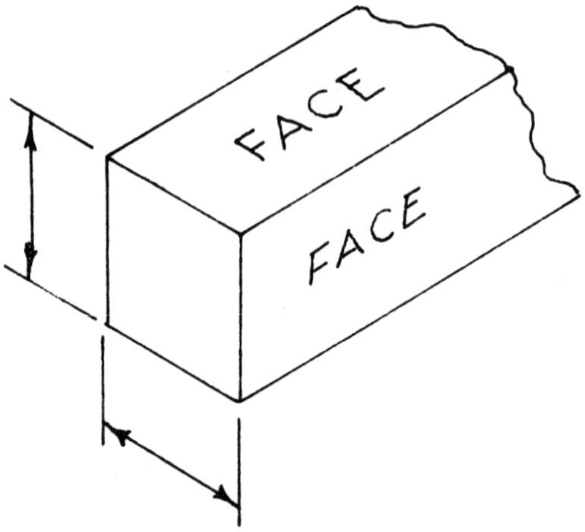

Fig. A.11. Square bar

		Sch. 40		E.H.		Dbl. E.H.	
Pipe size in in.	O.D. in in.	Wall in inches	Mass in pounds	Wall in inches	Mass in pounds	Wall in inches	Mass in pounds
1	1.32	0.13	1.68	0.18	2.17	0.36	3.66

Extra Heavy (Dbl. E.H.), and so forth, as shown above.

Round Bars are specified by diameter only, for example: ½ in. dia.

Square bars are specified by face-to-face measurements, for example ½ in. x ½ in.

Tee bars are specified by width of flange, depth of web, and thickness of web, for example:

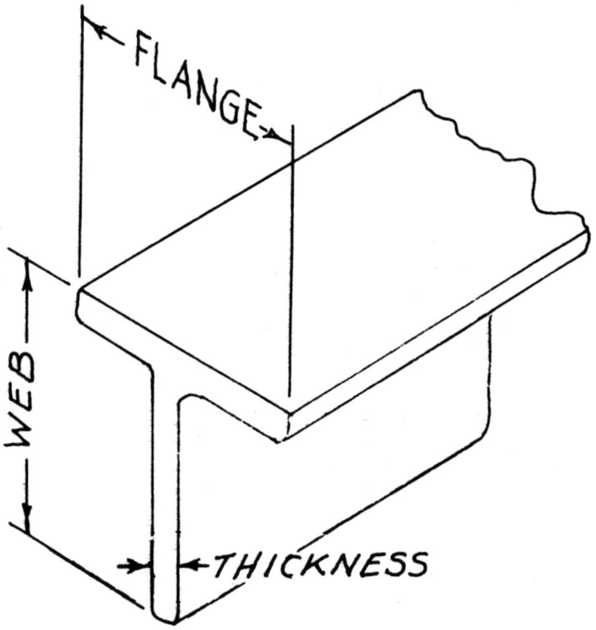

Fig. A.12. Tee bar

Flanges in in.	Depth of web in in.	Thickness of web in in.
4	4	½

Tubing is specified in O.D. size, wall thickness, and I.D., for example:

O.D. size in in.	Wall thickness in in.	I.D. size in in.
¾	0.13	0.48

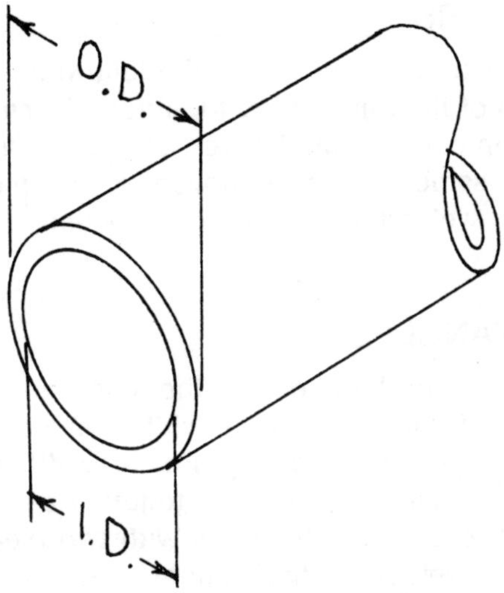

Fig. A.13. Tubing

Sometimes the wall thickness will be given in Birmingham wire gauge size; in that case, the example above would be #10 ga. Birmingham wire gauge is not an SI standard. At the time of publication no decision as to possible new standards has yet been made.

Zee bars are specified by a size which usually

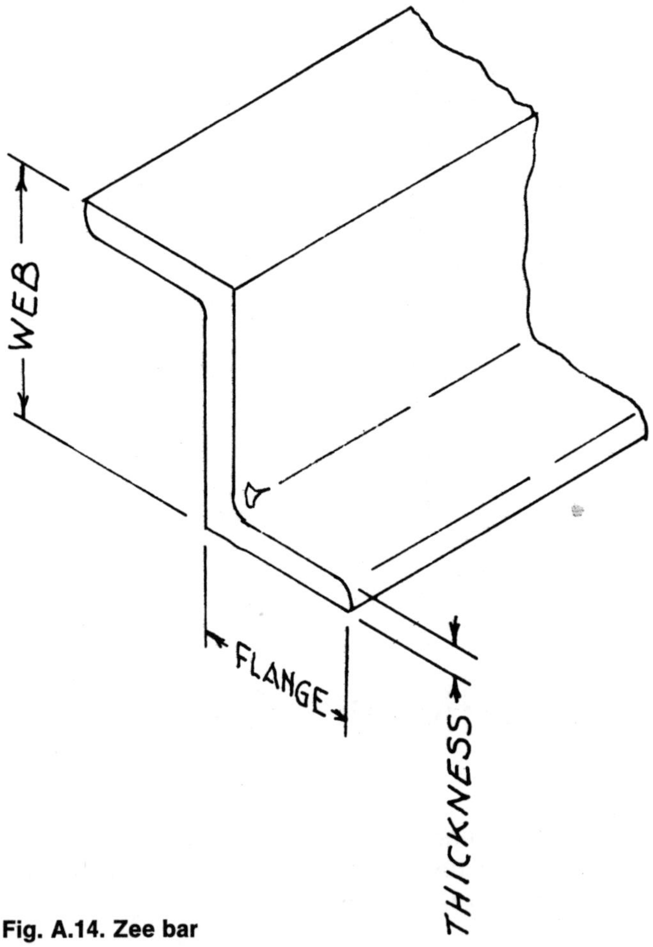

Fig. A.14. Zee bar

refers to the depth of web, or by the width and thickness of flange. For example:

Size	Depth of web in in.	Flange in in.	Thickness in in.
3	3	2$\frac{11}{16}$	¼

EXPANDED METAL

A popular form of metal, decorative for projects, is expanded, or perforated, metal. All types of metals can be expanded. Mild steel is the most

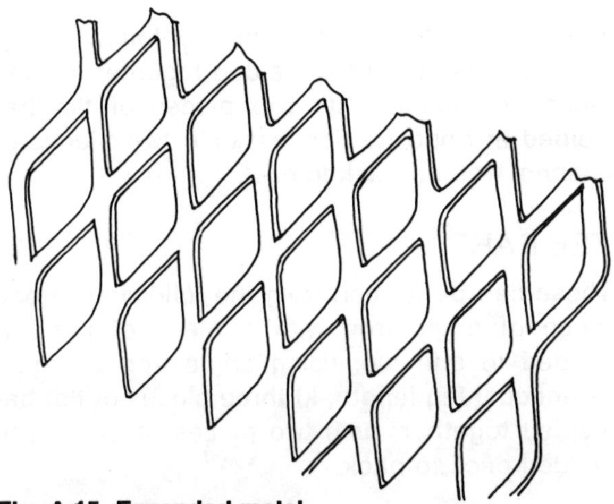

Fig. A.15. Expanded metal

common metal so used. On occasion aluminum and stainless steel are also expanded.

Basically, a solid sheet of metal is fed into a machine that perforates the metal in a variety of designs and patterns. When the metal emerges from the machine, it is called "standard expanded metal." When processed further by passage through a heavy rolling machine, it is then called "flattened expanded metal." This type is most often used for metal projects. Because it is made from one solid piece of metal with no joints that could separate, expanded metal can be cut and formed in any shape.

FORMING BASIC SHAPES

SECTION B

A few suggestions are offered here on constructing basic shapes from other components, whenever necessary.

SQUARES

These can be formed by welding together the following: a) two pieces of angle iron; b) four pieces of flat bar; or c) two pieces of channel.

ANGLE IRONS

These can be formed from the following: d) two pieces of flat bar welded together; e) one piece of zee bar with one leg removed; or f) one piece of channel cut to product angle iron with equal or unequal leg length.

BEAMS

These can be formed from the following: g) three pieces of flat bar welded together; h) one piece of channel with two pieces of flat bar welded at opposite flanges; or i) two pieces of channel welded back to back.

ZEE BARS

These can be formed from the following: j) one piece of angle iron with a piece of flat bar welded to one toe, using angle iron of equal or unequal leg length; k) three pieces of flat bar welded together; or l) two pieces of angle iron welded back to back.

CHANNELS

These can be formed from the following: m) three pieces of flat bar welded together; n) one piece of angle iron welded to one piece of flat bar; or o) one piece of square tubing cut to suit the depth of flange required on the channel.

TEE BARS

These can be formed from the following: p) two pieces of flat bar welded together; q) one piece of beam cut through the web so that the flanges on the same side are removed; or r) a piece of flat bar welded to the heel of a section of angle iron.

RECTANGLES

These can be formed from the following: s) two pieces of flat bar of equal width welded to two other pieces of equal but greater width; t) two pieces of angle iron welded together as shown; or u) two pieces of channel with two pieces of flat bar welded to the flanges and separating the channels.

Note. 1. Thicknesses and dimensions would depend on the type of assembly desired. 2. Welding time must be considered. 3. If the assembly is simply welded without any type of restraint being used, expansion and contraction may cause distortion. Welding procedures must be carefully planned beforehand.

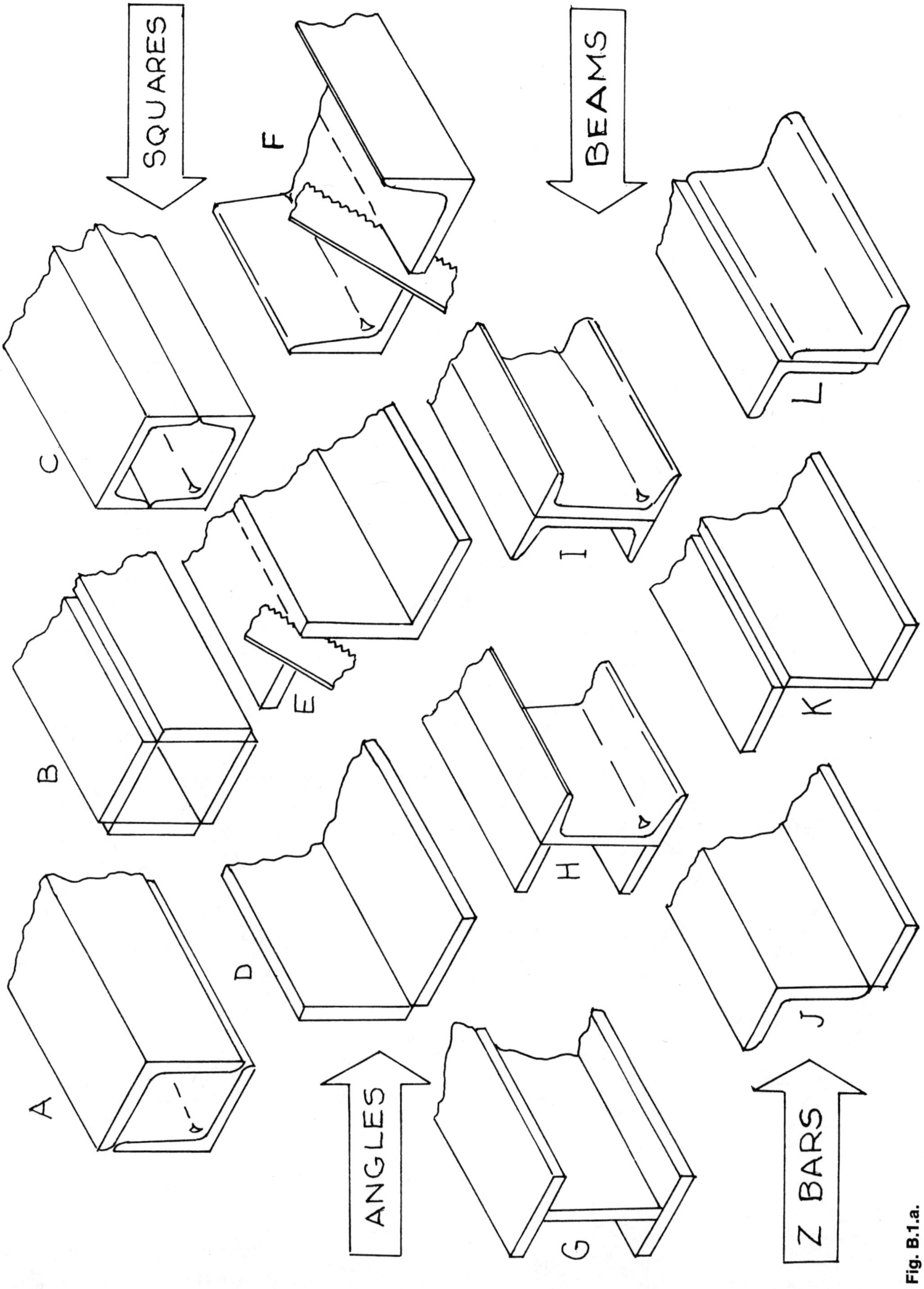

Fig. B.1.a.

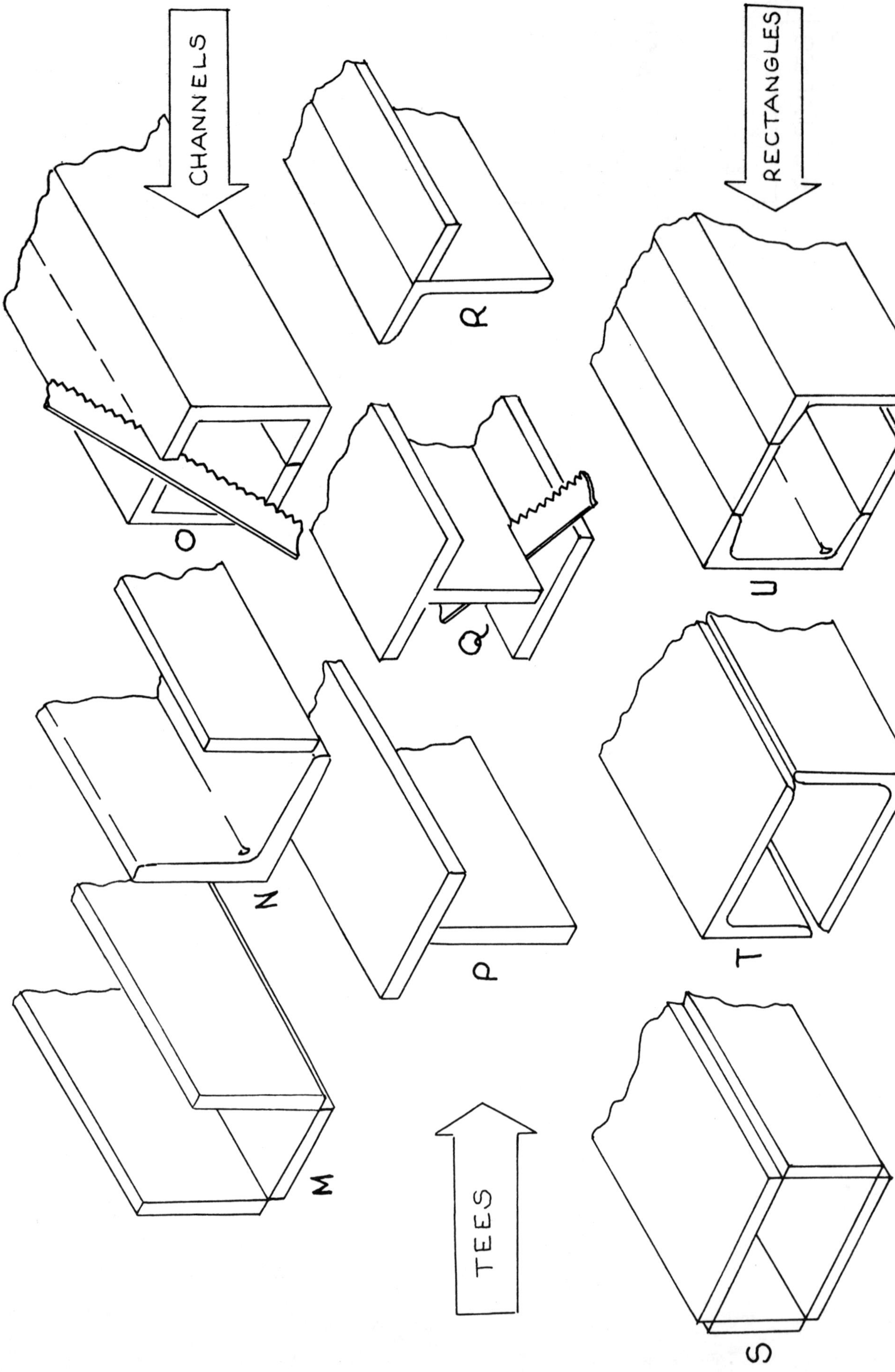

Fig. B.1.b.

MEASUREMENT & LAYOUT

To make parts fit properly, accuracy is essential in laying out and measuring projects. It is important to check all measurements, especially before cutting. Squareness should also be continually verified as many projects are cut accurately to size but assembled improperly.

Students should learn to read measuring instruments quickly and accurately. The most common instrument used is the rule (steel rule, tape measure, and so forth). The graduations could be marked on one, two, three, or four edges.

Steel rules are available in both metric and inch–pound systems. The metric rule will probably be the one used most often but the inch rule will also be necessary on some occasions. See Fig. C.1.

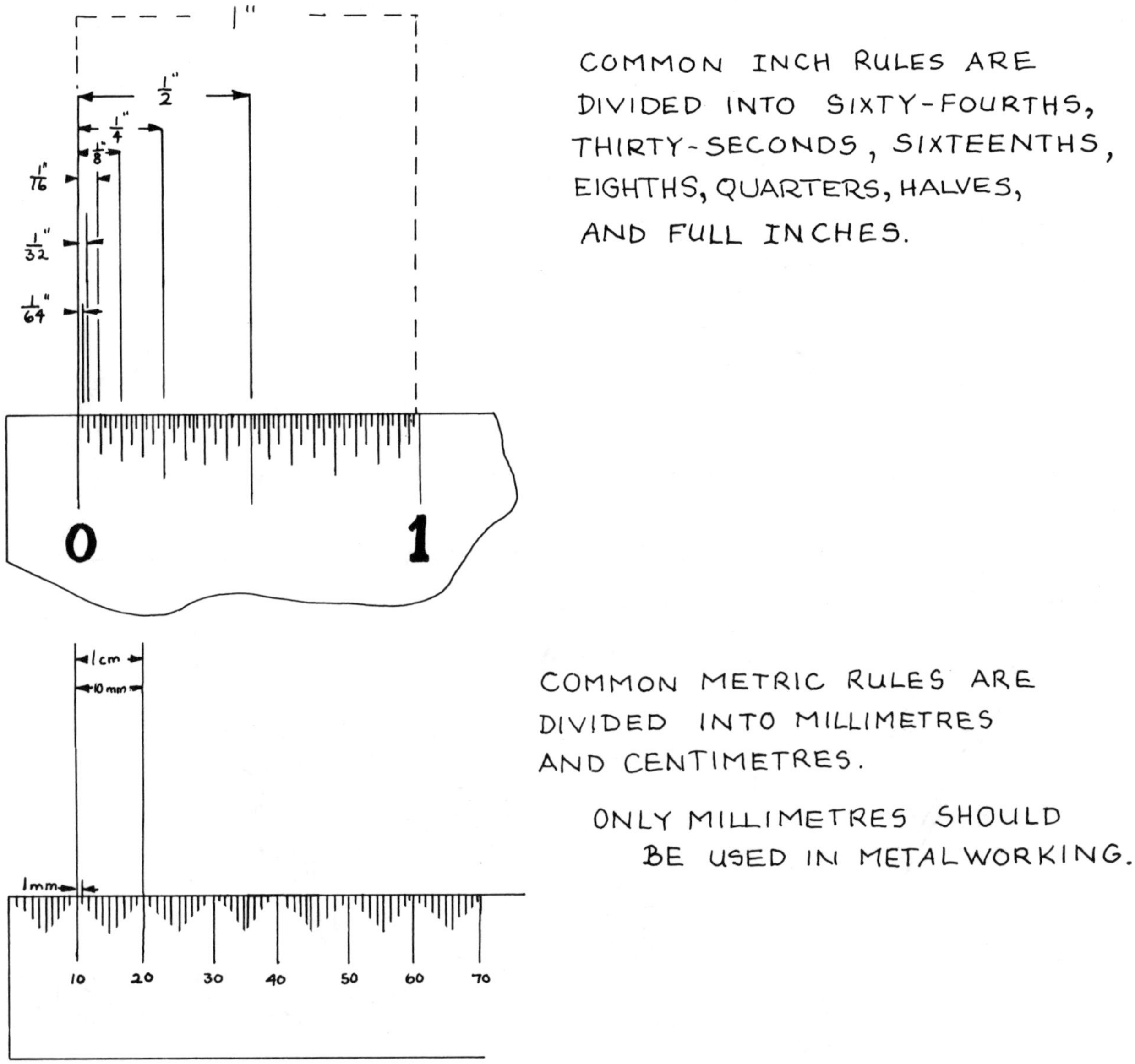

Fig. C.1. Inch and metric rules

Note. When the end of a rule or tape measure becomes worn with constant use, for accuracy it is better to start the measurement from the 10 mm (or 1 in.) mark rather than from 0. *Do not forget to subtract the 10 mm (or 1 in.) from the total measurement to compensate for this.* See Fig. C.2. (p. 10.)

The square is also used for measuring. As well, it is a tool for checking if two members are positioned at right angles to each other,

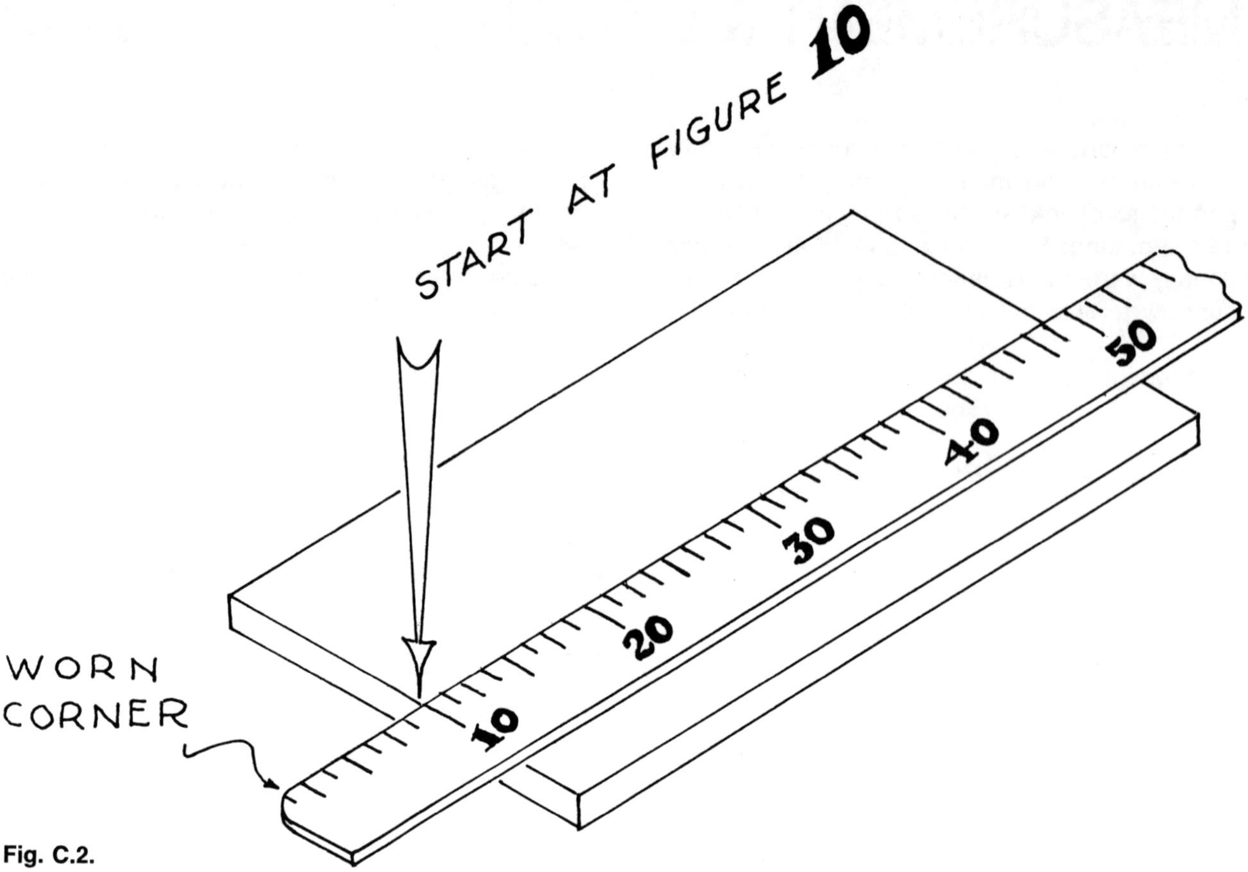
START AT FIGURE 10
WORN CORNER
10
20
30
40
50

Fig. C.2.

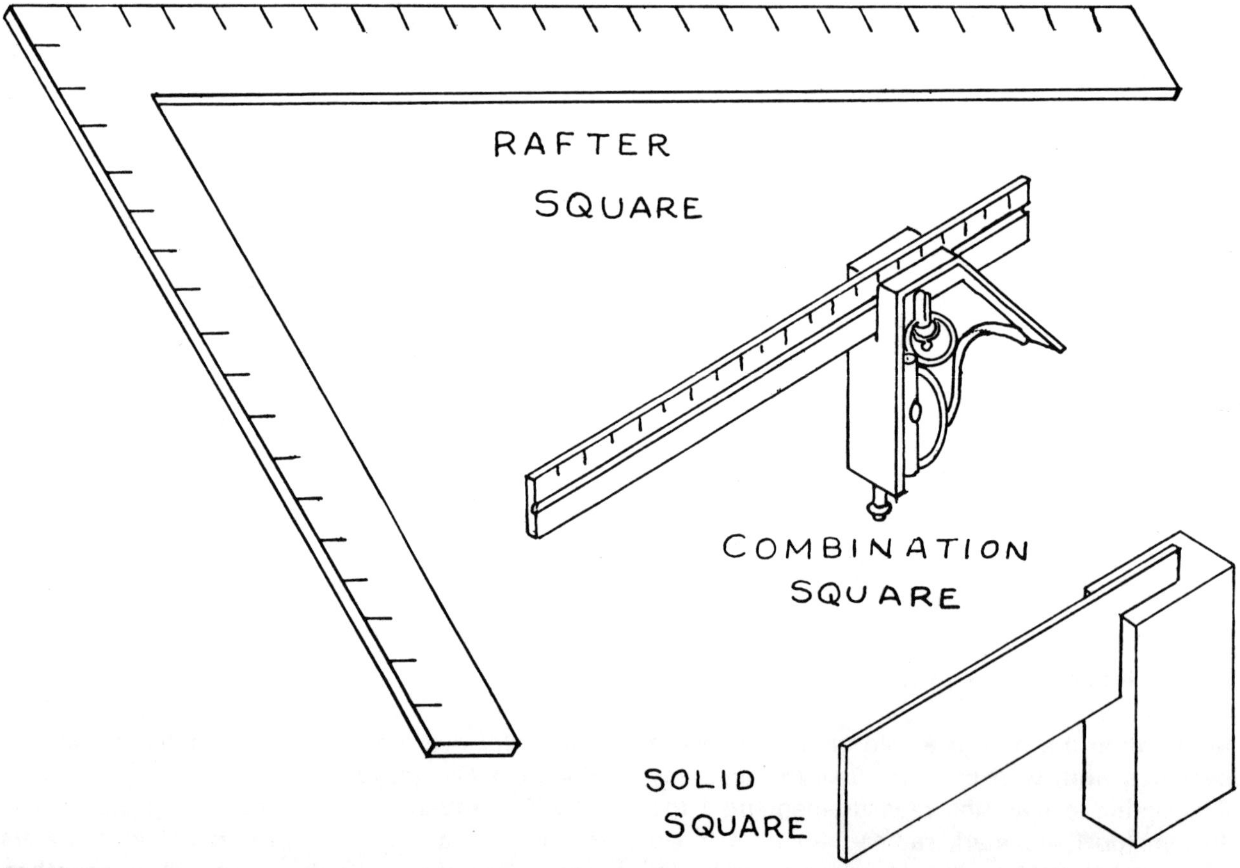
RAFTER SQUARE
COMBINATION SQUARE
SOLID SQUARE

Fig. C.3.

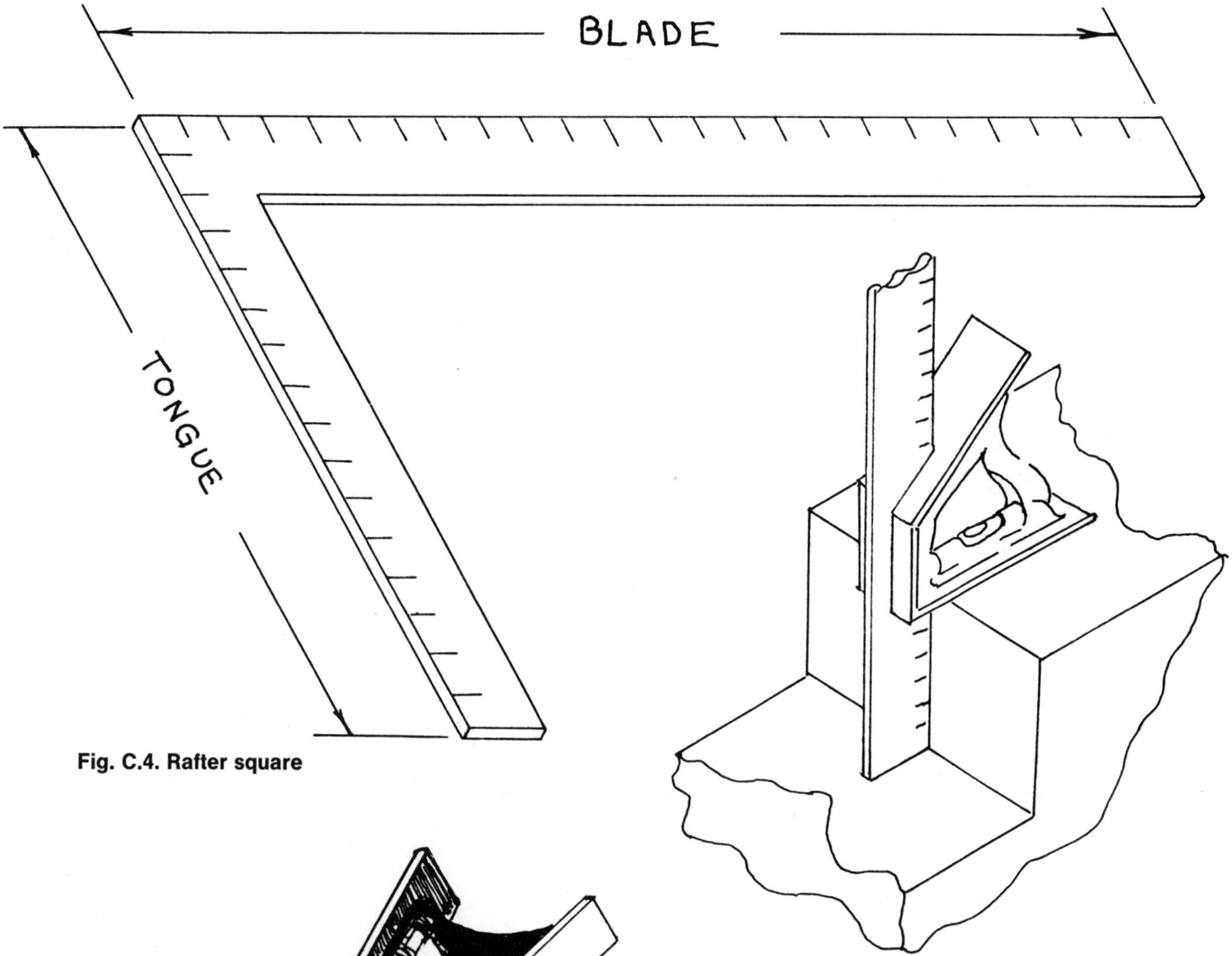

Fig. C.4. Rafter square

Fig. C.5.b. Combination square

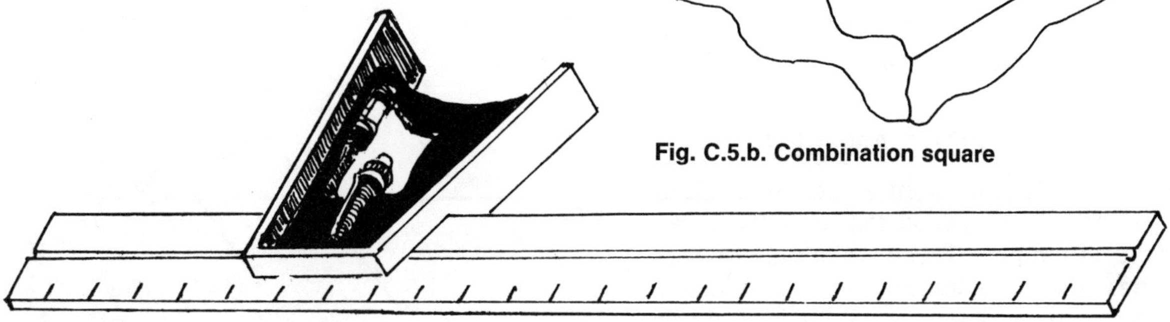

Fig. C.5.a. Combination square

for laying out lines, and, as a straight edge, for checking surfaces. The more common types of squares are shown in Fig. C.3.

The two most commonly used squares in metal project building are the combination square and the steel (rafter) square. (Fig. C.4.) The steel rafter square is a one piece instrument consisting of two arms, commonly called the blade and the tongue. The measurements and tables on the square are used mainly by carpenters for woodworking.

The combination square is a versatile instrument; the blade can be used as a rule and the body can be used to lay out or check 90° and 45° angles. The moveable body allows it to be used as a depth gauge. See Fig. C.5. and C.6.

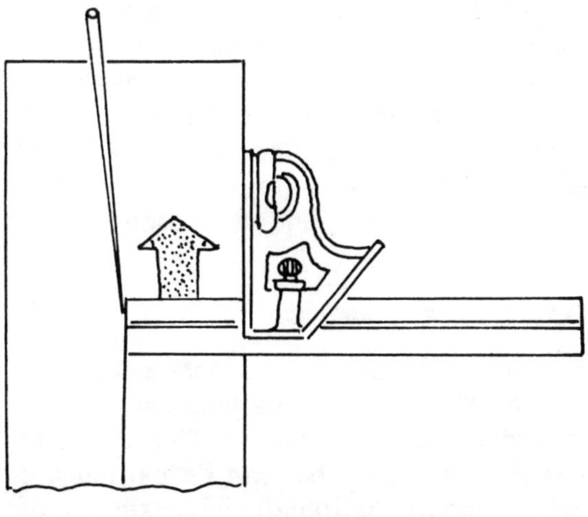

Fig. C.6. Combination square

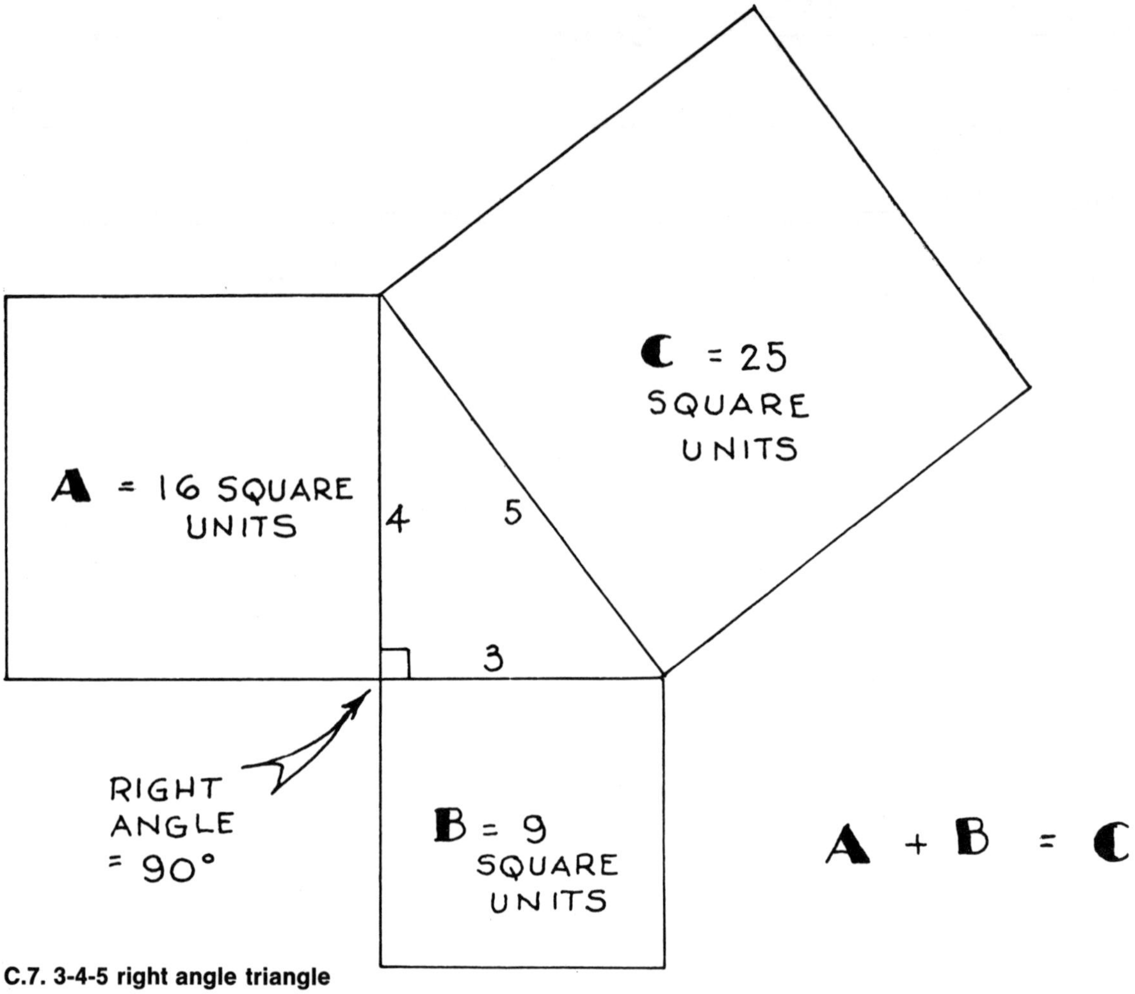

Fig. C.7. 3-4-5 right angle triangle

WORKING WITH RIGHT ANGLES

Squares are made from right angles — angles of 90° — and are based on the principle of a right–angle triangle. These triangles have three sides: base, altitude, and hypotenuse. The hypotenuse is always opposite the right angle and is the longest side.

During math classes, students have at some time probably used the Pythagorean theorem. This states that the square of the hypotenuse of a right–angle triangle is equal to the sum of the squares of the opposite two sides. The usual method of demonstrating this is the 3-4-5 right–angle triangle illustrated in Fig. C.7. Many projects — tables, chairs, trailers, and so forth — require 90° (square) and 45° corners.

TO MAKE A 45° ANGLE

1. From one end, measure and mark a distance equal to the total width of the material. 2. Connect the diagonal as shown in Fig. C.8. and cut across that mark. This can be carried out on tubing (square or round), pipe, square or round bar, and flat bar.

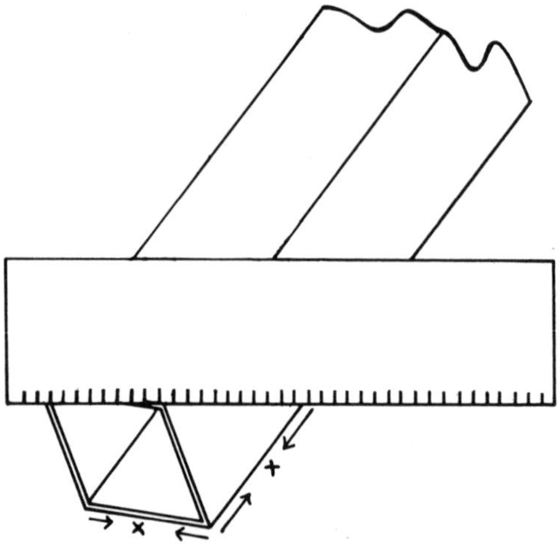

Fig. C.8. 45° angle

TO MAKE A 90° ANGLE

1. Measure the length required from one end of the material. 2. Measure back from that point a distance equal to the total width "w" of the material. 3. Connect the diagonal between these two marks, and cut along that diagonal to form two parts A and B as shown in Fig. C.9. 4. Turn piece B so that it is matched

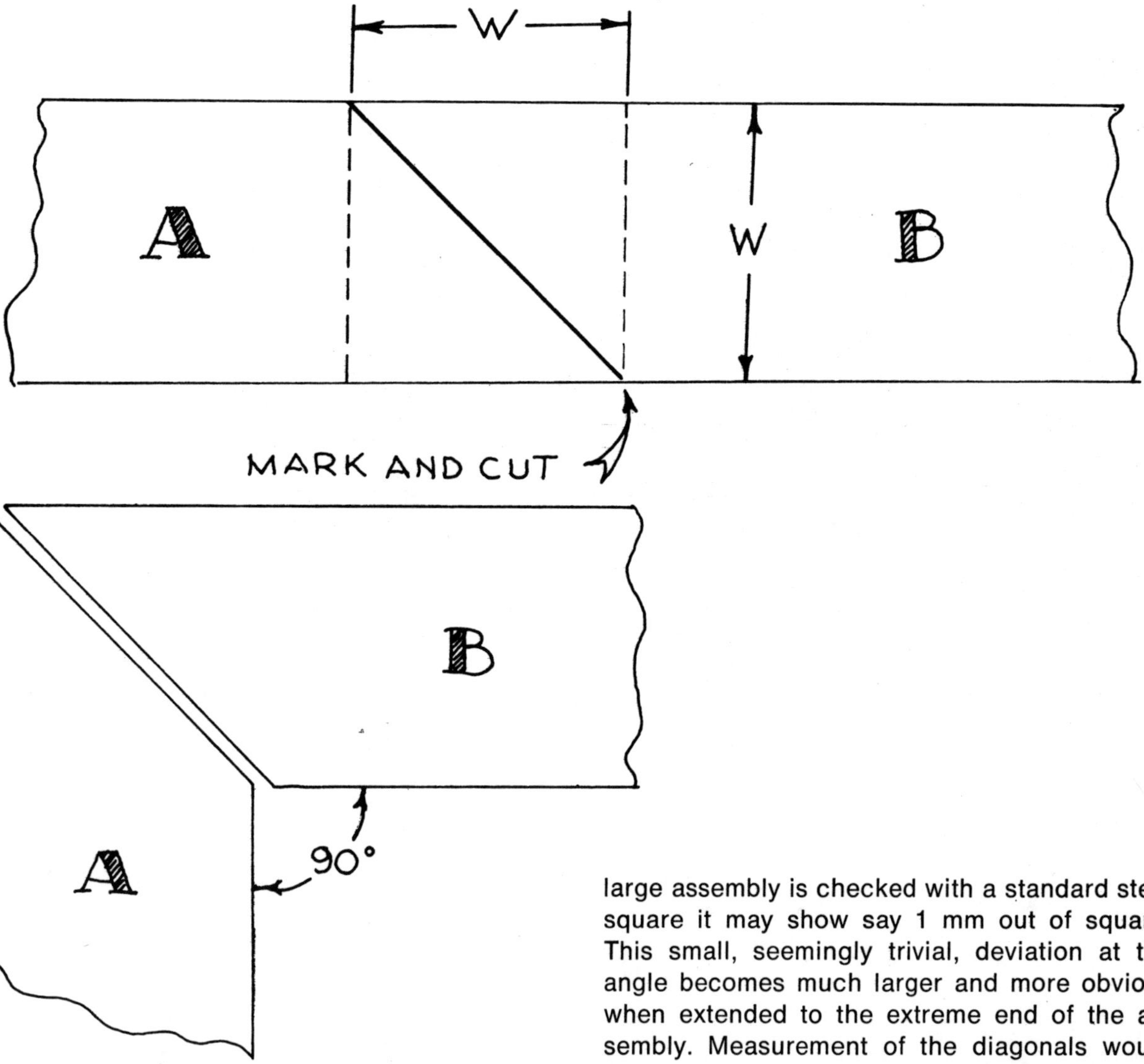

Fig. C.9. 90° angle

up as shown in Fig. C.9. 5. These pieces should match up perfectly and form an exact 90° angle. This can also be carried out on bar (flat, round, or square), pipe, and tubing (round or square). **Note.** This is simply an extension or example of the use of the hypotenuse theory mentioned at the beginning of the section. Instead of forming two 45° angles and matching them, it is easier to take one piece of metal and cut it as described to form two 45° angles at the same time.

This theorem has many practical applications, especially as a check on assemblies with right angles. For example, if you build a rectangular or square frame for a small table or chair, you could use a steel square to check the right angles for accuracy. More simply, you could measure across the diagonals: if they measure exactly the same, then the frame will be square. See Fig. C.10. (p. 14.) The larger the assembly the more important is the Pythagorean theorem in checking for squareness. When a large assembly is checked with a standard steel square it may show say 1 mm out of square. This small, seemingly trivial, deviation at the angle becomes much larger and more obvious when extended to the extreme end of the assembly. Measurement of the diagonals would prove a much more meaningful test of accuracy. See Fig. C.11. (p. 14.) The 3-4-5 ratio is also an excellent method of locating and aligning tongues and axles on boat or box trailers, as shown in Fig. C.12. (p. 15.)

In certain cases you can also find the optimum length of an unknown side. If the angle you are working around is 90° and you know the length of any two of its sides — the longest (hypotenuse), one side (the base), or the other (altitude) — you can use the Pythagorean theorem to calculate the length of the third side. This is the theorem expressed mathematically: $A^2 = B^2 + C^2$. A represents the hypotenuse (the side opposite the right angle), B represents the altitude, and C represents the base. To calculate A, find the square root of $B^2 + C^2$. To calculate B, find the square root of $A^2 - C^2$. To calculate C, find the square root of $A^2 - B^2$.

Note. Be sure to first square and then add or subtract the figures under the square root sign before you figure out the square root itself. Otherwise, your calculations will be wrong.

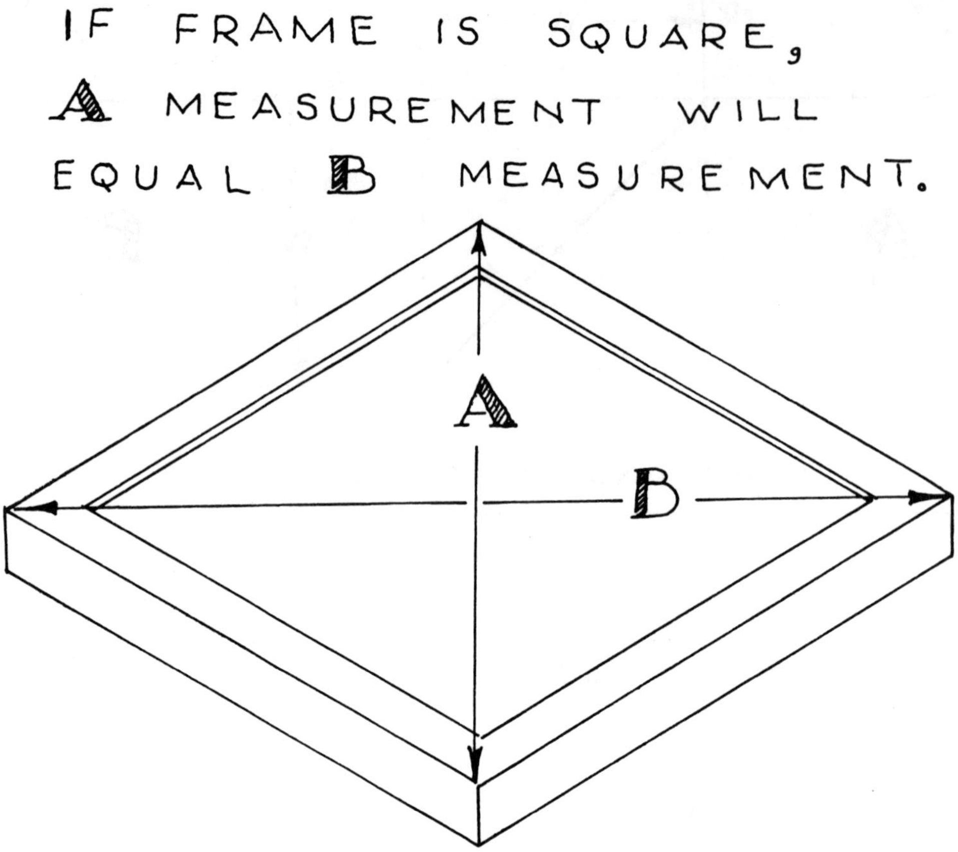

Fig. C.10.

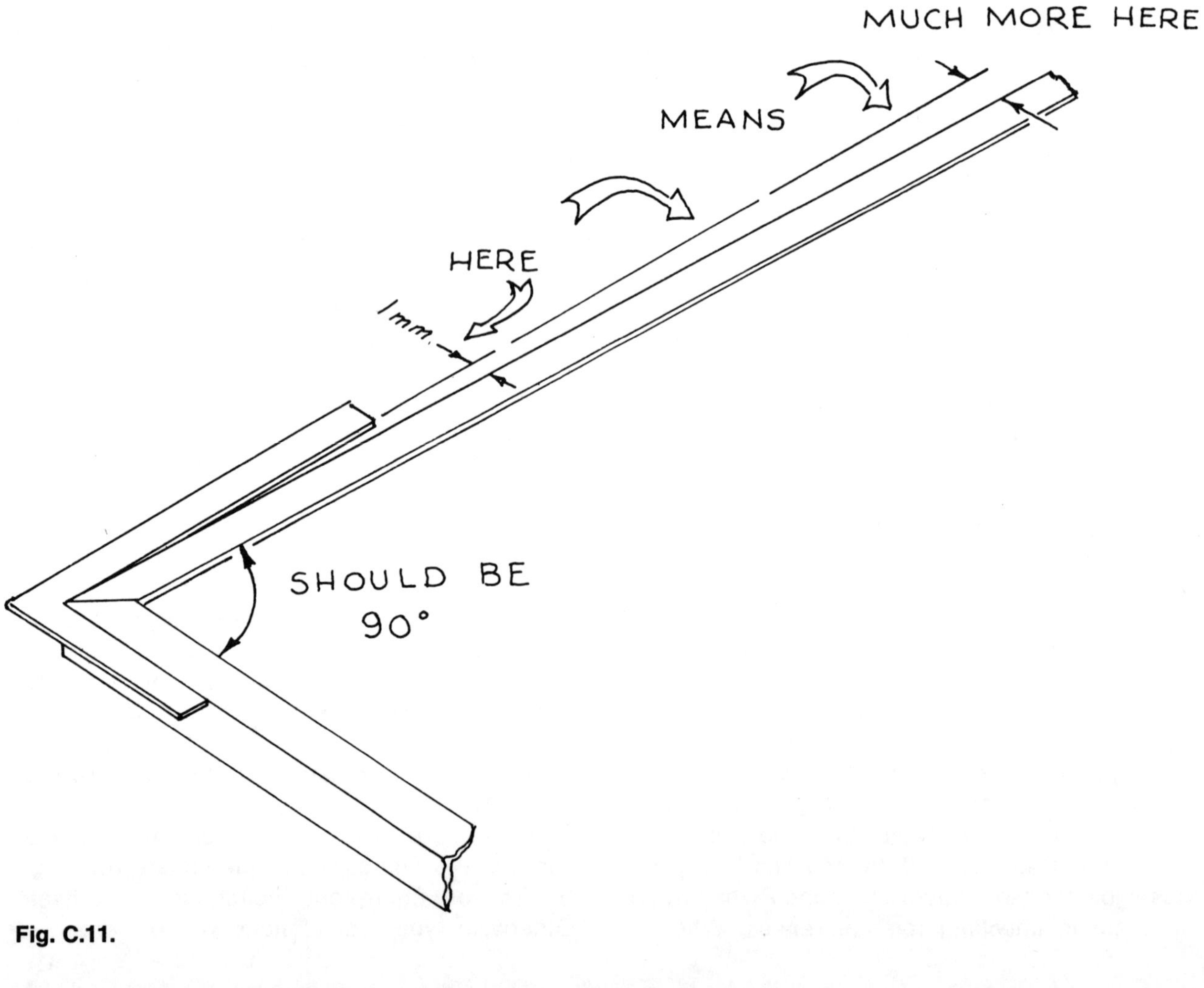

Fig. C.11.

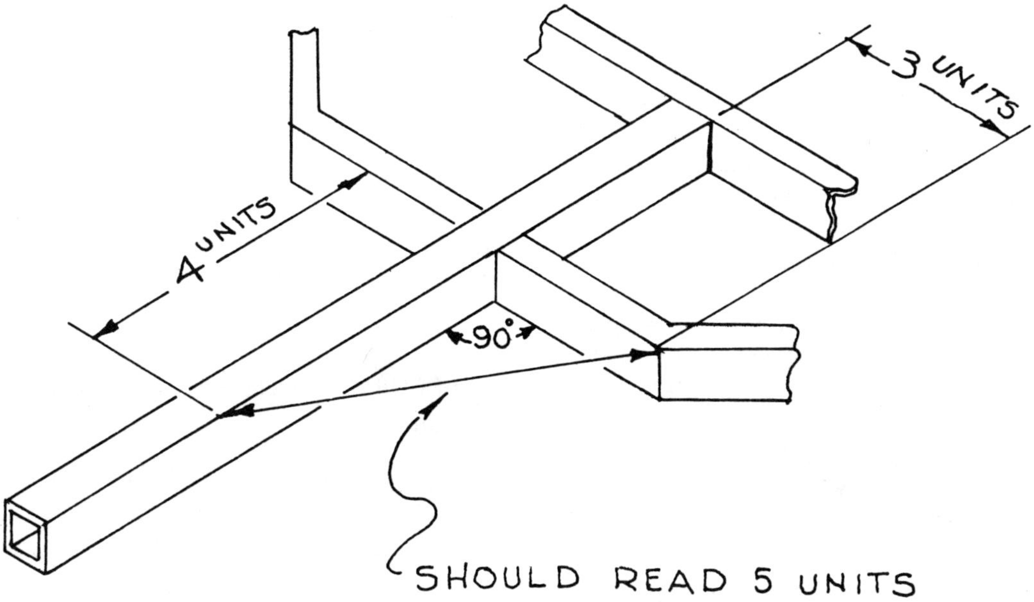

Fig. C.12.

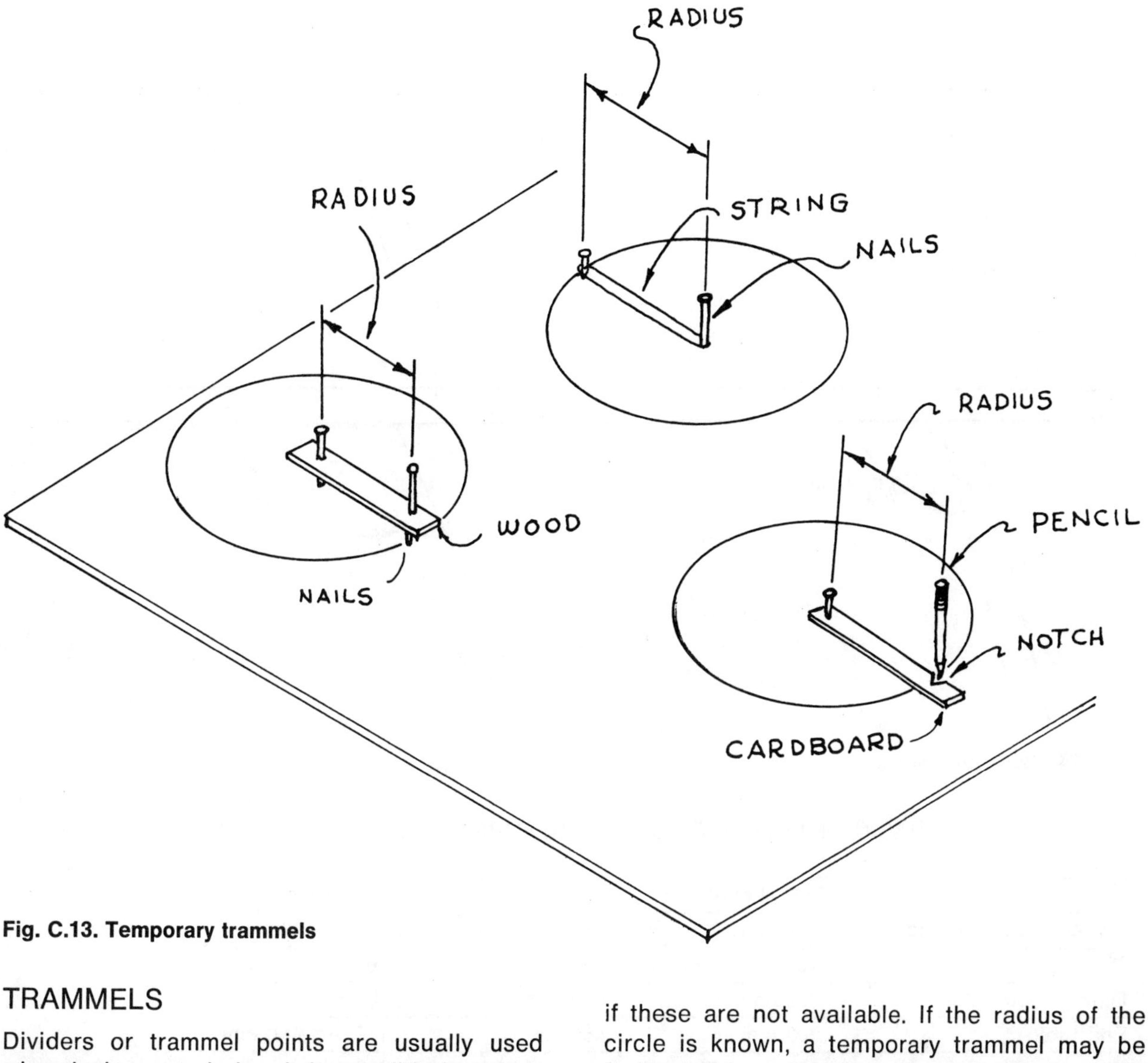

Fig. C.13. Temporary trammels

TRAMMELS

Dividers or trammel points are usually used when laying out circles. It is possible however, to lay out the circle with reasonable accuracy if these are not available. If the radius of the circle is known, a temporary trammel may be fashioned in a number of ways as shown in Fig. C.13.

DIMENSIONS

Personal choice and variety of designs and applications possible in a single project make it impractical to give specific dimensions in many cases. Each project is unique. You yourself must determine the measurements according to its use and design. A dining room table and chair set could be built, for example, so that the height of the tables in relation to the chair would not allow for comfortable seating. These heights should complement each other and care must be taken to ensure this.

Students should measure heights and dimensions of various pieces of furniture to familiarize themselves with basic dimensions. The following list of *average* dimensions may prove helpful in constructing projects; *remember they are to be used as a guide only.*

Note. The great variety of styles available makes it impossible to give exact dimensions for chairs. However, the average seat height of most chairs is 410 to 460 mm. The back support from the seat up should reach at least as high as the middle of a person's back. A sheet such as the one shown below may prove helpful in sorting out this information before beginning the drawing or actual project.

SPECIFICATIONS	MATERIALS
Height.	Accessories.
Length.	Back.
Width.	Braces.
	Frame.
	Legs.

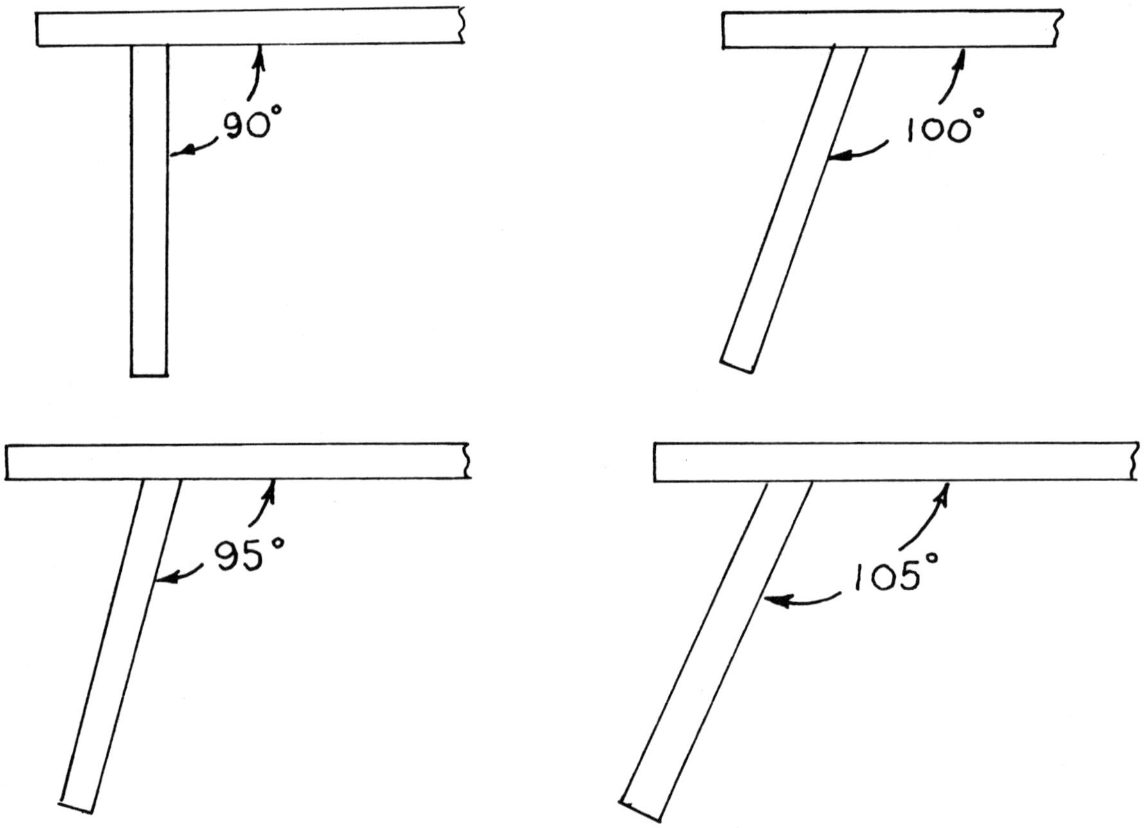

Fig. C.14. Table leg angles

BASIC DIMENSIONS FOR METAL FURNITURE

Project	Length	Width	Height
Coffee table (rectangular)	900 mm-1220 mm	460 mm	410 mm-460 mm
Desks	760 mm-1065 mm	510 mm-610 mm	760 mm
Dining room table	1220 mm	810 mm	760 mm
End table	900 mm	460 mm	410 mm-460 mm
Foot stools	380 mm-460 mm	380 mm-460 mm	300 mm
Lamp table	750 mm	460 mm	560 mm

Sometimes setting a table leg at an angle greater than 90° can enhance the beauty and design of a table. However, too great an angle can be as unattractive as too little an angle. There is no hard and fast rule. Generally it will be found that the angle increases as the height of the table decreases. For example, a large or high table should have an angle of 95°, a medium height table an angle of 100°, and finally a small table should have an angle of 105°, as shown in Fig. C.14. (p. 14.)

Note. At no time should the leg be allowed to extend beyond the boundaries of the table itself. Remember if the legs are placed at an angle as suggested above that the bottom of the leg must also be adjusted to make the table level and stable. See Fig. C.15. Remember also to trim any cross members or braces if you adjust the legs to other than 90°. See Fig. C.16.

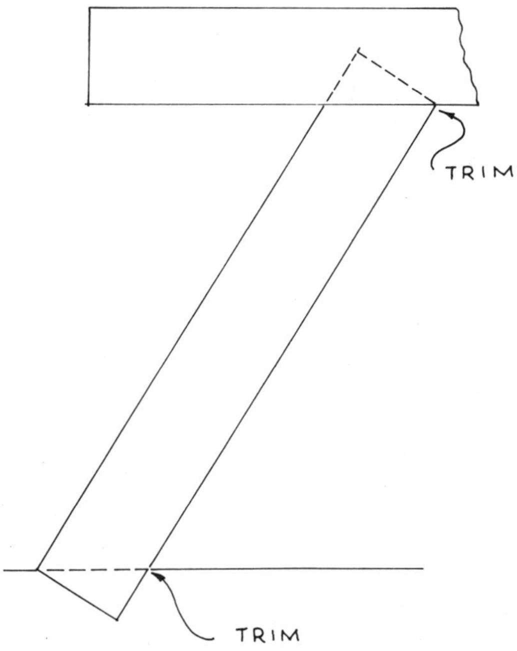

Fig. C.15. Trim leg

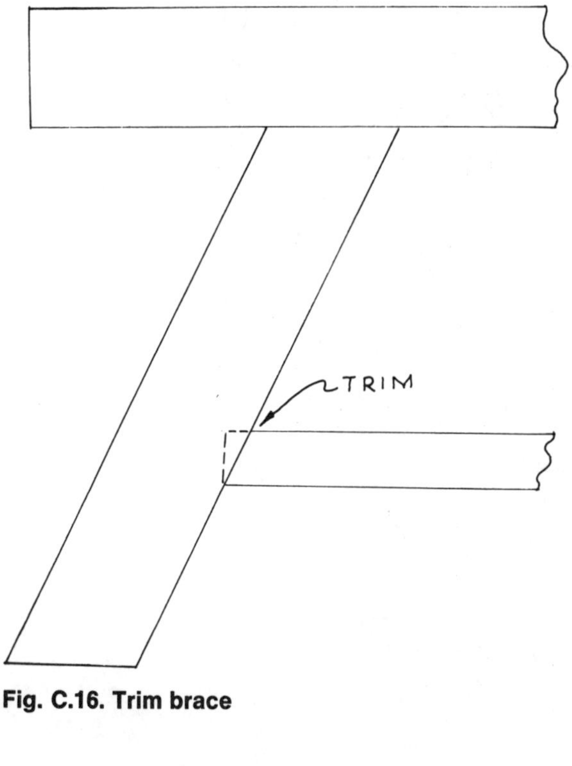

Fig. C.16. Trim brace

LAYING OUT DUPLICATE PARTS

METHOD 1

If a piece of metal from a full size pattern or original project is to be duplicated and it involves irregular shapes, e.g., the leg of a chair as in Fig. C.17., a piece of wire can be bent to the exact shape. Having bent the wire to shape, you can use it as a model to duplicate a number of parts.

Note. To determine the length of the above chair leg or any object that has already been formed, a piece of soft soldering wire can be bent as described above, straightened, and the length measured. If a small spool of this soldering wire is kept handy and not cut, it can be used over and over again.

METHOD 2

Another method would be to trace the shape of the object on a piece of plate or a large bench with soapstone. This sketch could be used then to check each piece for uniformity.

Note. The chalk lines would have to be carefully preserved or marked with a centre punch. See Fig. C.18. (p. 18.)

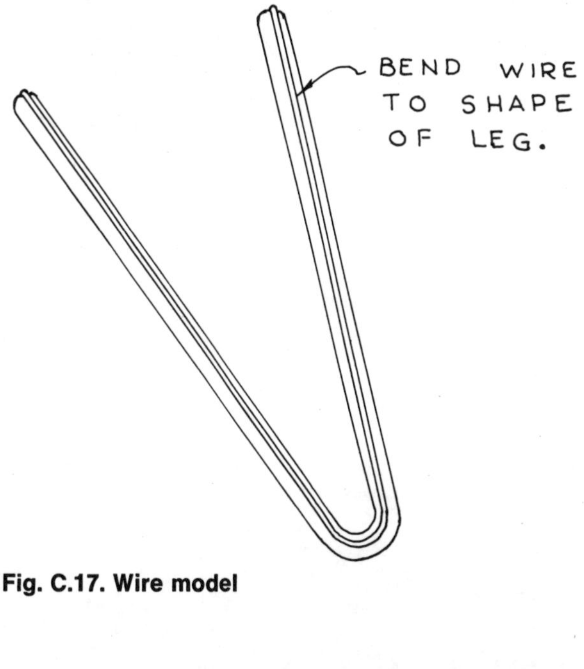

Fig. C.17. Wire model

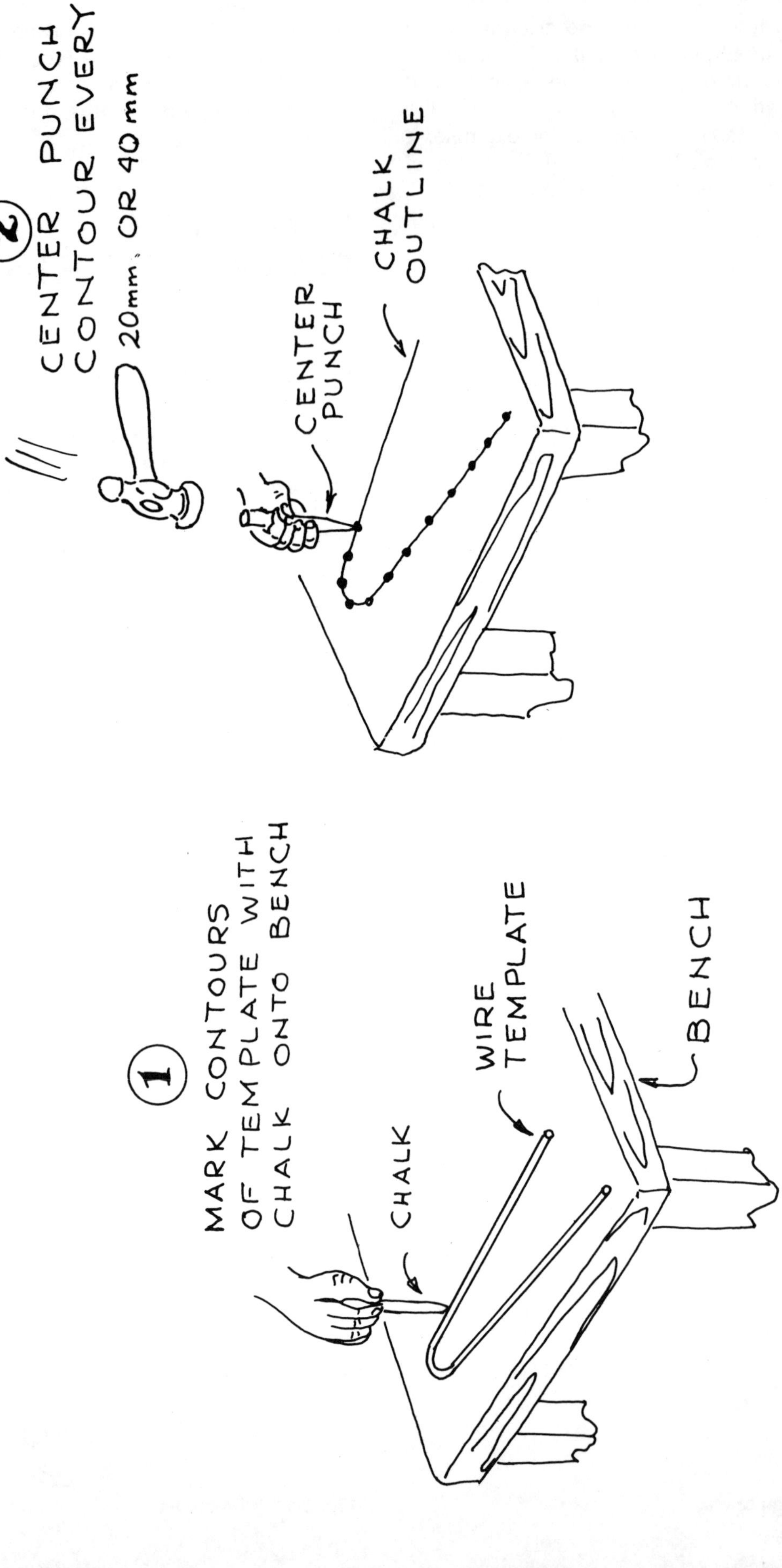
2
CENTER PUNCH CONTOUR EVERY
20mm, OR 40 mm
CENTER PUNCH
CHALK OUTLINE
1
MARK CONTOURS OF TEMPLATE WITH CHALK ONTO BENCH
CHALK
WIRE TEMPLATE
BENCH

Fig. C.18.

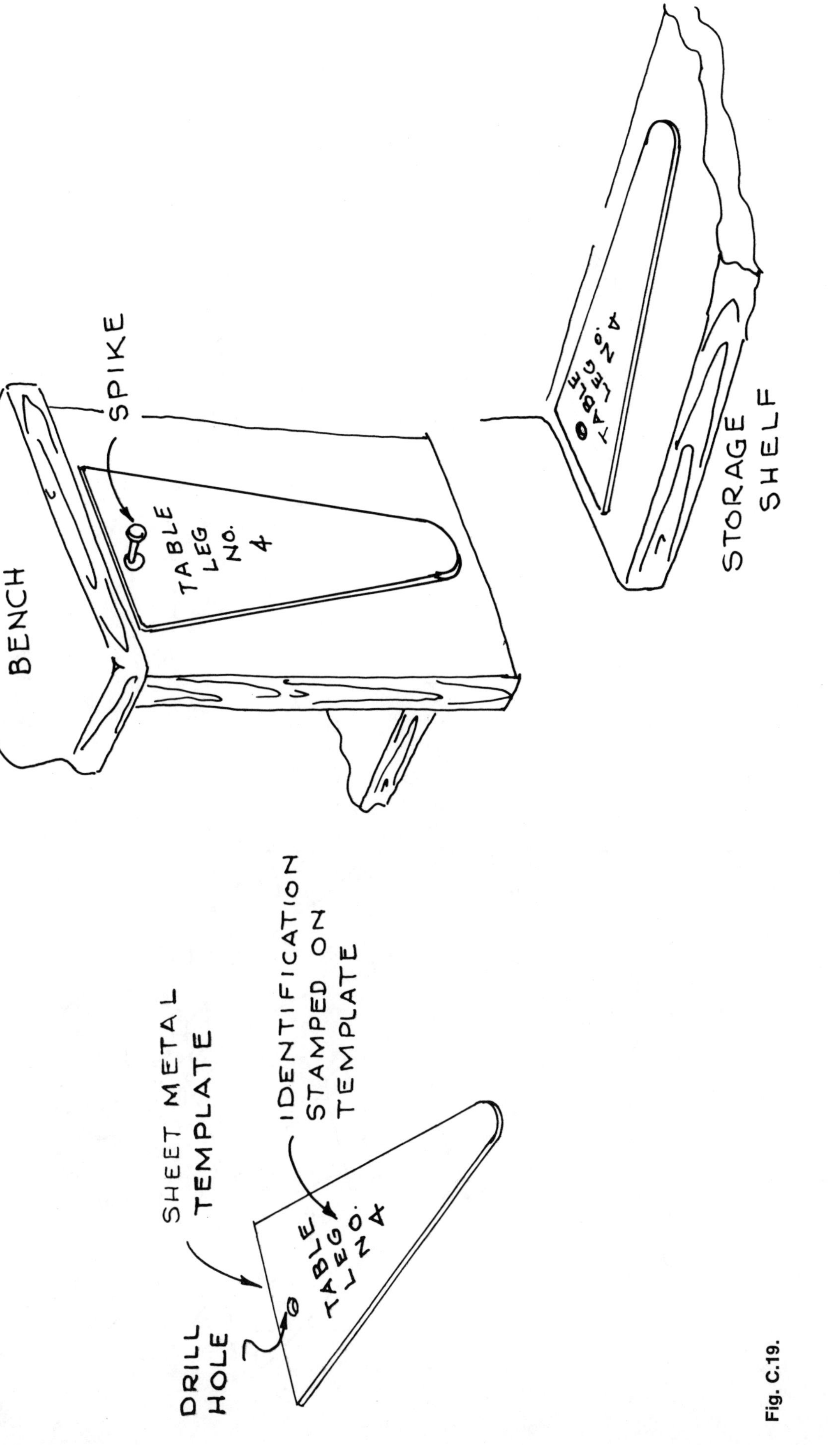
DRILL HOLE
SHEET METAL TEMPLATE
IDENTIFICATION STAMPED ON TEMPLATE
TABLE LEG NO. 4
BENCH
SPIKE
TABLE LEG NO. 4
STORAGE SHELF
TABLE LEG NO. 4

Fig. C.19.

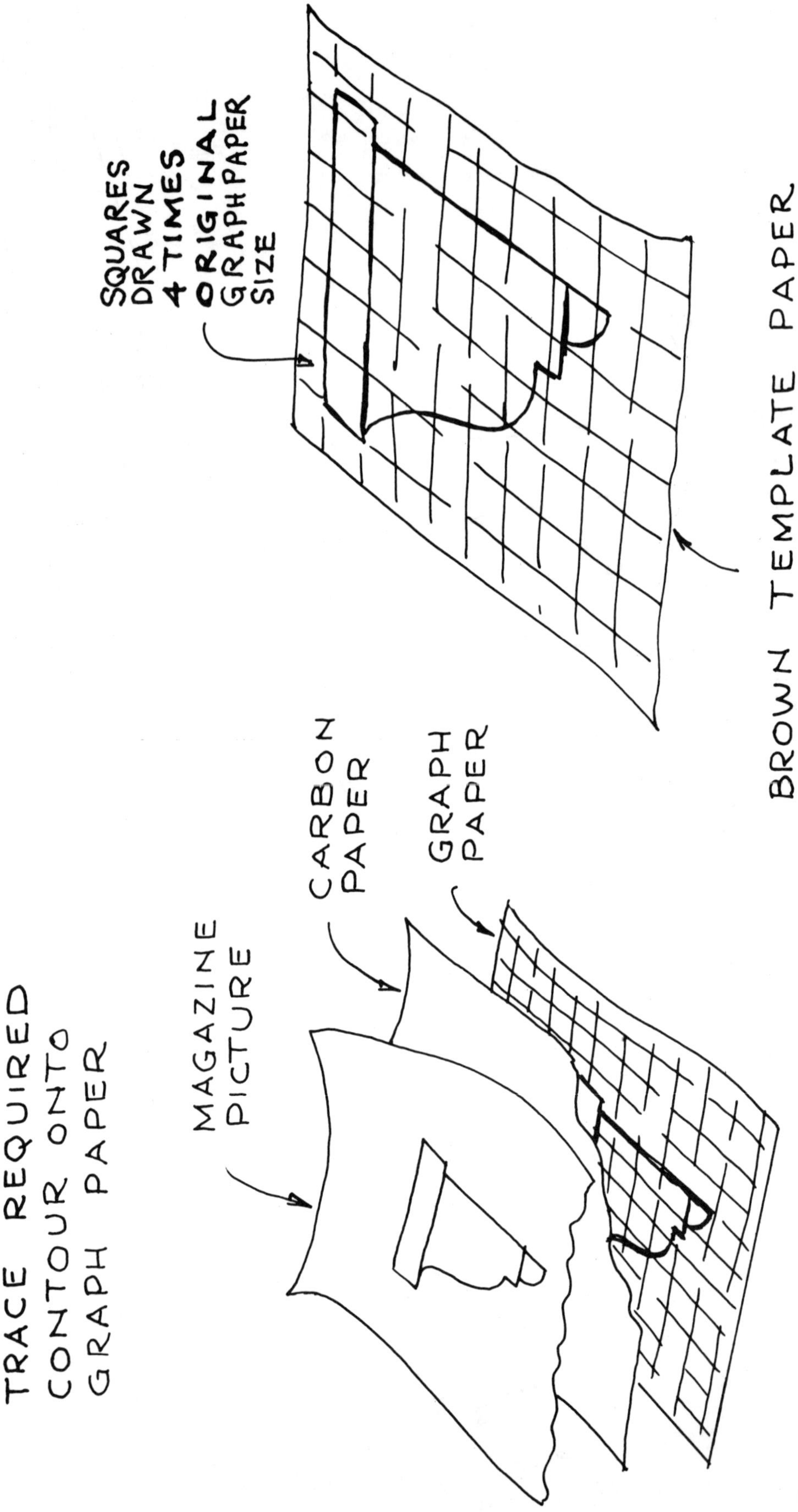
TRACE REQUIRED
CONTOUR ONTO
GRAPH PAPER
MAGAZINE
PICTURE
CARBON
PAPER
GRAPH
PAPER
SQUARES
DRAWN
4 TIMES
ORIGINAL
GRAPH PAPER
SIZE
BROWN TEMPLATE PAPER

Fig. C.20.

METHOD 3

If a large number of pieces are to be duplicated, a metal template could be used instead of the previous two methods. This template could be identified in some way and stored for future use. See Fig. C.19. (p. 19.)

Note. A large number of these sheet metal templates would require a very small amount of storage space.

METHOD 4

Sometimes projects appear in magazines and books, in a reduced size. A full size pattern can be made of the project in the following manner. Place a piece of graph paper under the magazine page and insert a piece of carbon paper between the two. Trace over the object on the page so that it is transferred onto the graph paper.

Next, take a large sheet of heavy brown paper (the kind that is usually found in sheet metal shops for layout, or in auto body shops for masking) and lay it out like the graph paper but with squares four or more times as large.

Start with a point on the small graph where the design crosses the lines of the squares and locate and mark it on the large pattern. Continue in this manner until all the points are marked. Connect all the dots or points, transferring the design to a full-scale working drawing or pattern. See Fig. C.20. (p. 20.)

Note. If you decide to make a large number of duplicate pieces after completing the initial project, you may want to make a copy of the pattern in metal.

TECHNICAL DRAWINGS: THE ALPHABET OF LINES

This short section is not intended as a course in technical drawing or blueprint reading. Depending on the student's background it is simply a refresher or introduction to blueprint fundamentals. Lines, dimensions and figures on blueprints are universal symbols, each with a specific meaning. An "alphabet of lines" is used by draftspersons; a few of the more common lines are listed below.

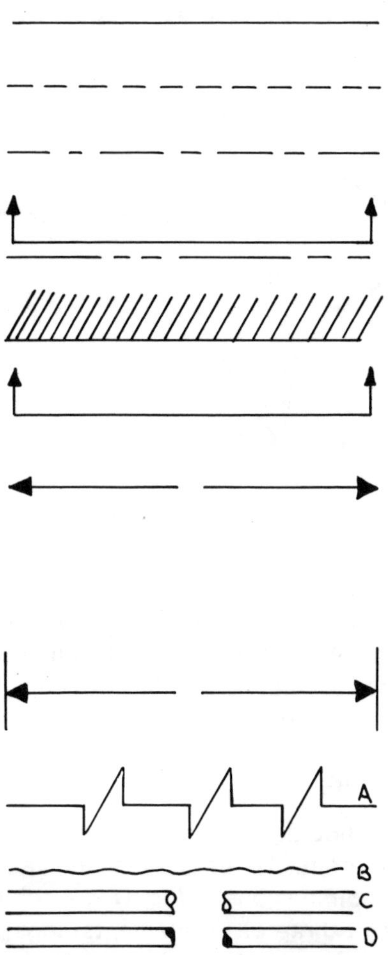

The object line is used to show the visible edges of an object.

The hidden line is used to show the edges that cannot be seen.

The centre line is used to show the centre of the object.

The cutting plane line is used to show where the object is cut to expose a section for viewing.

Section lines are used to show the area that has been exposed by the cutting plane line.

The viewing plane line is used to show the direction that the section area should be viewed from.

The dimension line is used to show a dimension or measurement. The ends of the dimension line are completed by arrow heads, and the dimension line itself has a break in the centre where the dimensions or measurements are shown.

The extension lines are used in conjunction with the dimension line and are used to show the boundaries or extremities of the dimension.

The break line is used to show an object that is uniform in length but too long to be drawn on a blueprint. (A) is used to indicate long breaks. (B) is used to indicate short breaks. (C) is used to indicate round bars that are solid. (D) is used to indicate round bars that are hollow, e.g., pipe.

DESIGN CONCEPTS

There is a current trend away from the repetitive mass–produced articles that symbolize our technological society. More and more people are appreciating individual crafting as shown by the rise of small stores dealing in hand-crafted items, do-it-yourself kits, and evening courses in assorted arts and crafts. Perhaps this is because people have more leisure time today, because many manufactured items are becoming so costly, or because individualized items, things made and used with self-expression in mind, are becoming more popular. People appreciate things designed individually and with common sense.

The word "design" has many definitions and explanations. Three main ideas prevail in all of them. 1. Plans are made mentally. 2. Plans are detailed physically as drawings, sketches, patterns, and so forth. 3. The actual parts are constructed to make a tasteful finished piece. With these three points in mind we in this book would prefer not to present the student with a set of standard projects, but to help develop his or her own ability to observe and to design individual projects. There is no limit to the scope, design, and possibilities of a project.

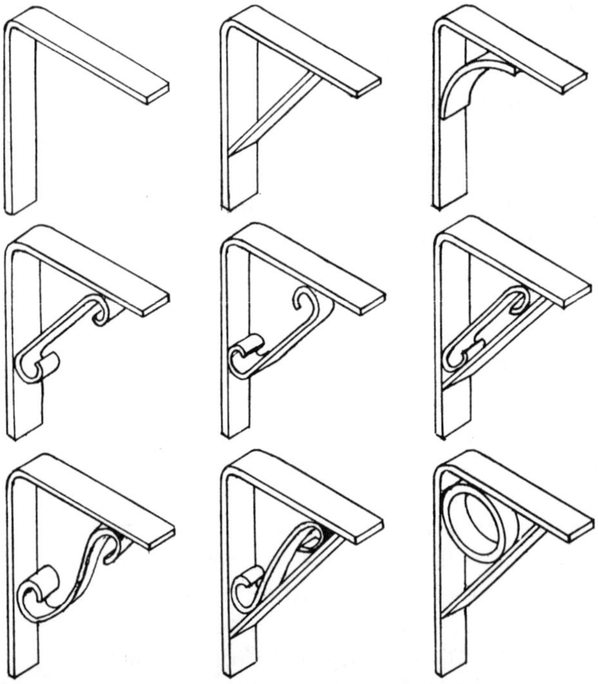

Fig. D.1. Shelf brackets

A shelf bracket as shown in Fig. D.1. can be more than just a piece of metal bent at right angles. By the addition, rearrangement or deletion of other pieces of metal, many designs are possible.

With this in mind, the student should consider the following ten points before starting any project. 1. Decide on what you want, on the overall function of your project. 2. Choose a project to match that function. If it is to be a table, as shown in Fig. D.2., will it be a dining room, coffee, occasional, step, or utility table? Stores, books, magazines, and newspapers are valuable sources of ideas. Make sure the project will serve the intended purpose. 3. Study alternative methods and metals that might be used. Select the best based on available equipment, fixtures that may have to be built, cost, and so forth. 4. Study the design and think of possible improvements that might be made. Would round legs look better than square legs, for example? Would a circle look better than a diamond design on your particular table? 5. Use sketches for changes. 6. Now consult the instructor with your plan. Be open to sensible suggestions. 7. Make a drawing and estimate the amount and cost of material

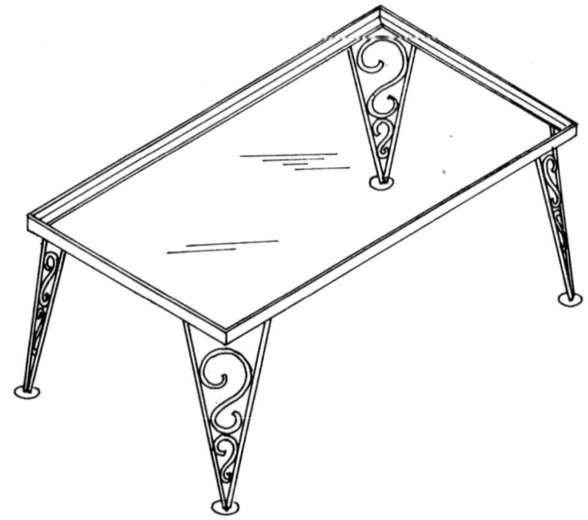

Fig. D.2. Sample table

needed. 8. Make templates and patterns, if these are necessary. 9. Decide on the method and type of finishing. 10. Proceed with the building of the project.

PLANNING

Projects should be carefully planned. To illustrate this point, let us look at the table of unspecified dimensions in Fig. D.2. A table usually consists of three basic elements: legs, top, and ornamentation and/or supports. Tables may be built to any dimension, but before students begin the project there are certain questions they must ask themselves. 1. Is the size and shape correct for the area the table is to be used in? 2. Is the height correct? For example, if it is a dining room table will it be comfortable to sit at? 3. What type and style of leg complements the overall design? 4. If I use angle iron for the top, should the frame be turned up or down? What are the advantages and disadvantages of doing it either way? 5. Will I use glass, wood, or mesh (expanded metal) for the top? 6. What type of ornamentation, if any, should I use? 7. What type of finish will I use? Will it be a natural, painted, or hammered finish? These and many other questions will be illustrated and explained throughout the text so that the student will have a clearer understanding of design concepts.

In summary, the basic steps for any project are as follows. They are to be done in order, one at a time. 1. A choice of project is made. 2. Design alternatives are carefully considered. 3. Procedure is planned in detail. A little time now may save a lot of time and waste later. 4. Layout involves whatever full size or scale patterns, jigs, and fixtures you will need. 5. Materials are selected. Mark, check, and double-check everything before you actually cut your materials. 6. Assembling, checking for square and accuracy where necessary, and then welding are next. 7. Finishing is the final step and includes the preparation by grinding, filing, and so forth, for the finished surface.

COST ESTIMATING

SECTION E

Ordering metal or estimating costs for project materials can be confusing. With the exception of a few shapes such as pipe and tubing, which are sold by the linear foot, steel suppliers presently charge for steel in cents per pound, or dollars per 100 lbs. The student, however, usually deals in the total length of material needed per project. For example, a student builds a project which involves the use of 3 mm of 1 in. x 1 in. x ⅛ in. H.R.S. angle iron, 6.10 mm of ⅛ in. x 2 in. H.R.S. flat bar, and 3 mm of ½ in. x ½ in. square bar. He or she is then confronted by the information that the angle iron costs $14.00 per 100 lbs., the flat bar costs $18.00 per 100 lbs., and the square bar costs $12.50 per 100 lbs.

The student now has to find out the mass per metre of material needed. The price per metre is found by converting from the price per foot. (See Section F — "The Changeover Period".) He or she could possibly determine the mass of the material and work out the price, but shops do not normally have scales of the type required. Similarly, the student ordering 100 lbs. of 1 in. x 1 in. x ⅛ in. angle iron will not know how many metres this will measure.

In the following section we have tried to make this a little easier by introducing some tables of the more commonly used metal shapes and their approximate masses and lengths. We have also included a typical bill of material that will help in estimating overall costs of a project. Since H.R.S. is the most commonly used metal for these types of projects, tables will refer to that material only, and although the prices may vary, the mass should remain the same.

Description in in.	Lbs. per ft.	Lbs. per 20 ft.	Ft. per 100 lbs.	20-ft. lengths per 100 lbs.
Angle iron.				
½ x ½ x ⅛	0.38	7.60	263	13
¾ x ¾ x ⅛	0.59	11.80	170	8½
1 x 1 x ⅛	0.80	16.00	125	6
1½ x 1½ x ⅛	1.23	24.60	80	4
2 x 2 x ⅛	1.65	33.00	61	3
3 x 2 x ½	4.10	82.00	25	1½
3 x 3 x ½	4.90	98.00	21	1
Round bar.				
⅛	0.04	0.84	2381	119
¼	0.17	3.34	588	29
5/16	0.26	5.22	385	18
⅜	0.38	7.52	263	13
½	0.67	13.36	150	7½
Square bar.				
⅛	0.05	1.06	2000	100
3/16	0.12	2.40	833	41½
¼	0.21	4.26	476	23½
5/16	0.33	6.64	303	15
⅜	0.48	9.56	209	10½
½	0.85	17.00	118	6
Flat bar.				
⅛ x ½	0.21	4.26	476	23½
⅛ x ¾	0.32	6.38	313	15½
⅛ x 1	0.43	8.50	234	11¾
⅛ x 1½	0.64	12.76	157	8
⅛ x 2	0.85	17.00	118	6
3/16 x ½	0.32	6.38	313	15½
3/16 x ¾	0.48	9.56	209	10½
3/16 x 1	0.64	12.76	157	8
¼ x 1	0.85	17.00	118	6
¼ x 1½	1.28	25.50	79	4
¼ x 2	1.70	34.00	59	3
I-beams.				
3	5.70	114.00	18	0
4	7.70	154.00	13	0
Channel beams.				
¾ x 5/16 x ⅛	0.50	10.00	200	10
1 x ½ x ⅛	0.84	16.80	119	60
1½ x ¾ x ⅛	1.17	23.40	86	4
2 x 1 x ⅛	1.78	35.60	59	3
Tee bars.				
¾ x ¾ x ⅛	0.62	12.40	162	8
1 x 1 x ⅛	0.85	17.00	118	6
1½ x 1½ x 3/16	1.90	38.00	53	2¾
1½ x 1½ x ¼	2.43	48.60	41	2
2 x 2 x ¼	3.62	72.40	29	1½
Zee bars.				
1¼ x 1¾ x 1¾	2.79	55.80	36	1¾

Note. As steel suppliers do not normally send bars of fractional length, the figures in the last column would be rounded to the nearest whole number.

Since tubing and pipe are usually sold by the linear foot there is really no problem in calculating the cost after conversion. However, there are many types of tubing and pipe; the most often used are listed below with the mass per foot included should mass calculation be necessary. It will be noted that pipe is much heavier than tubing of the same size and this should be taken into consideration when planning a project.

TUBING — PIPE

		Seamless mechanical					Welded mechanical							
		Gauge					Gauge					A.S.A. schedule		
Size		18	16	14	12	10	18	16	14	12	10	10	40	80
⅜ in.	R.				0.31	0.35	0.17	0.22	0.26			0.42	0.57	0.74
	S.						0.22	0.27						
½ in.	R.	0.24	0.30	0.37	0.46	0.52	0.24	0.30	0.37	0.46		0.67	0.85	1.09
	S.						0.30	0.38						
¾ in.	R.	0.37	0.48	0.59	0.75	0.88	0.37	0.48	0.59	0.75		0.86	1.13	1.47
	S.						0.47	0.61	0.95	1.03				
1 in.	R.	0.50	0.65	0.82	1.04	1.24	0.50	0.81	1.04			1.40	1.68	2.17
	S.						0.63	0.83	1.04	1.32	1.58			
1¼ in.	R.	0.63	0.82	1.03	1.33	1.60	0.63	0.82	1.03	1.33	1.60	1.81	2.27	3.00
	S.						0.80	1.05	1.32	1.69	2.03			
1½ in.	R.	0.76	1.00	1.26	1.62	1.96	0.76	1.00	1.26	1.62	1.96	2.09	2.72	3.63
	S.						0.97	1.27	1.80	2.06	2.49			
2 in.	R.	1.02	1.34	1.70	2.20	2.67	1.02	1.34	1.70	2.20	2.67	2.64	3.65	5.02
	S.						1.30	1.71	2.16	2.46	3.40			

Note. "R" denotes round tubing and pipe; "S" denotes square tubing and pipe. All masses are shown as pounds (or part) per linear foot.

Students should have some experience in estimating the cost of an item. It would give an accurate cost of an individual project before it was started; it would show students the importance of cost estimating should they someday wish to start a small business of their own; and it would encourage economy in the choice of method and material. A bill of materials (Fig. E.1.) may be of help.

WORK RECORD CARDS

Another way of estimating costs is to have all of the materials for a project preprinted on one side of a card or sheet, leaving a few blank spaces for extra or miscellaneous items. See Fig. E.2. (p. 26.) The reverse side would be used to sketch and show the dimensions of the project.

A well detailed card could act as each or all of the following: a sort of blueprint, to be consulted throughout fabrication; a ready check by the teacher or supervisor; and a reference

SAMPLE BILL OF MATERIAL

Item No.	Quantity	Size	Description	Unit cost	Item price
1.	3 m	1 in. x 1 in. x ⅛ in.	Angle iron H.R.S.	$14.00/100 lbs.	$1.12
2.	6.1 m	2 in. x ⅛ in.	Flat bar H.R.S.	$18.00/100 lbs.	$1.53
3.	3 m	½ in. x ½ in.	Square bar H.R.S.	$12.50/100 lbs.	$1.06
Total cost					$3.71

Fig. E.1.

PROJECT WORK RECORD

WELDING DEPARTMENT

Date Started.................. **Student's Name**.................................... **Class**.................. **Date Completed**..................

Material List	Feet	Price	Paid	Material List	Feet	Price	Paid
1/2" Round				Furnace Chain			
3/8" Round							
5/16" Round							
1/4" Round							
3/16" Round							
1/2" Angle							
3/4" Angle							
1 1/2" Angle							
1" x 1/8" Flat Bar							
3/4" x 1/8" Flat Bar							
5/8" x 1/8" Flat Bar							
1/2" x 1/8" Flat Bar							
1/2" x 1/2" Square Bar							
Total				**Total**			

Fig. E.2. Work record card

source to be filed away with other cards for future students wishing to build a similar project.

To further help understand cost estimation and to provide a more accurate picture of it, it is suggested that a further card or record be made out by the student to cover the hidden costs in building projects. For example, the cost of the following items should be estimated: amount of oxygen gas used; amount of acetylene gas used; welding rods (and flux if needed) used; grinding discs and cleaning materials used; paint; labour (total periods on job including preparation time and any drawing or sketching time); and overhead (depreciation of equipment, hydro, and so forth.) Profit (reasonable return over and above actual cost) should also be considered.

Gas consumption, in most cases, can be recorded directly from the cylinder content gauge.

Welding rods are currently sold by the pound; it will be necessary to establish how many rods are in one pound.

Cost of grinding discs, cleaning materials, and paint should not be too difficult to estimate, prices can be taken from requisitions or suppliers' catalogues.

Labour in this case can only be estimated since it is still a school project. It is suggested that you find out the average wage paid to apprentices and journey welders in the area.

Overhead will be difficult to determine but some attempt should be made to do so. Shop records of all expenses incurred over a period of time, say one year, should be kept. This should then be broken down into cost per school day or better still cost per class period, and this amount added for each period worked on the project.

Profit can be calculated by visiting local stores to see the price of similar manufactured items, and subtracting the total cost of your project from that price. Accuracy in cost calculations will come with experience.

Many schools are getting involved in small repair jobs for the local neighbourhood or community that are not profitable for a welding repair shop. Schools might do this for free as public relations or establish a small minimum charge for the service performed.

Note. The materials listed in the example above are the materials mentioned at the beginning of the section, and are calculated in the following manner.

ANGLE IRON

The price mentioned was \$14.00 per 100 lbs. = 14¢ per lb. From the table the mass is 0.80 lbs. per foot, therefore 3 m $\simeq$ 10 ft. = 8 lbs. The cost then is 14¢ × 81 lbs. = \$1.12. Similarly, *the flat bar* cost is estimated at 18¢ × 8.5 lbs. = \$1.53. *The square bar* cost is estimated at 12.5¢ × 8.5 lbs. = \$1.06.

SI (METRIC) CONVERSION

SECTION F

Over the next few years North America will use, more and more, Système International (SI) units of measurement instead of units from the "inch-pound" system. Students using this book will be using SI in the classroom. SI conversion does not happen overnight. During the transition period, while both measurement systems are used, certain problems arise. The following notes explain these problems so that you can cope with them; they also give some general hints about how to begin using SI units, especially as they apply to the welding topics covered in this book.

SI UNITS TO BE USED IN METALWORKING

Before beginning to use SI in the workshop it would be wise to be familiar with the more common units and prefixes such as kilo-, centi-, and milli-. In basic metalworking we are almost exclusively concerned with only a few quantities: length, mass, temperature, and sometimes area and volume.

For these quantities the SI units most likely to be used in the classroom are as follows: for length, the metre (m) and the millimetre (mm); for mass, the kilogram (kg) and the gram (g); for area, the square metre (m^2) and the square millimetre (mm^2); and for volume, the cubic metre (m^3) and litre. You will soon learn to estimate lengths in millimetres and will find that it makes sense to try to work to the nearest 5 mm because your tape or rule is graduated in groups of 5 mm.

RECOGNITION POINTS

The average height of a chair seat is 460 mm. The average table height is 740 mm. The average height of a door is 2 m. The diameter of a 5¢ piece is 20 mm.

Identify the dimensions in SI units of other common objects around you. Workshop tools are especially good. Try to think of recurring dimensions, that is, measurements that you see again and again, and how they compare with the common objects you have identified. It will seldom be necessary for you to convert from "inch-pound" measurements as you will quickly and easily learn to "think SI." Centimetres are quite unnecessary in the workshop and would cause serious mistakes if confused with millimetres. Therefore always use millimetres for all measurements. For dimensions of 1000 mm and above use metres expressed to 3 places of decimals, e.g. 1.655 m. In this way 1 mm accuracy is maintained, as no rounding of numbers is necessary.

THE CHANGEOVER PERIOD

At the present time almost all the materials in the school workshop are made to "inch-pound" standards and therefore come in "inch-pound" units — ½ in. rod, ⅛ in. x 1 in. strip, and so forth. The steel industries are now working on new metric standards for certain products — standards which may be available from the Canadian Standards Association by the time this text is published. They are CSA 312.1 — preferred dimensions of flat metal products — and CSA 312.2 — preferred dimensions for round, square, rectangular, and hexagonal metal products. These standards do not cover angled stock.

The dimensions of the products covered by these standards will be, for the most part, in whole millimetres. Some of these SI products will have dimensions very close to existing ones. For instance, the replacement for ⅛ in. x 1 in. strip may be 3 mm x 25 mm — close to the original, but not exactly the same. It is impossible to forecast when these SI products will be readily available and it is almost certain that after they appear "inch-pound" products will still be available for some time. While the dimensional differences between SI and "inch-pound" products are unlikely to be particularly significant, in most school metalworking projects it is still very important not to confuse the two types of product. The only way to accurately differentiate between them and avoid confusion is to go on calling the "inch-pound" products by the "inch-pound" measurement unit names.

In the changeover period, call products built to "inch-pound" standards by "inch-pound" units and products built to SI standards by SI units. Never call a product built to one system's standard by the units of another system. The dimensions in this text reflect this rule.

MEASURING LENGTH

Throughout this text non-critical dimensions (generally length) are given in SI units. The seat of a chair does not have to be exactly 18 in. (457.2 mm) from the ground. It is therefore possible to work in metric units for most length dimensions, using stock dimensions in customary units, rounding off lengths sensibly. But, if you change an existing inch drawing to millimetres change *all* the relevant dimensions. It is a big mistake to mix millimetres and inches on the same job. Work either in inches or in millimetres — do not mix them. Designing a job from scratch in metric measurements presents no problem. How should you convert existing designs? Here are some rough equivalents.

36 in. is about 900 mm.	¼ in. is about 6 mm.
12 in. is about 300 mm.	⅛ in. is about 3 mm.
1 in. is about 25 mm.	¹⁄₁₆ in. is about 1.5 mm.
½ in. is about 12 mm.	¹⁄₃₂ in. is about 1 mm.

However, there is an error in all these values which can create problems. For instance, a design may specify a component 10 ft. long. To call this 3 m introduces an error of 48 mm (almost 2 in.) If this is within the tolerance for the job, well and good. If not, a rounding to 3.05 m might be more acceptable. Similarly if a design calls for 20 pieces of 6 in. bar and you decide to make them 155 mm or even 160 mm instead of 150 mm, remember the material list will have to be increased from its original 10 ft. (3.05 m) to accommodate the extra material.

For greater accuracy apply the exact factor 1 in. = 25.4 mm and round off to two decimal places or less, as appropriate (see chart below.)

Examples: to change 4 in. to millimetres, $4 \times 25.4 = 101.6$ mm
to change 1¼ in. to millimetres, $1.25 \times 25.4 = 31.75 \simeq 31.8$ mm or 32 mm
to change 12 in. to metres, 12 in. $\times 25.4 = 304.8$ mm $= 0.3048$ m $\simeq 0.30$ m

TOOLS

Metric rules and tapes are readily available now. Metric drills are available but are not likely to be necessary until metric rods and fasteners come on the market. This will probably not occur for some time yet.

FRACTIONS OF AN INCH WITH APPROXIMATE DECIMAL AND MILLIMETRE EQUIVALENTS

Inch fractions			Approximate Equivalents	
1/16	1/32	1/64	Decimal	Metric mm
		1	0.02	0.40
	1		0.03	0.79
1			0.05	1.19
		3	0.06	1.59
		5	0.08	2.00
	3		0.09	2.38
		7	0.11	2.78
2			0.13	3.18
		9	0.14	3.57
	5		0.15	3.97
		11	0.17	4.36
3			0.19	4.76
		13	0.20	5.16
	7		0.22	5.56
		15	0.23	5.95
4			0.25	6.35
		17	0.27	6.75
	9		0.28	7.14
		19	0.30	7.54
5			0.31	7.94
		21	0.33	8.33
	11		0.34	8.73
		23	0.36	9.13
6			0.38	9.53
		25	0.39	9.92
	13		0.41	10.32
		27	0.42	10.71
7			0.44	11.11
		29	0.45	11.51
	15		0.47	11.91
		31	0.48	12.30
8			0.50	12.70

Inch fractions			Approximate Equivalents	
1/16	1/32	1/64	Decimal	Metric mm
		33	0.52	13.10
	17		0.53	13.49
		35	0.55	13.89
9			0.56	14.29
		37	0.58	14.68
	19		0.59	15.08
		39	0.61	15.48
10			0.64	15.88
		41	0.65	16.27
	21		0.66	16.67
		43	0.67	17.06
11			0.69	17.46
		45	0.70	17.86
	23		0.72	18.25
		47	0.73	18.65
12			0.75	19.05
		49	0.77	19.45
	25		0.78	19.84
		51	0.80	20.24
13			0.81	20.64
		53	0.83	21.03
	27		0.84	21.43
		55	0.86	21.83
14			0.88	22.23
		57	0.89	22.62
	29		0.91	23.02
		59	0.92	23.41
15			0.94	23.81
		61	0.95	24.21
	31		0.97	24.61
		63	0.98	25.00
16	32	64	1.00	25.40

MASS

Students may have to calculate mass during project building, especially if they are required to order metals. Steel manufacturers presently sell metal by its mass measured in pounds or tons. In SI, the units of mass of particular interest to the student will be the gram, kilogram, and the metric tonne.

The relationship between these units is as follows:

Units	g	kg	t
g	1	1000	1 000 000
kg		1	1000
t			1

Students may well talk of the heaviness or lightness of objects in such units. It is not yet certain however whether steel manufacturers will still base their prices on mass under the new SI standards being devised. Students should therefore be careful to stick to the pound or ton when talking of ordering steel supplies until such time as new standards become available and are put into common practice.

MASS VS. WEIGHT

The word "weight" has been often confused with "mass." Mass refers to the molecular makeup of an object whereas weight refers only to the force of gravity acting on some mass. A steel bar in outer space would have mass but no weight. On earth or on the moon it would have both mass and weight, but on the moon the bar would only weigh 0.17 of what it would on earth since the moon's gravitational force is only 0.17 of that of the earth's. On earth it is often difficult for us to quickly distinguish between mass and weight. The best word to use is mass, since it is almost always more accurate and less confusing. Never use the word weight except when you are certain its use will refer only to the force of gravity. Example: the crane picked up the heavy mass and lifted it straight into the air before the cable broke, causing the mass to fall to the ground because of its weight.

TEMPERATURES

Since the melting points of metals are important in project building, it would seem appropriate at this point to include at least a table of the more common metals and their approximate melting points.

MELTING POINTS OF METALS

Metal	Approximate melting point, °C
Tin.	230
Lead.	340
Zinc.	420
Aluminum.	620-650
Bronze.	880-915
Brass.	925-980
Silver.	960
Copper.	1080
Cast iron.	1230
Monel metal.	1340
High carbon steel.	1230
Medium carbon steel.	1425
Stainless steel.	1425
Low carbon steel.	1510
Tungsten.	3410

There are various crayons, pellets, and liquids on the market which, when applied to a piece of metal, will indicate the temperature to which the metal has been heated. There are also thermometers that can be placed directly on the hot metal and will indicate the exact temperature of the metal. However, it is faster and more economical if the student learns this simple method of recognising approximate temperatures by judging the colours associated with the forging and joining of steel. The table opposite should be helpful for this purpose.

Note.

1. Welding usually begins at the lower end of the range.
2. Forging usually begins at the high end of the range.
3. The opposite colours are as seen with the naked eye, *but, coloured safety lens should always be worn when welding or forging* and some adjustment should be made in distinguishing these colours.

Conversions to Fahrenheit will not normally be needed but if an unusual occasion arises, use the formula $C = 5/9\,(F - 32)$, where C and F are the temperatures in Celsius and Fahrenheit, respectively.

Example: $420° = 5/9\,(F - 32)$

$9/5 \times 420°/1 = 9/5 \times 5/9\,(F - 32)$

$756 = F - 32$

$756 + 32 = F$

$788° = F$

HEAT COLOUR RANGE

Colour Range.	Temp. °C	Remarks.
White.	1200-1540	Welding range.
Yellow/orange.	980-1200	Between range.
Red.	370- 980	Forging range.
Blue/black.	0- 370	Preheat range (welding).

ASSEMBLING ANGLE IRON

SECTION G

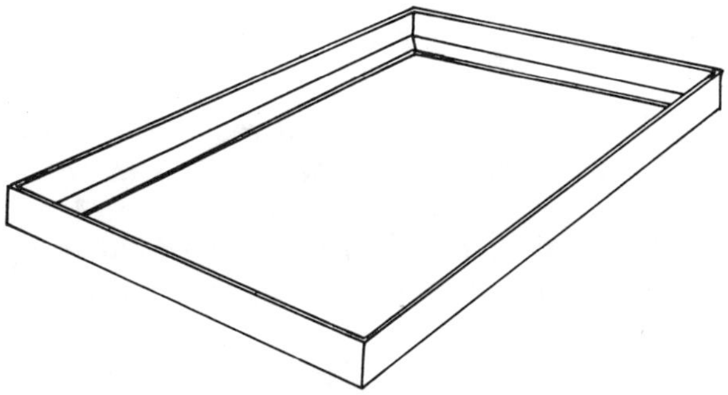

Fig. G.1. Angle iron frame

A large number of projects require the forming of angle iron frames. These include table tops, seats for chairs, and, on a larger scale, boat and utility trailers. Regardless of the size of the project, the following methods are used to form the corners. Study them carefully. There are slight variations in each one which could affect the overall design of the project.

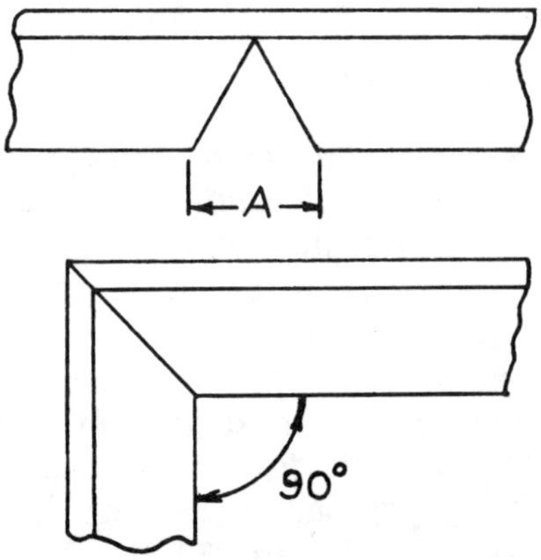

Fig. G.2. Method #1

Method #1, shown in Fig. G.2., will produce a rounded corner and will require the least welding. The dimension "A" will be equal to the total width of the angle iron from toe to heel.

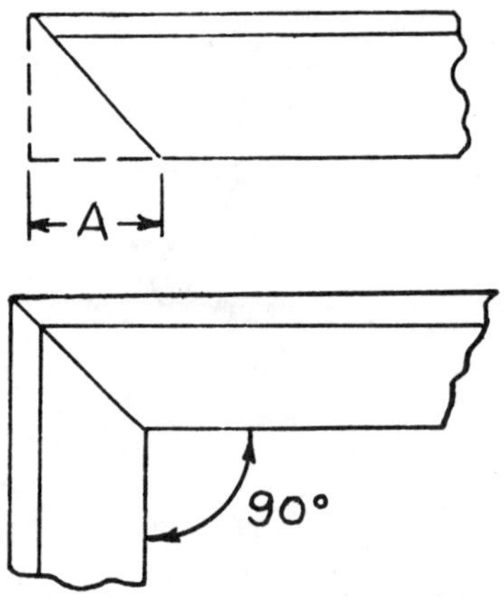

Fig. G.3. Method #2

Method #2, shown in Fig. G.3., entails more fitting and welding, but will produce a sharper corner than method #1. The angle "A" should be 45° and the distance "A" is determined by measuring the total width of the angle iron from toe to heel.

Method #3, shown in Fig. G.4. (p. 32.), requires the same amount of fitting as method #2 and more welding. However, it will produce a neater and better fit. The dimensions for the section to be cut out (section A) are found by measuring the total width of the angle iron from toe to heel.

Methods #1, #2, and #3 show the forming

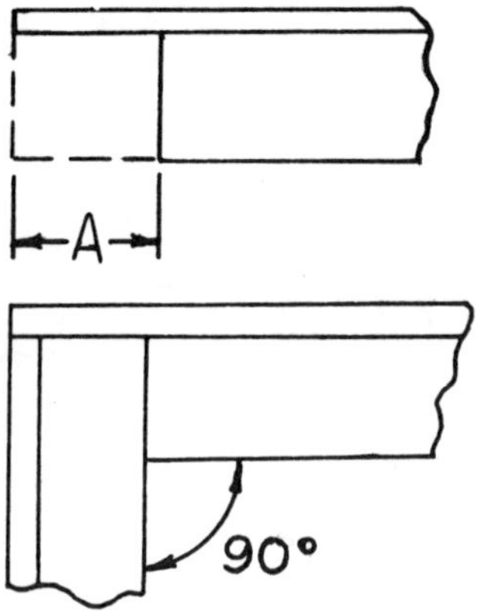

Fig. G.4. Method #3

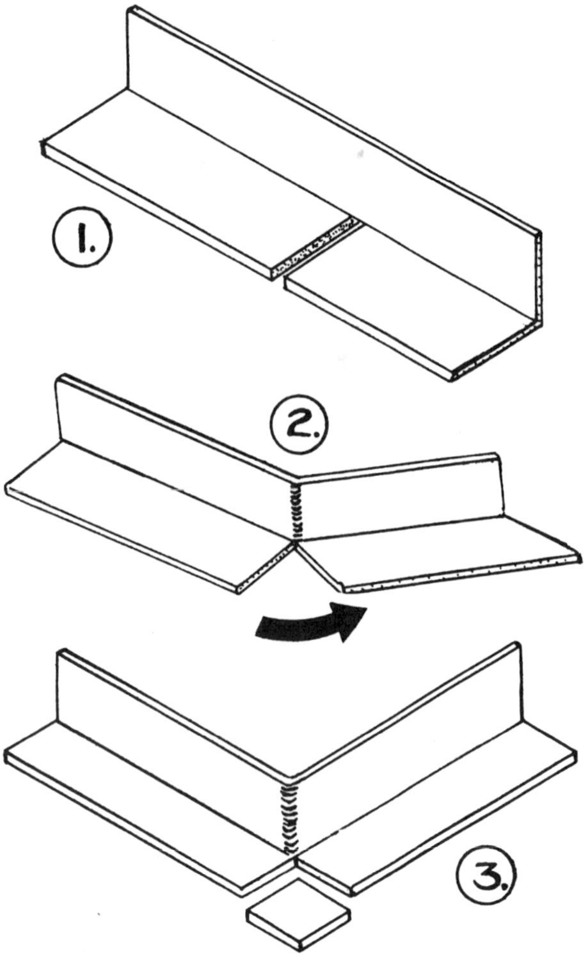

Fig. G.5. Method #4

of what are commonly called inside corners. Sometimes it is desirable to form an outside corner for various reasons, and this method #4 is shown in Fig. G.5.

In **Method #4,** a cut is made at the point where the corner is to be formed. The angle is bent out in the opposite direction from the previous methods shown. After bending to the desired degree of angle, a piece of metal of the same thickness as the angle iron is cut to shape and inserted in the opening. It is welded in place to close the corner and add strength to the assembly.

WELDING TIPS

When welding angle iron corners, the most successful method of maintaining the correct degree of angle is to weld from the heel toward the toe as shown in Fig. G.6.

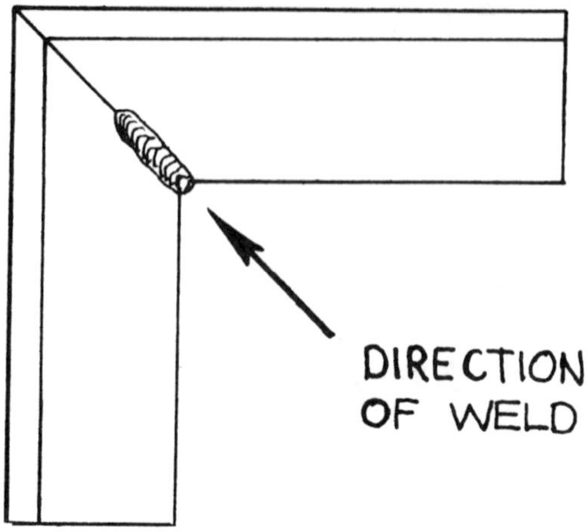

Fig. G.6.

If possible, some type of restraint, such as "C" clamps, should be used to hold the angle iron in place during welding and to help prevent distortion. The joint should be welded and left to cool before removing the clamps. See Fig. G.7.

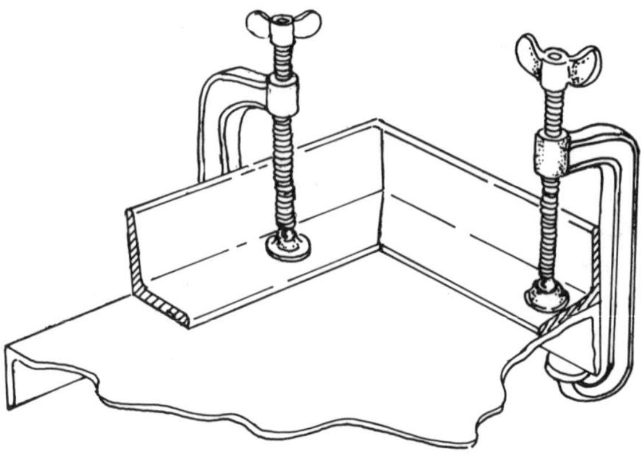

Fig. G.7. "C" clamps

If the metal expands to more than the desired degree of angle during welding, the only solution is to cut the weld through, reset the degree of angle, and reweld, as shown in Fig. G.8. (p. 33.)

If, however, the metal contracts, and forms less than the desired degree of angle, it is sometimes possible to hammer the corner on an anvil and spread it back to the desired degree, as in Fig. G.9. (p. 33.)

Welds inside angle iron corners can cause

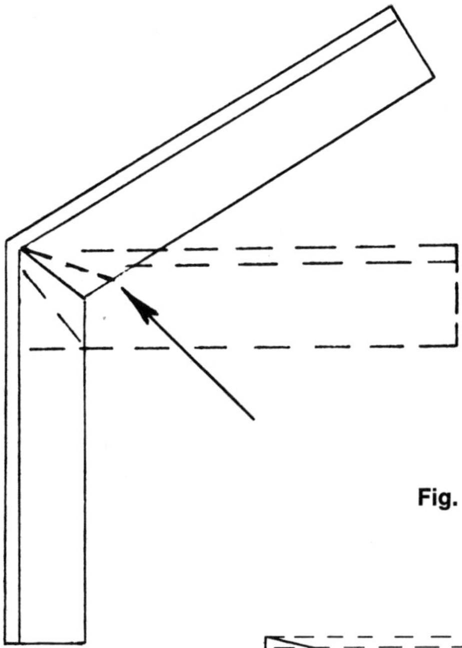

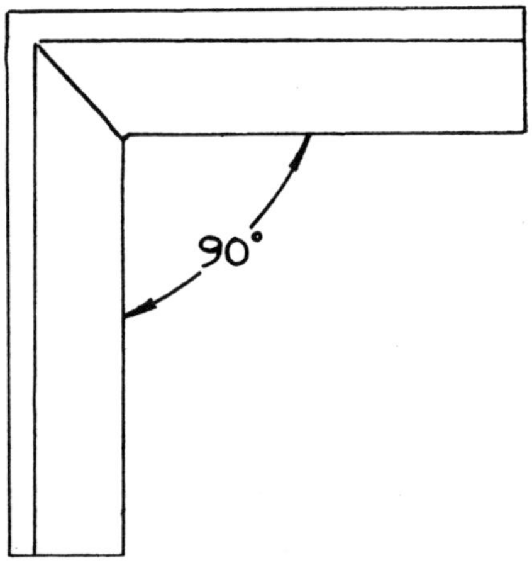

Fig. G.8. If metal expands

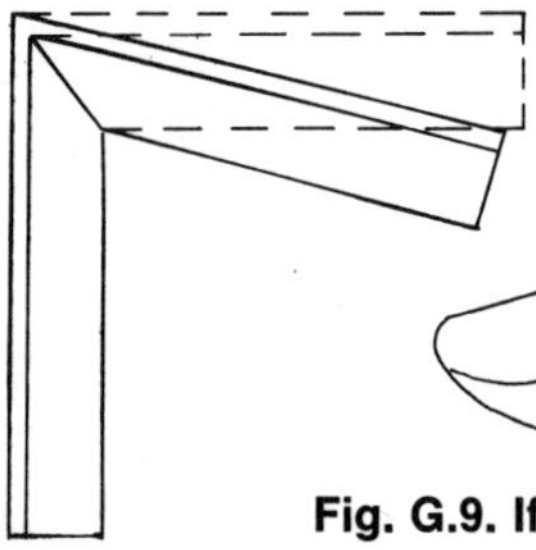

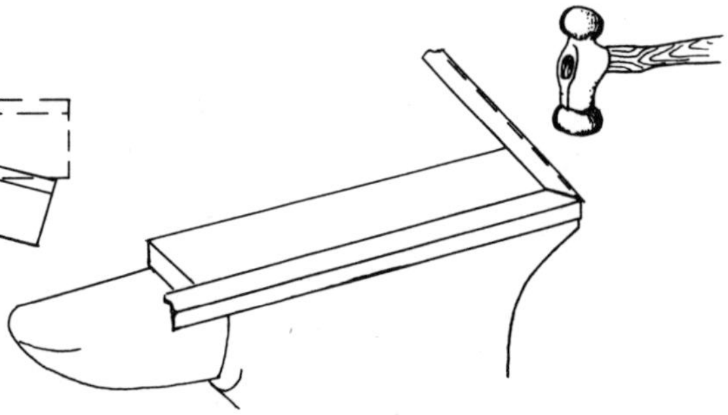

Fig. G.9. If metal contracts

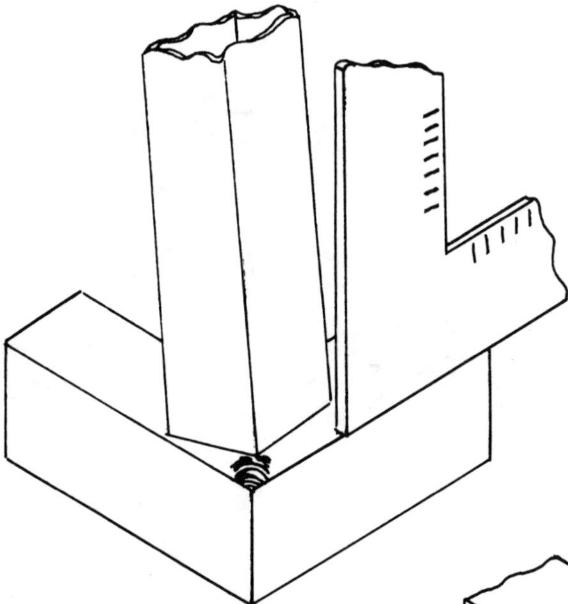

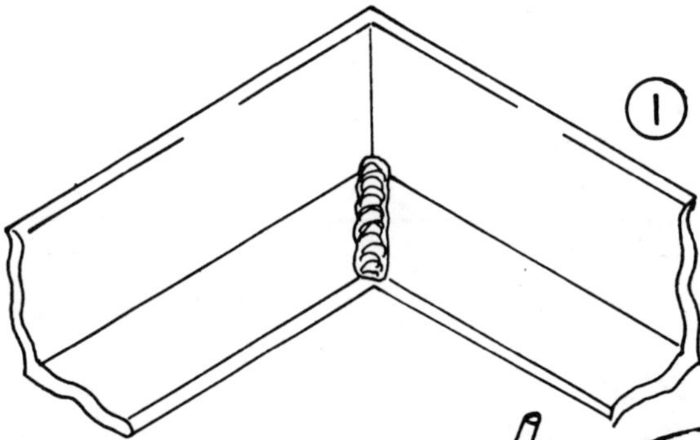

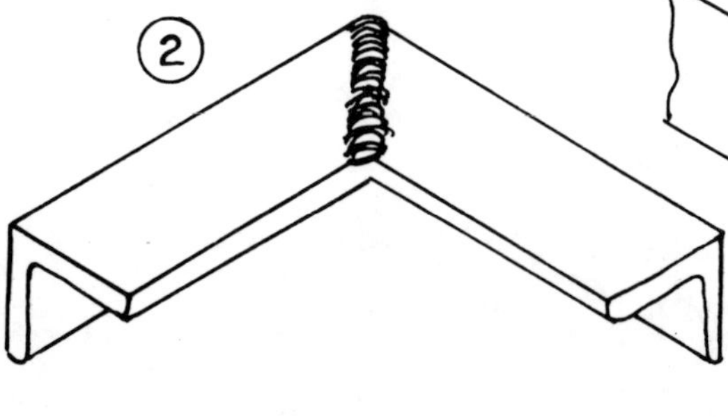

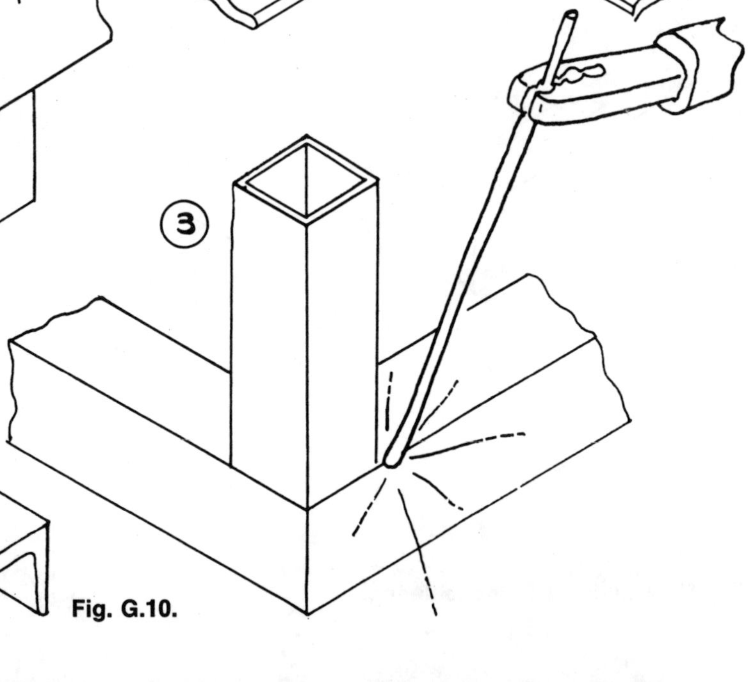

Fig. G.10.

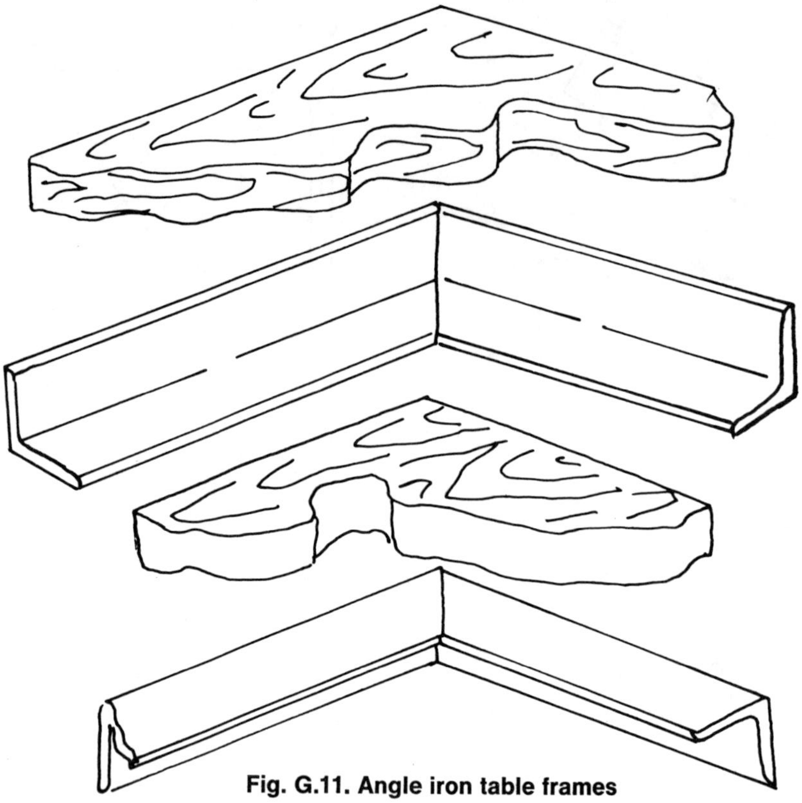

Fig. G.11. Angle iron table frames

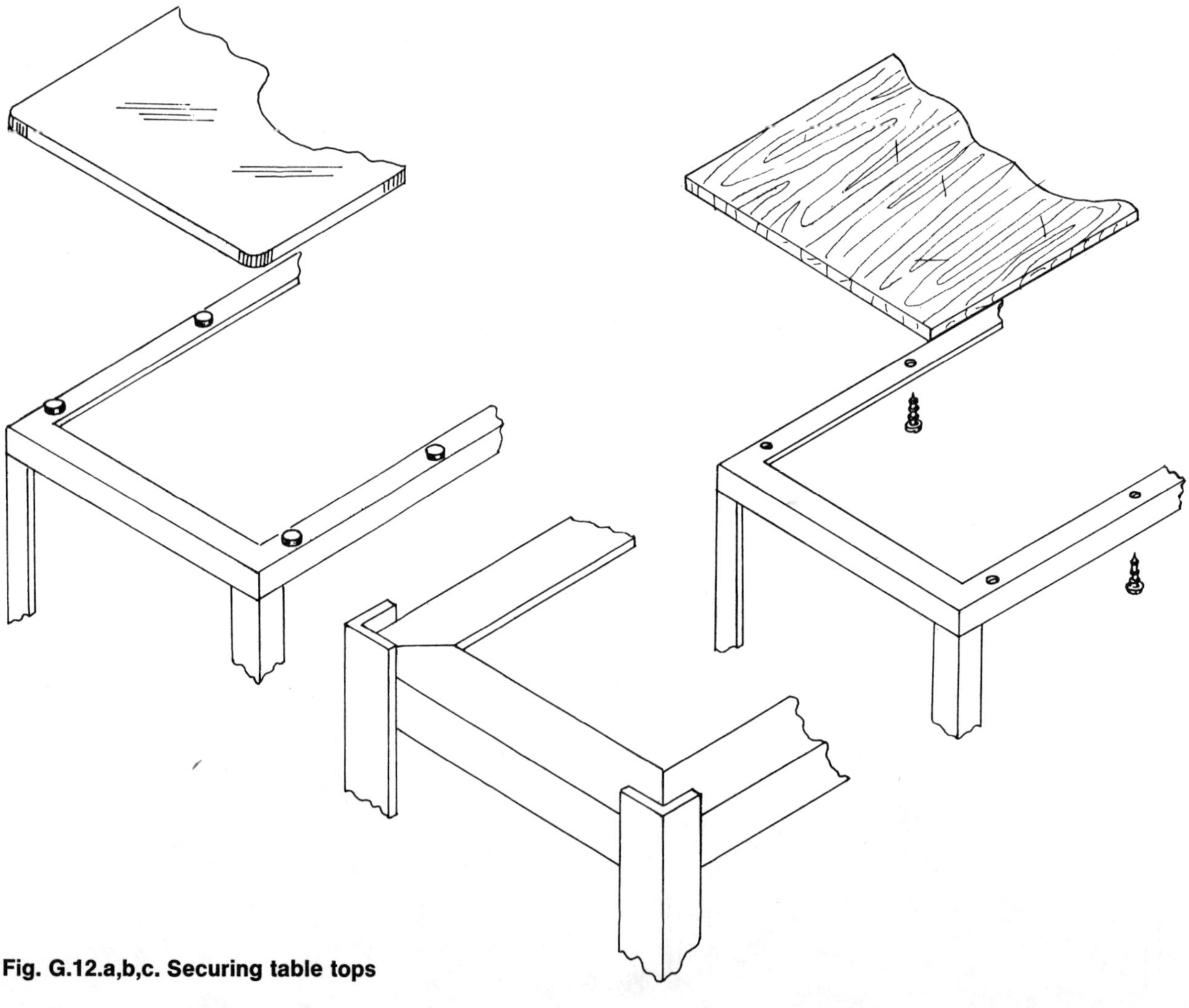

Fig. G.12.a,b,c. Securing table tops

problems, especially if a frame is being formed for angle iron tables. If the weld is too high, the top will not fit properly, and it is almost impossible to remove the excess weld metal. Since in the building of tables strength is not a major factor, these corners should be welded on the outside only. If the leg is to be welded directly on to the outside at the corner, then the outside corner weld will have to be ground down to make it level. In this case the additional welding of the leg at the corner will reinforce that area. The usual procedure is to tack weld the angle iron corner, apply the leg, and then weld the complete assembly all at once. See Fig. G.10. (p. 33.)

Before assembling angle iron frames for coffee, dining room, and other tables, thought must be given to the positioning of the top. A frame formed with the angle iron facing up will provide a natural ledge for the top to fit into. However, unless the top is made to completely cover the frame, the edge will be a trap for crumbs, dirt, and dust.

If the frame is formed with the angle iron facing down, this problem will be eliminated. Some method of securing the top must be found, however, or the top could easily slip off of the frame. In this case the edges of the top will require finishing since they will now be exposed. To secure the table top, a number of methods can be used. A variety of adhesives are available or screw nails could be used from underneath. Be sure the screws do not penetrate the surface. Also, angle iron legs may be allowed to project at the corners, thus securing the top. See Fig. G.12. A, B, C. (p. 34.) Remember, you must consider all these methods and decide which way you will have the table top secured *before* you start the project.

BENDING METAL

SECTION H

"To bend" is defined as *"to direct or turn in a certain direction"*. Most projects require that the metal components be bent and formed into various shapes before they are joined. There are a number of machines on the market that will perform these operations by the simple assembly of dies, pins, and blocks. These machines will bend ferrous and non-ferrous metals, ranging from small rod (round or square) through pipe and tubing to flat stock, angle iron, tees, and so forth.

Fig. H.1.a.

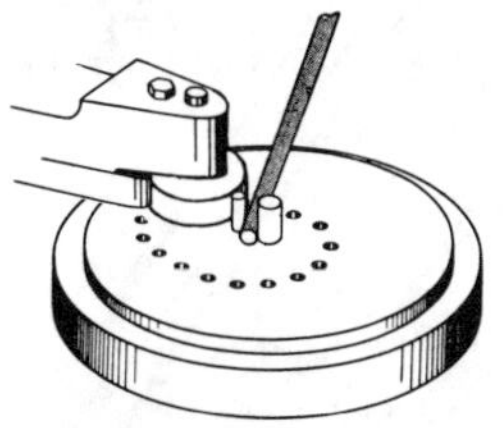

1. To reduce work marking and ease bending effort install a Forming Roller. Standard Forming Nose can also be used for this operation.

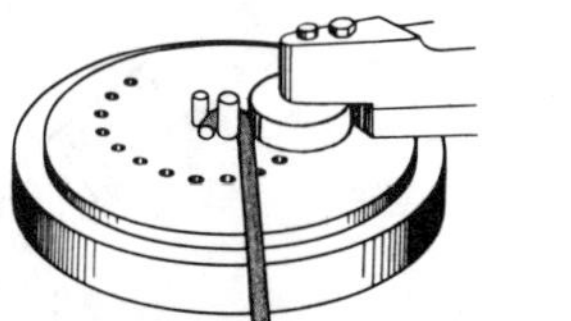

2. Insert material between Locking Pin and Radius Pin and set Forming Roller against material. Lock material against Radius Pin as illustrated in figure 1. Advance Operating Arm until it strikes Angle Gauge.

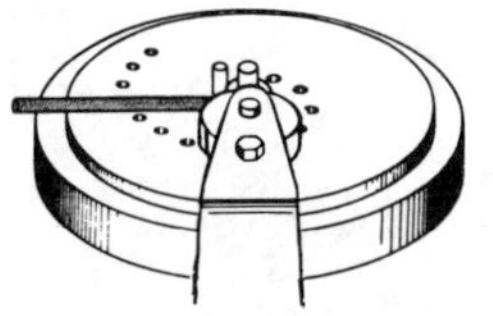

3. An off-center eye has now been completed in one operation and additional parts can be rapidly and exactly duplicated by following the same procedure. This same shape can be readily formed in all types of solid materials.

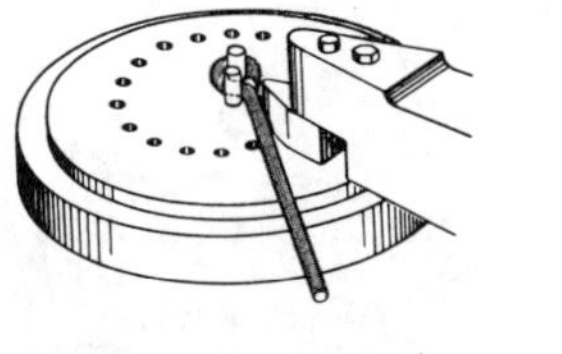

4. Centering Eye in Second Operation

When the material is too heavy to be formed into a centered eye in one operation, an off-center eye should first be produced as shown above in figures 1 to 3. The eye should then be placed over the Locking Pin and the centering bend made around a Radius Pin of the desired size as illustrated.

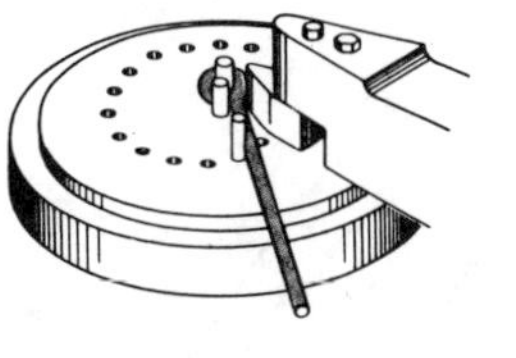

5. Centering Eye Using Two Locking pins

It is sometimes possible to form a centered eye in heavy materials in one operation by following the procedure in figures 1 to 3. The centering bend is obtained by forcing the material against a pin remotely located in a Mounting Plate hole as shown. If necessary, this pin can be mounted from the Bender Base.

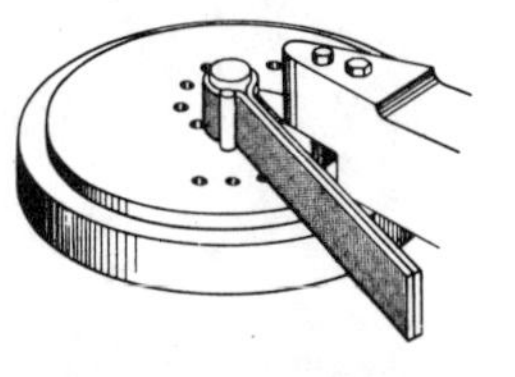

6. Forming Pipe Strap in One Operation

A pipe strap containing three individual bends can be formed in one operating cycle by following the procedure outlined on page 6. The only difference is that the material must be gauged so that both ends will extend beyond the Locking Pin and thereby be formed during the centering operation.

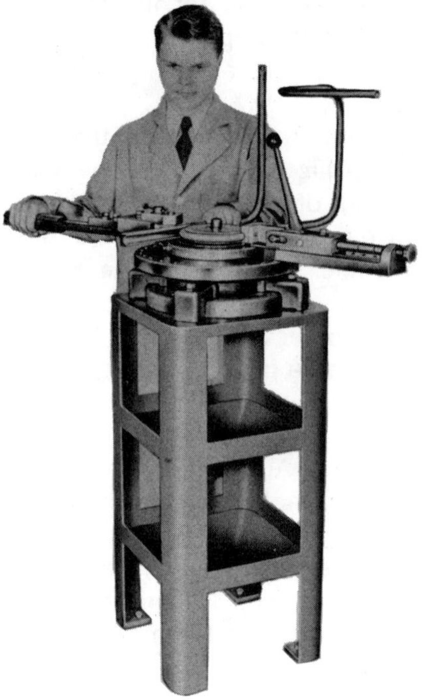

Fig. H.1.b.

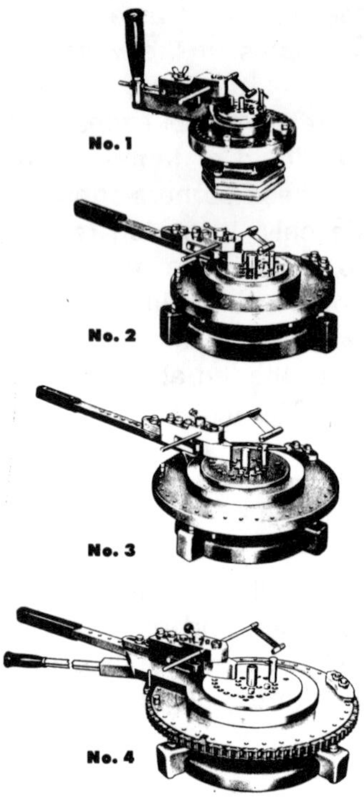

Fig. H.1.c. Di-acro benders

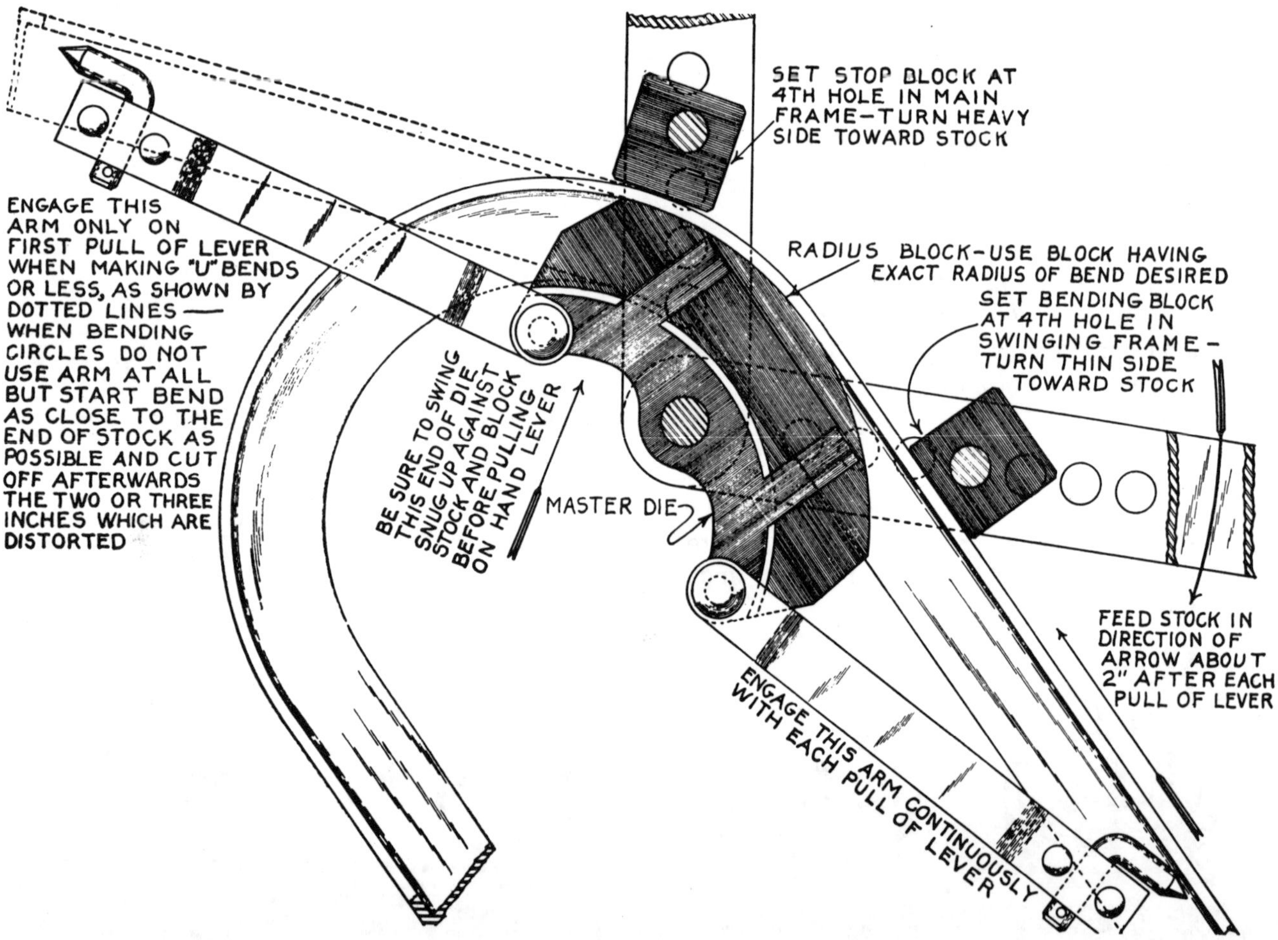

Fig. H.2.a. Bending angle iron flange inward

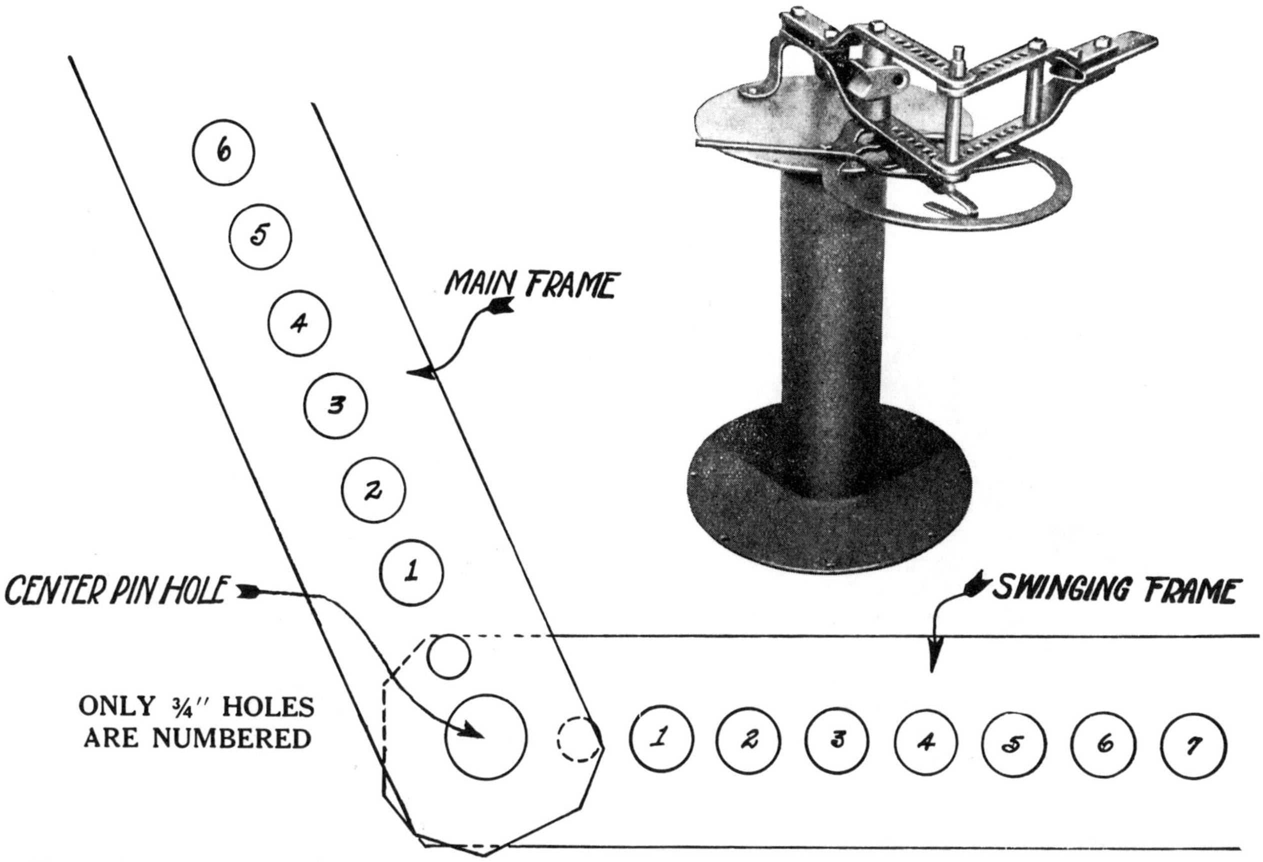

Fig. H.2.b. Mounting Hossfeld bender

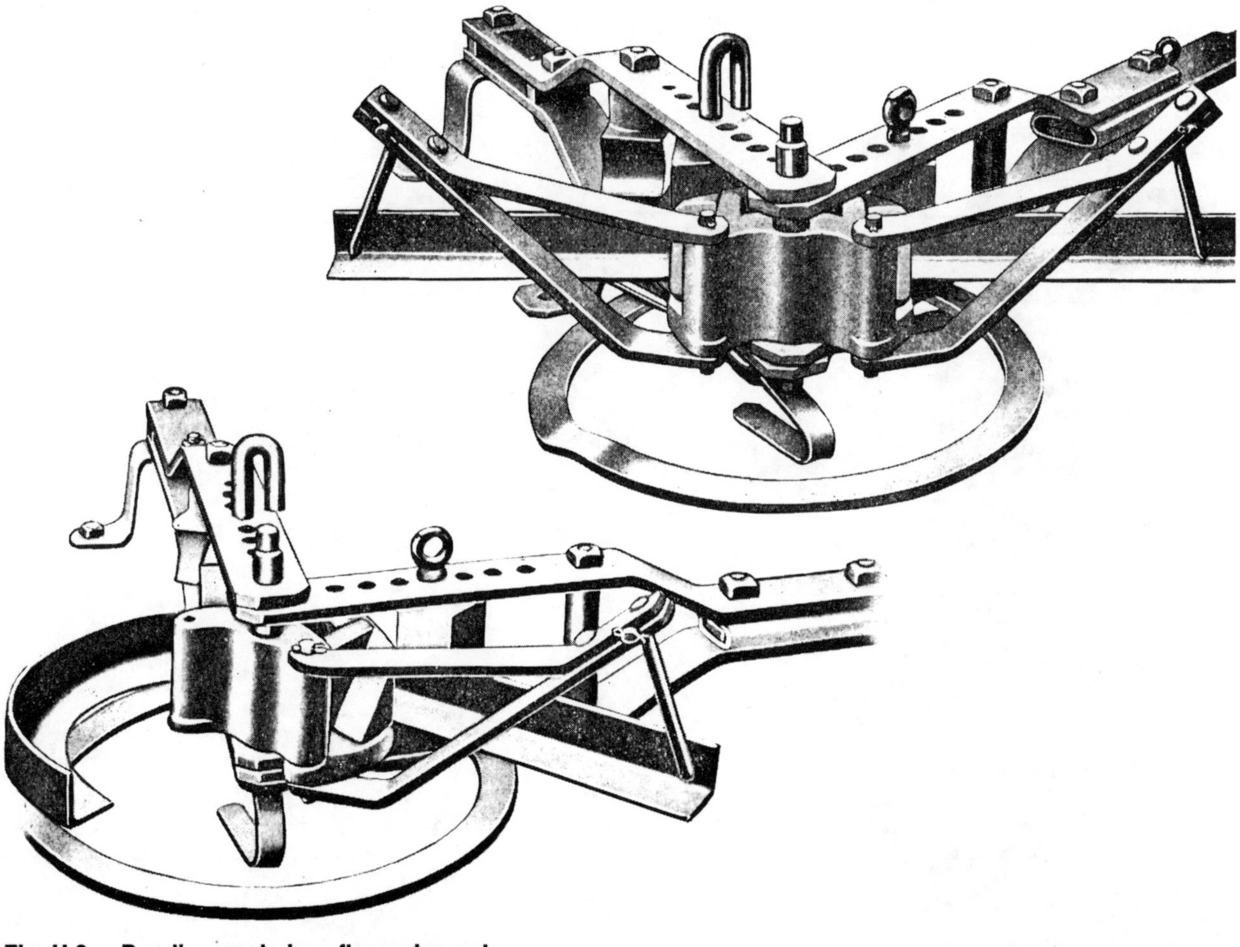

Fig. H.2.c. Bending angle iron flange inward

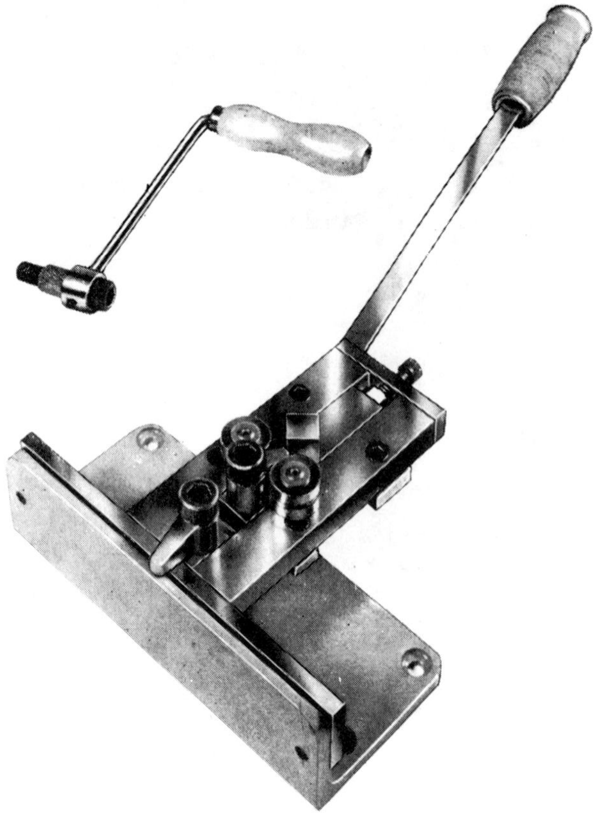

Fig. H.3. Riveting, bending, rolling tool

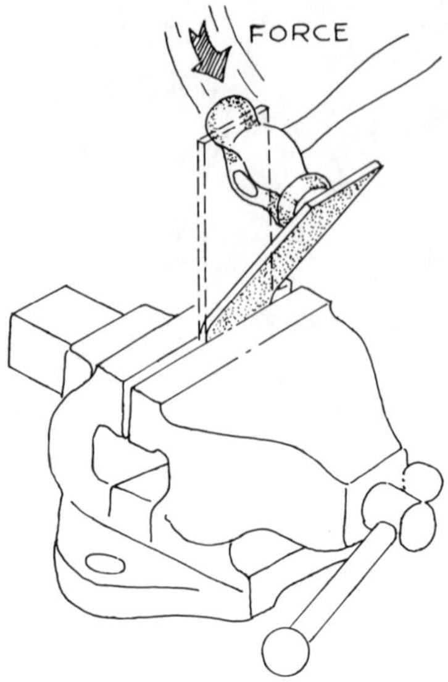

Fig. H.4. Applying force

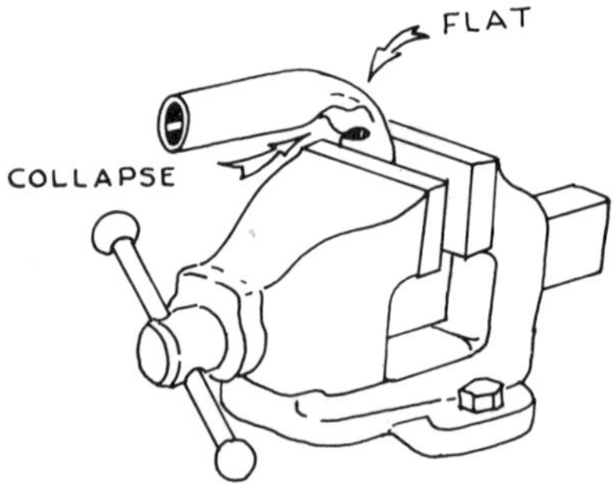

Fig. H.5. Wrinkling at bend in metal

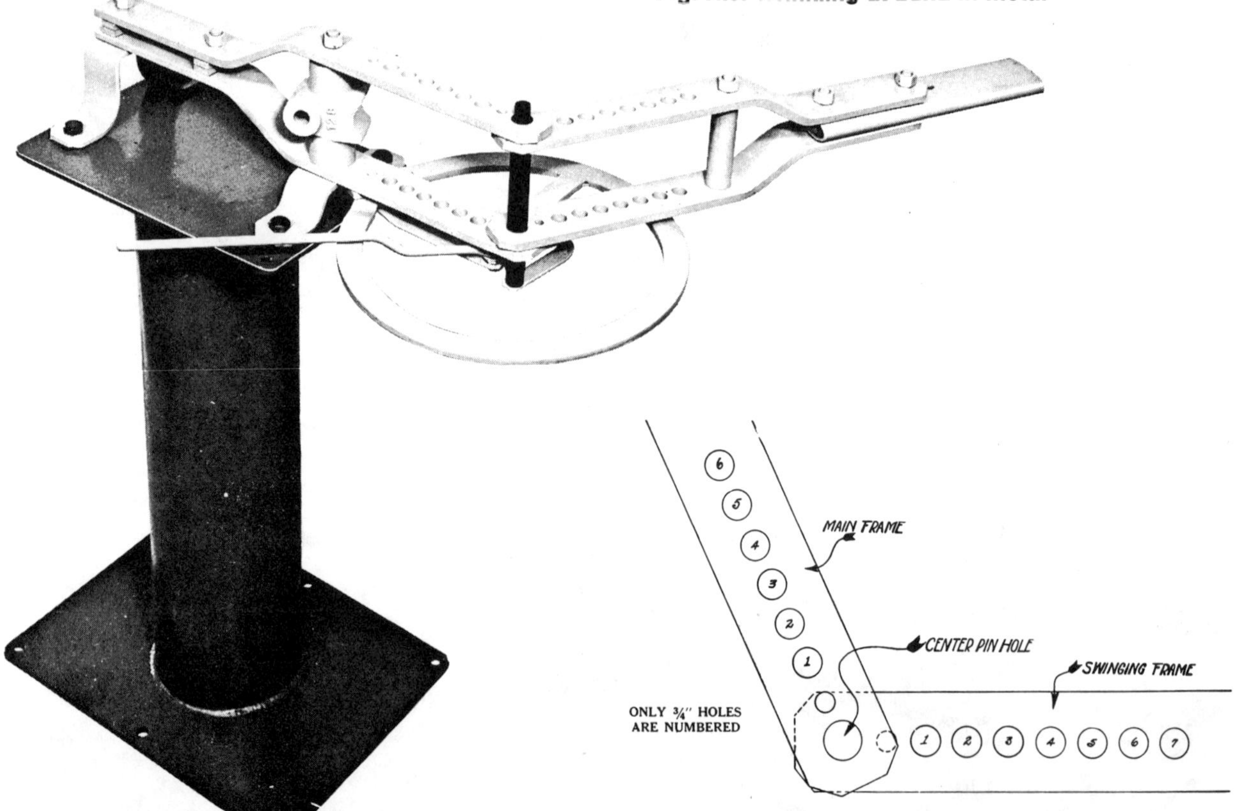

Fig. H.6. Hossfeld bender

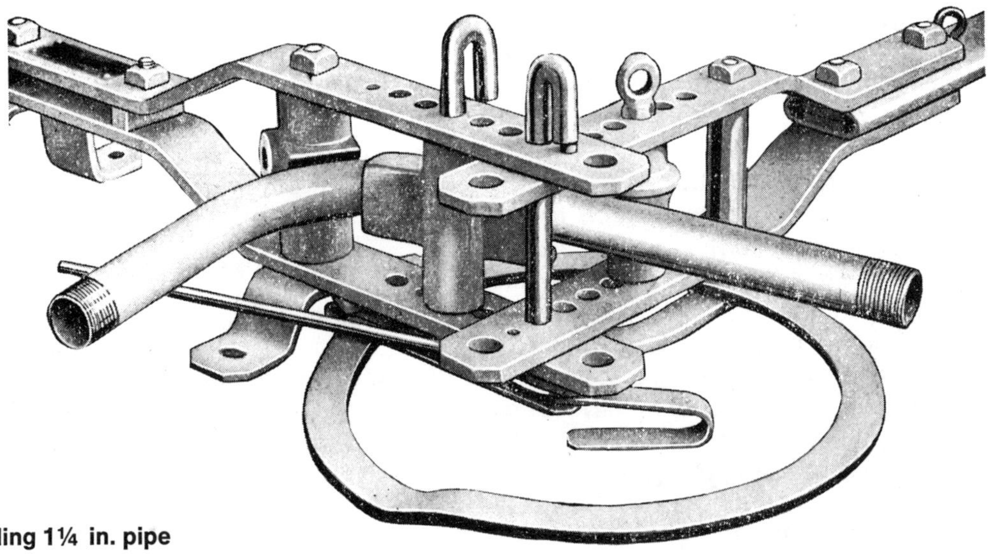

Fig. H.7. Bending 1¼ in. pipe

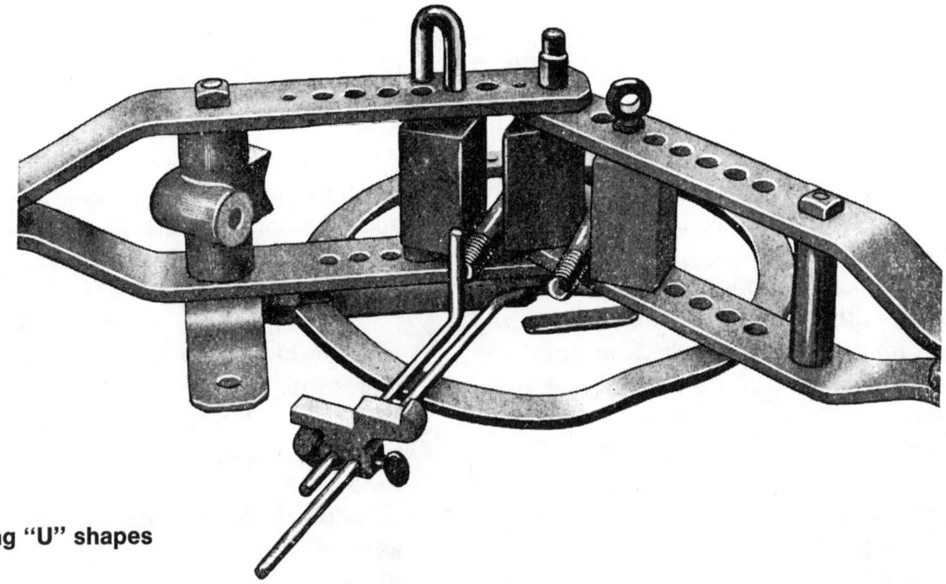

Fig. H.8. Bending "U" shapes

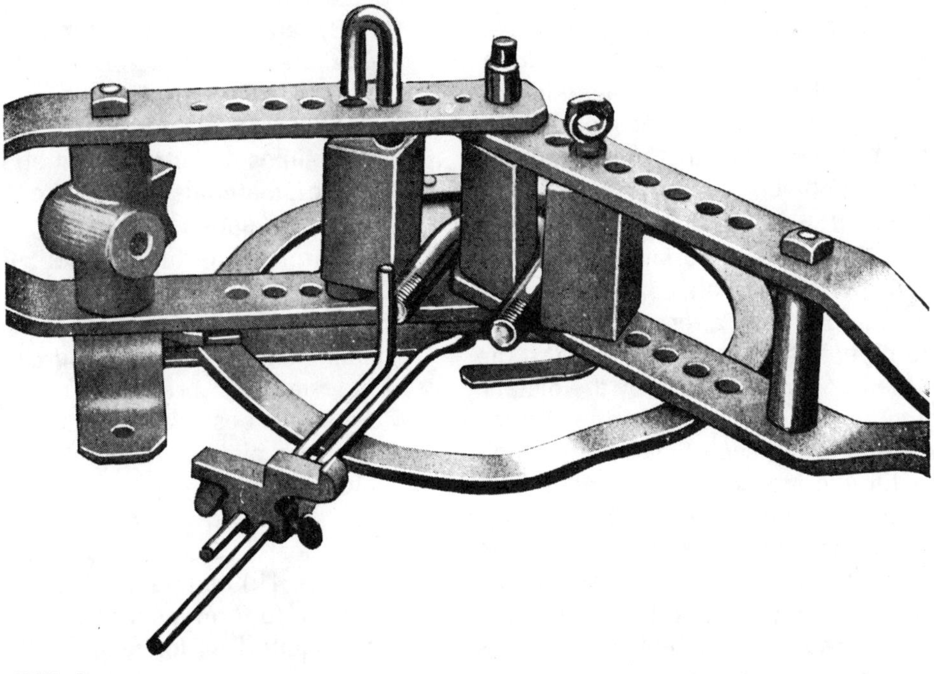

Fig. H.9. Bending "U" shapes

All of the above machines follow this basic principle: when a piece of metal is secured at one end and force is applied to the free end, then the piece will diverge or turn away from a direct line. This is easily demonstrated by clamping a thin piece of metal or wire in a vise and pushing, pulling, or applying force to the free end. See Fig. H.4. (p. 38.)

An understanding of this basic principle plus a knowledge of allowances and tolerances is all that is required to bend or form metal to the desired shape. There are also a number of techniques and procedures that have been developed over the years, which will be illustrated or explained where applicable.

BENDING PIPE OR TUBING

One of the main problems in bending these materials is the tendency for wrinkling or buckling to occur at the point where the bend is made. See Fig. H.5. (p. 38.)

This occurs because the metal on the outside wall is being stretched while the metal on the inside is being compressed. If the bend is too sharp, the metal may even rupture.

If a sharp bend is desired, the pipe or tube should be blocked at one end and the pipe or tube filled with wet sand or lead. The sand should be rammed in and packed as tightly as possible. The open end of the pipe or tube must now also be plugged, and the bending operation carried out carefully and slowly. Check continually for any rupturing. After the bend is completed, the sand or lead is removed, the latter by melting.

Note. When working with molten metal (lead) great care must be taken as with any hot metal against burns or spilling. Moisture should not be present as it will cause the hot metal to erupt from the tube or pipe and splash out. Before using lead, check with the instructor and wear proper safety equipment.

The Hossfeld bending machine shown in Fig. H.6. (p. 38.) features forming dies of the "feed-along type." The bend is formed by pulling the swinging frame after each feed (the amount the pipe or tube is moved into the die). The arc of the bend is controlled by the distance the swinging frame is pulled and the distance the pipe or tube is fed each time into the die. If the pipe or tube is moved a small amount each time and the swinging frame is pulled hard, a small radius bend will be produced. Similarly, if the pipe or tube is moved a large amount each time and the swinging frame is pulled gently, a large radius bend will be produced. Therefore, to obtain a smooth, even radius, the pipe or tube must be fed the same distance each time and the pressure or pull on the swinging frame must be consistent.

MINIMUM RECOMMENDED RADIUS
FOR STANDARD PIPE

Pipe size	Minimum radius
1/4 in.	65 mm
3/8 in.	70 mm
1/2 in.	75 mm
3/4 in.	115 mm
1 in.	130 mm
1 1/4 in.	205 mm
1 1/2 in.	230 mm
2 in.	255 mm

With the Di-Acro bending machine, the tube or pipe is clamped at one end and the moveable arm is rotated until the desired degree of bend is obtained.

The forming roller method of tube bending as shown in Fig. H.8. (p. 39.) is recommended for all large bends where the centreline radius is at least four times the outside diameter of the tubing. It can also be successfully used for bending heavy wall tubing or small diameter pipe to smaller radii and for bending very small diameter tubing.

Note. The forming roller and the matching radius collar must be grooved to exactly fit the tube or pipe, and the tube or pipe must not be allowed to slip during the bending operation as even a slight amount of slippage will cause distortion.

By supporting the tubing or pipe over a wide area using a follow block as shown in Fig. H.9. (p. 39.), thin wall tubing can be satisfactorily formed to a centreline radius as small as 2.5 times its outside diameter without the use of filler materials inside the tube. The follow block also eliminates marring of the metal as it rolls around the tube without friction or sliding action.

Note. The manufacturer's specification as to minimum radius size should always be followed. The smallest practical radius that can be formed is best determined by trial and error on scrap pieces of metal before a project is actually started.

If the shop is not equipped with bending machines, a bending hickey as shown in Fig. H.10. (p. 41.) can be used to form radii. It is also possible to form a set or sets of dies, which will serve quite well for many of the projects under construction.

The die shown in Fig. H.11. could be made in the carpentry shop from a piece of hardwood, or in the machine shop from a piece of steel.

Note. The groove in the bending die must equal the radius of the pipe or tube that is to be bent.

For a radius other than 90°, a circular die should be made. The base plate beneath can be marked off in degrees to facilitate bending.

Fig. H.13. (p. 42.) shows a very basic bending die or fixture in the form of a hardwood or steel block which may be secured in a vise. The holes drilled in the block should be of varying diameters and should be drilled very slightly larger than the outside diameter of the pipe or tube that is to be bent in the fixture.

Note. If the angle was to be less than 90° and a number of pieces were to be bent, a small hole could be drilled in the block (in line with, but a short distance away from, the centreline of the holes). A piece of strong wire or rod bent to the correct angle and inserted in this hole would speed the operation up considerably and result in more uniform bends.

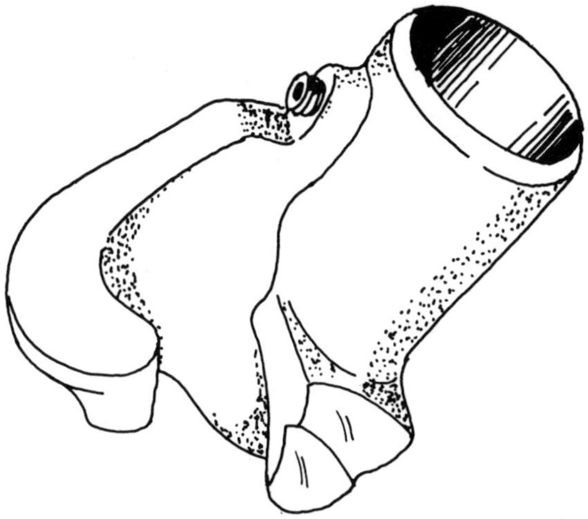

Fig. H.10. Bending hickey

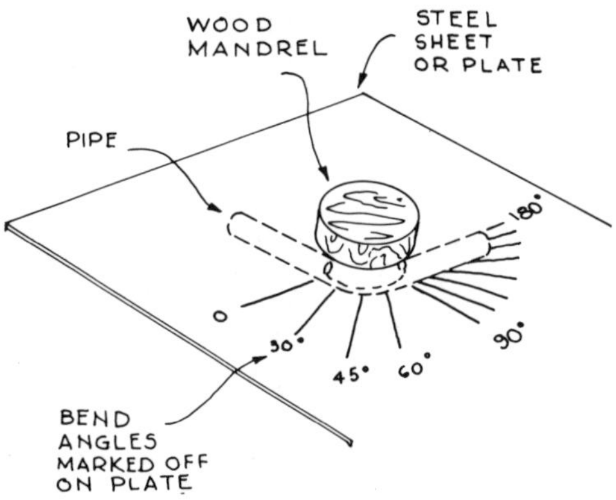

Fig. H.12.

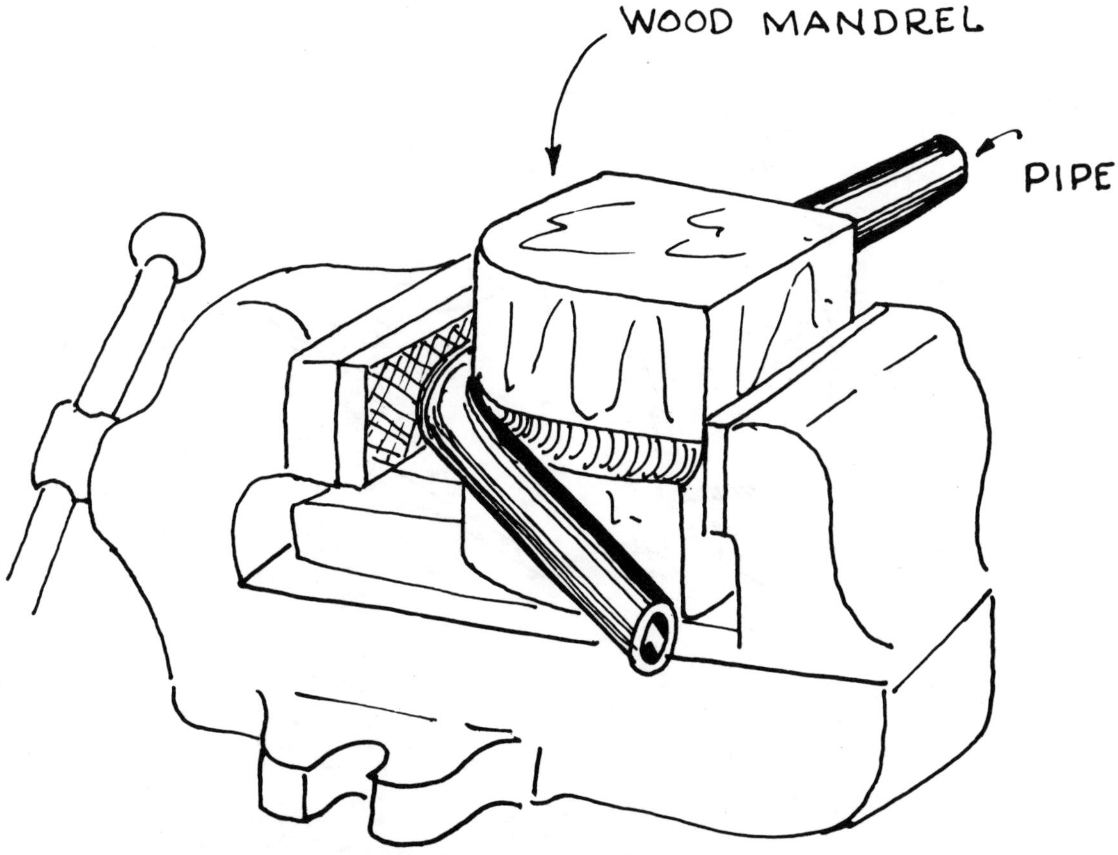

Fig. H.11. Bending die

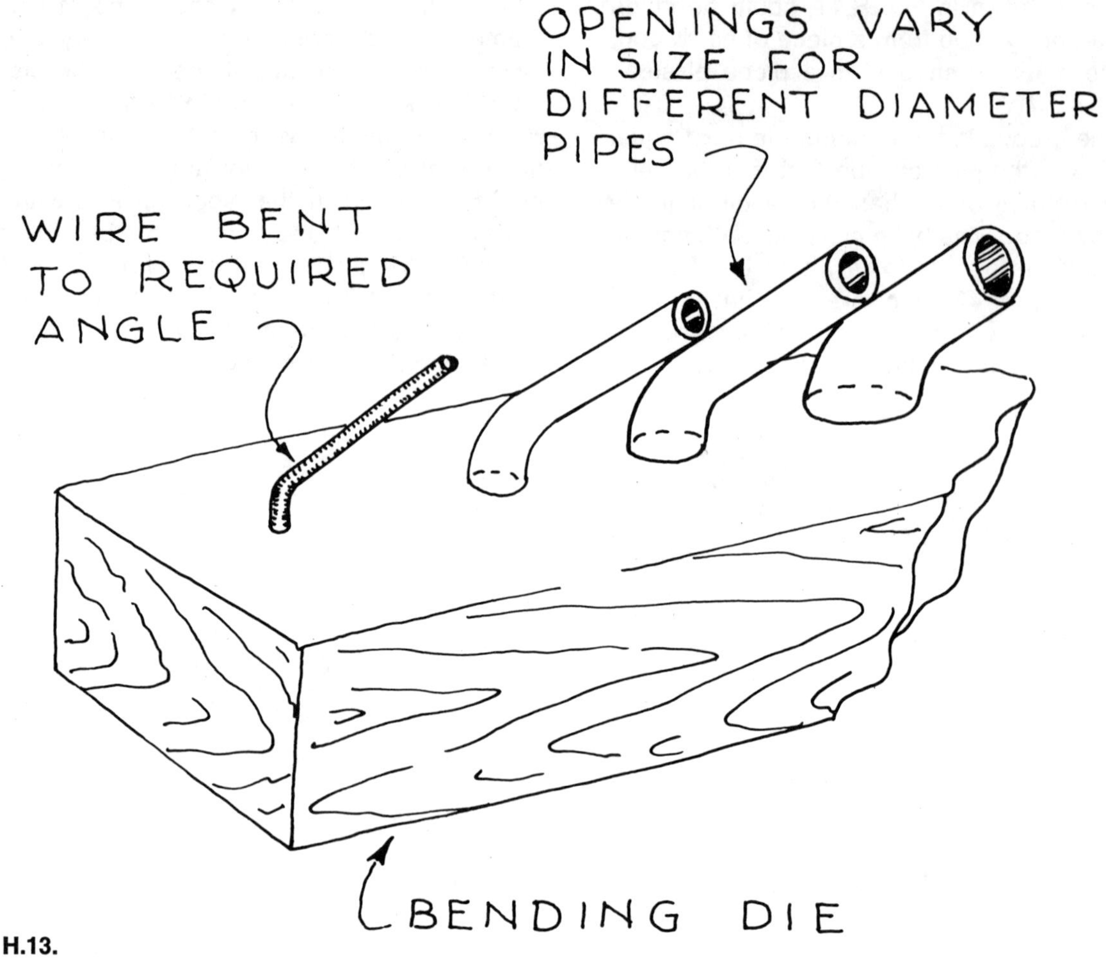

Fig. H.13.

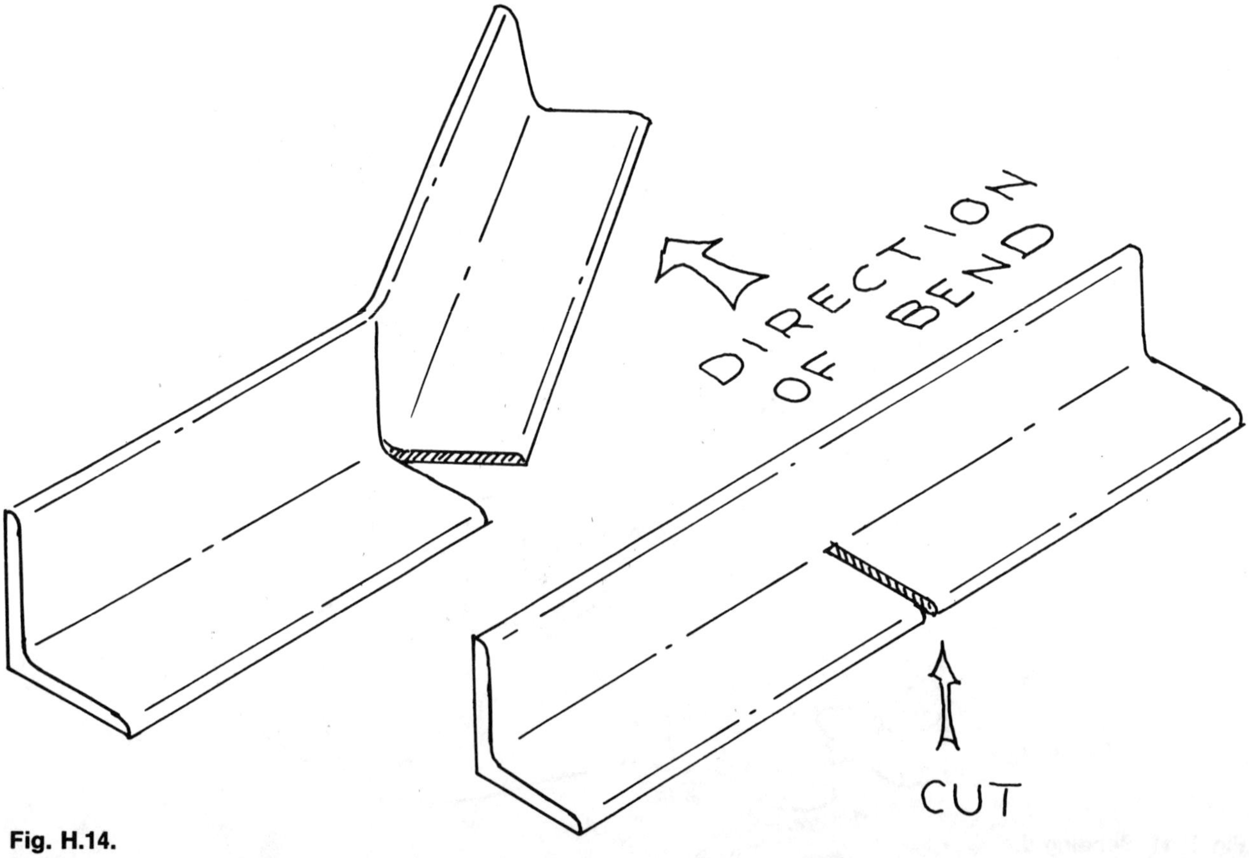

Fig. H.14.

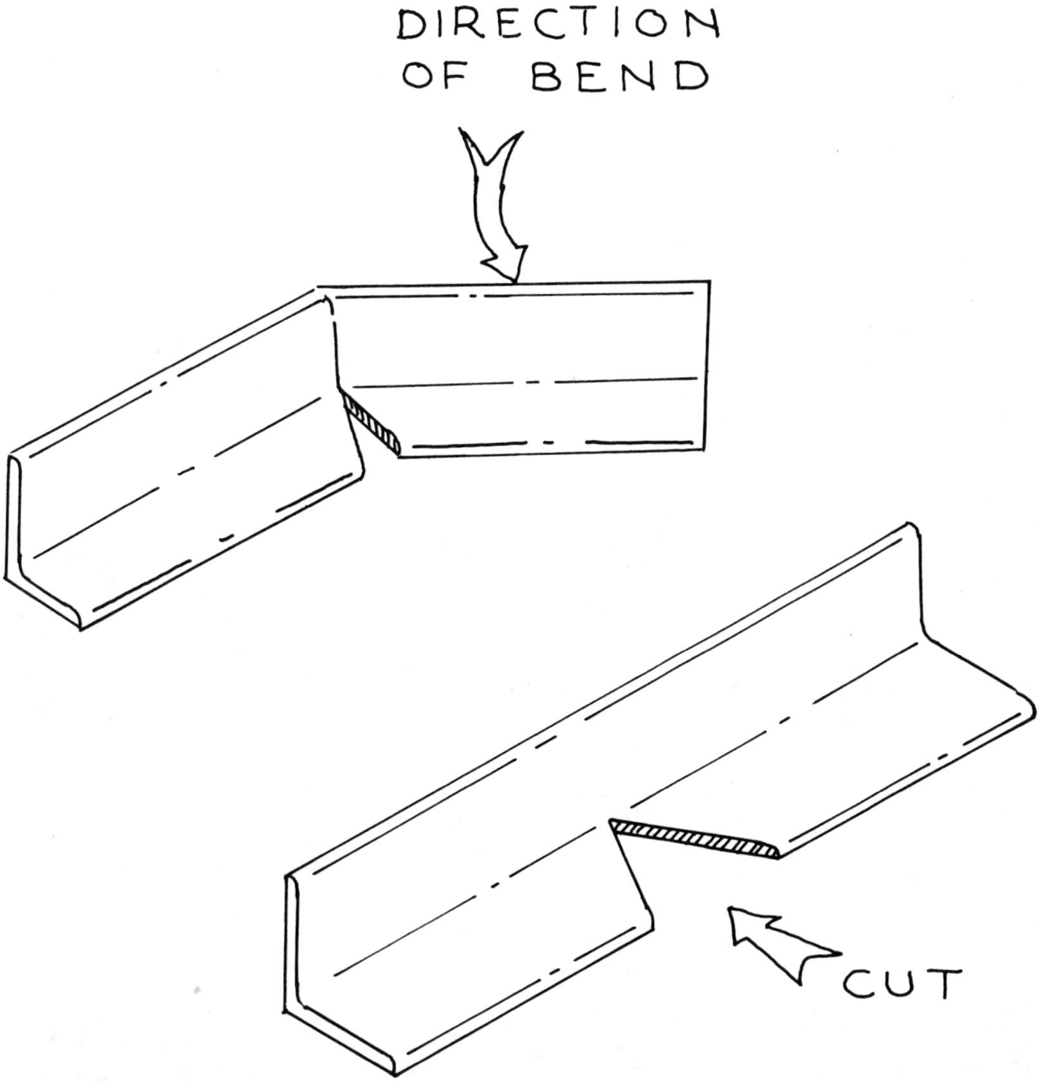

Fig. H.15.

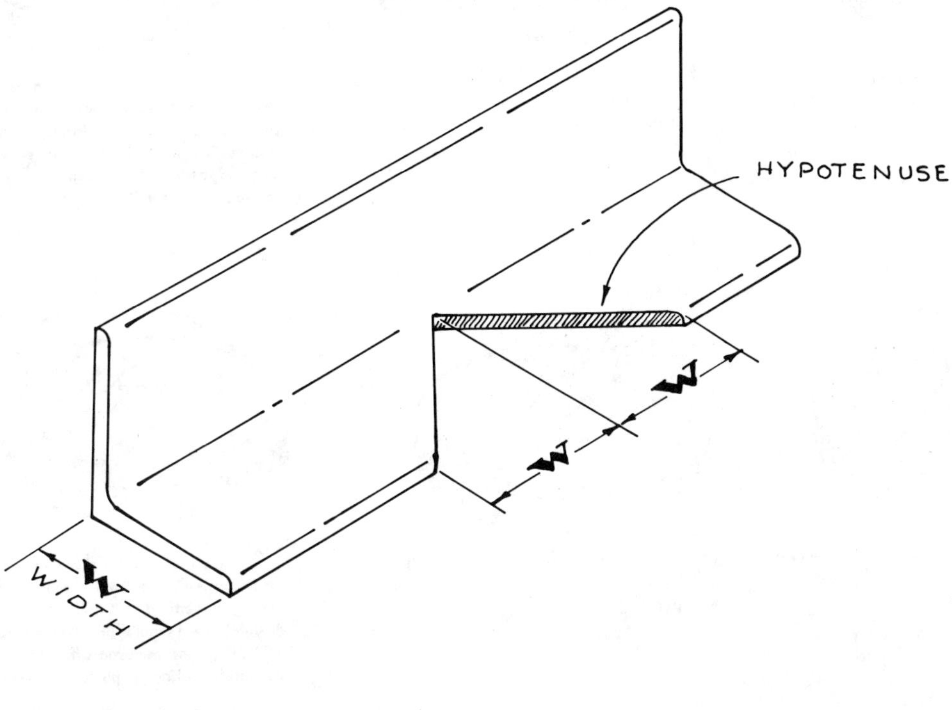

Fig. H.16.

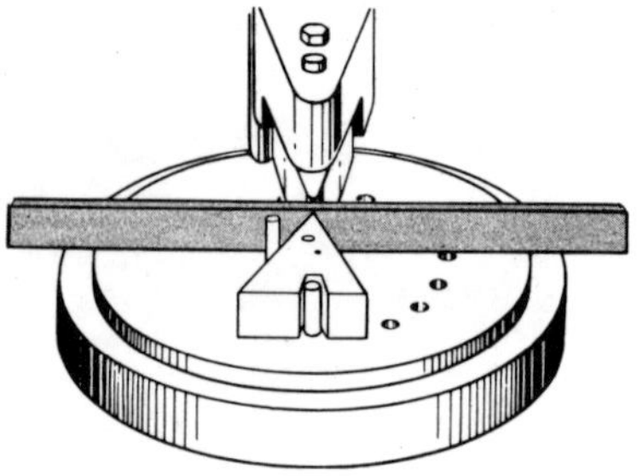

1. Adjust Forming Nose so material will fit snugly between Nose and apex of Zero Radius Block.

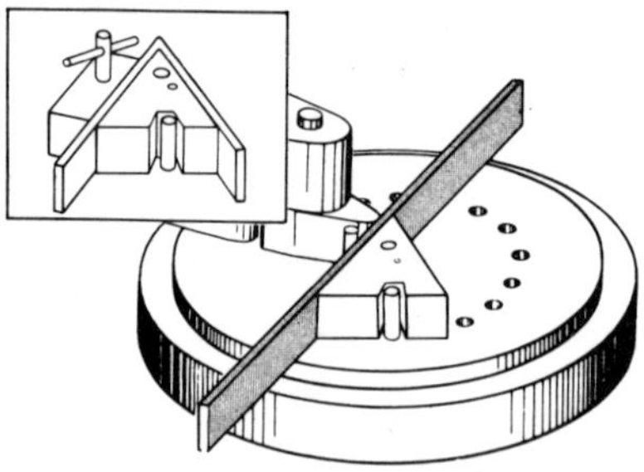

2. Clamp material close to bending edge using Locking Pin or Holding Block illustrated.

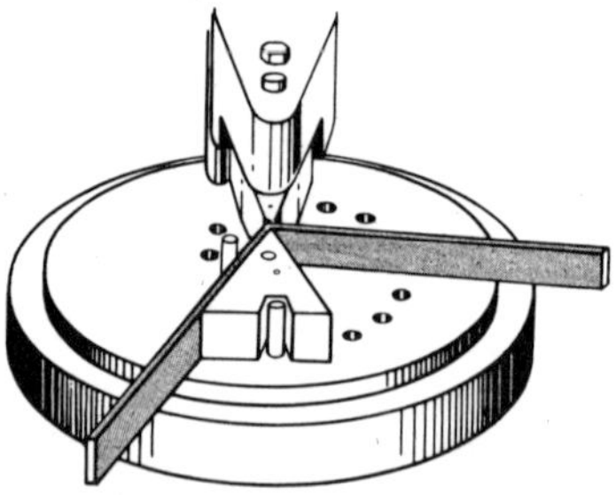

3. Advance Operating Arm until it strikes Angle Gauge, thereby establishing exact degree of bend.

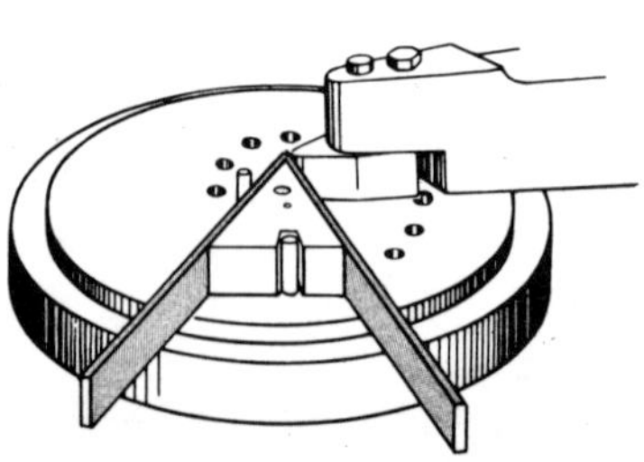

4. Bend is now completed and additional parts can be exactly duplicated by repeating cycle.

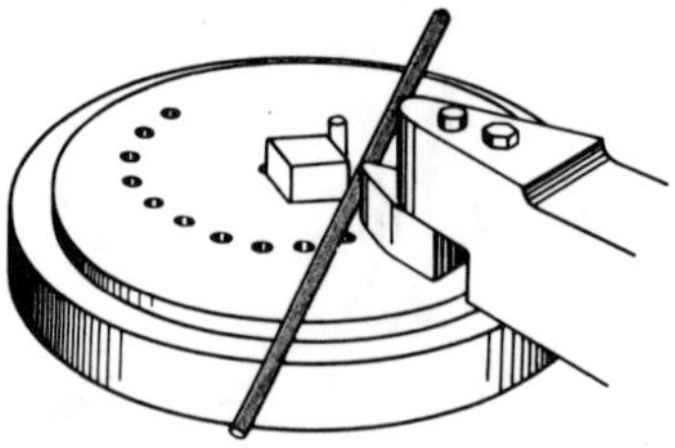

1. Adjust Forming Nose so material will fit snugly between Nose and any edge of Square Block.

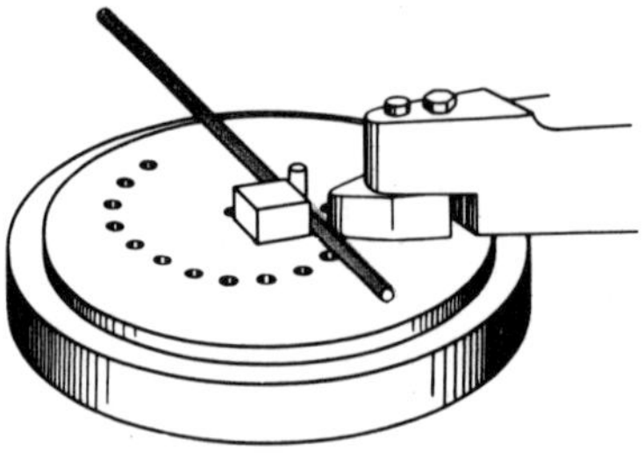

2. Clamp material between Locking Pin and Square Block and then advance Operating Arm.

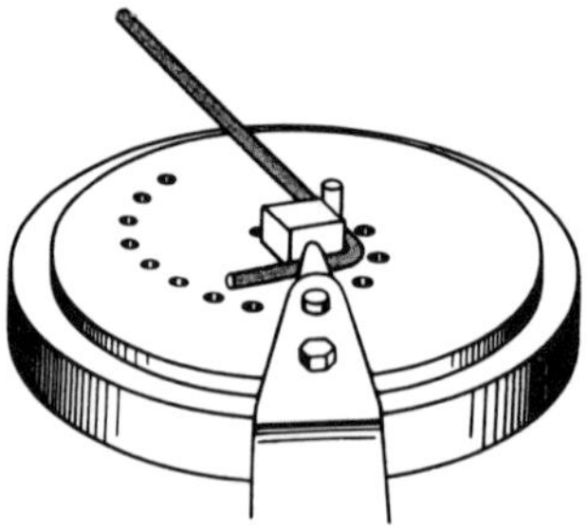

3. Note how material remains straight between corners of block as Forming Nose moves into position for second bend.

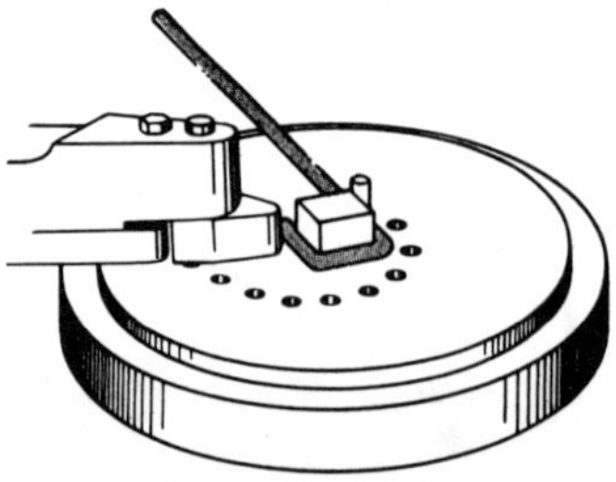

4. Two bends have now been completed. A third bend can be obtained by merely advancing the Operating Arm.

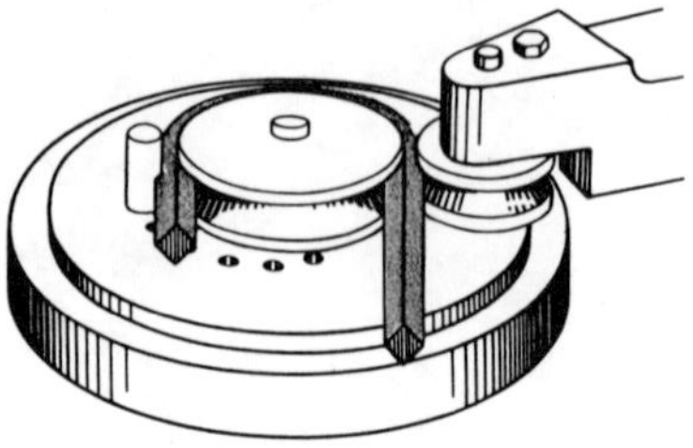

1. To form square material on edge it is only necessary to machine a 90° V groove in the Radius Collar and Forming Roller to eliminate twisting or marking of the material.

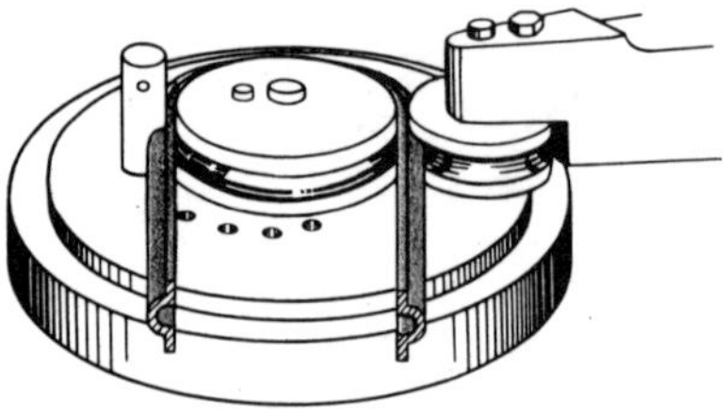

2. A convex shape has been turned on the Radius Collar while a concave groove has been machined in the Forming Roller to adequately support the contour of this extrusion.

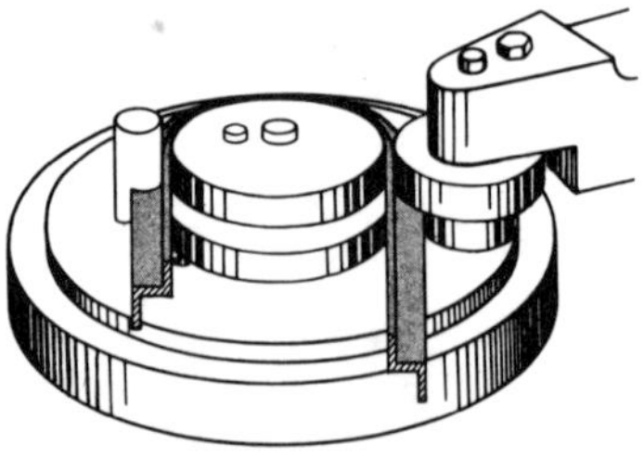

3. By turning shoulders on both the Radius Collar and Forming Roller it is possible to apply bending pressure against both vertical legs of this part and also confine the horizontal member which must be bent edgewise. The Locking Pin should also be shaped to fit the material.

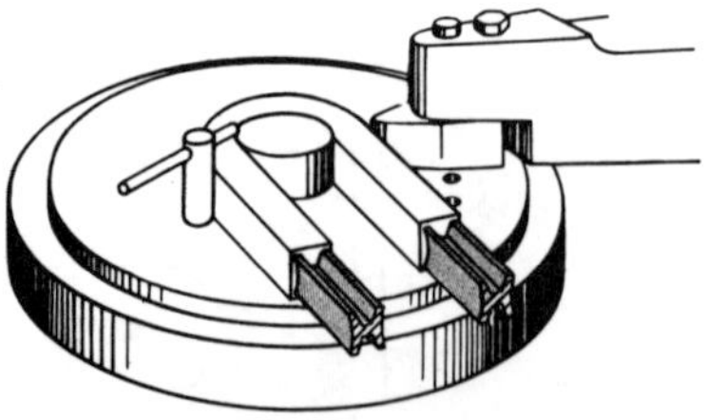

4. The contour of this aluminum extrusion cannot be supported or confined and has therefore been entirely imbedded in Cerrobend and formed as though it were a solid bar. As Cerrobend melts at 158°, it can be removed after the part is formed by merely placing the piece in hot water.

Fig. H.17.

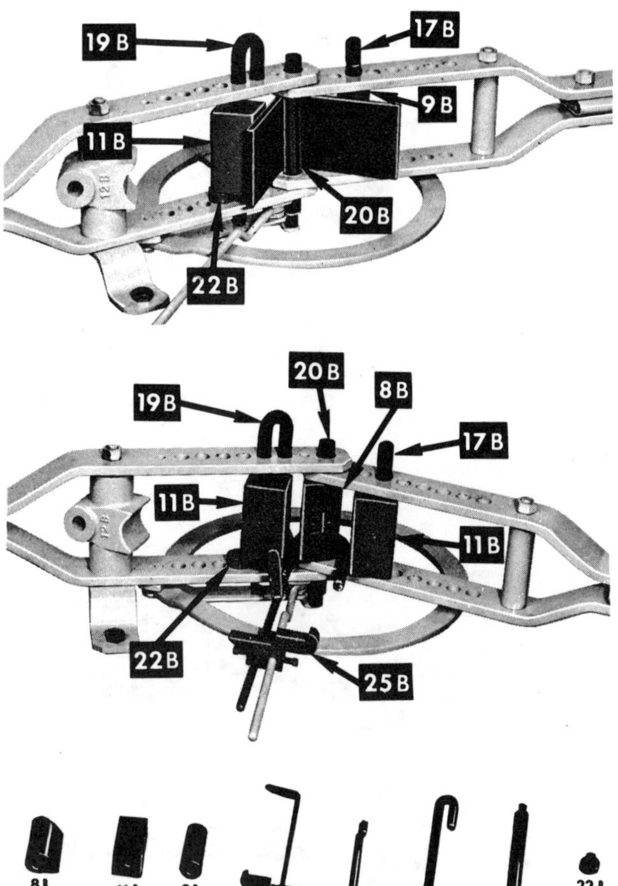

Fig. H.18.a. Bending heavy stock cold and bending with cam die

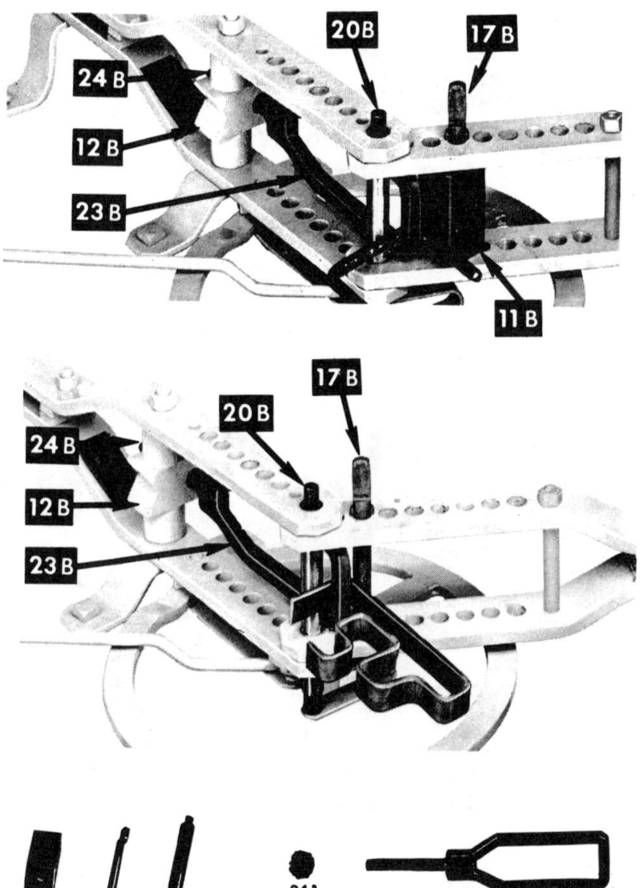

Fig. H.18.b. Bending sharp, square bends with bending yoke

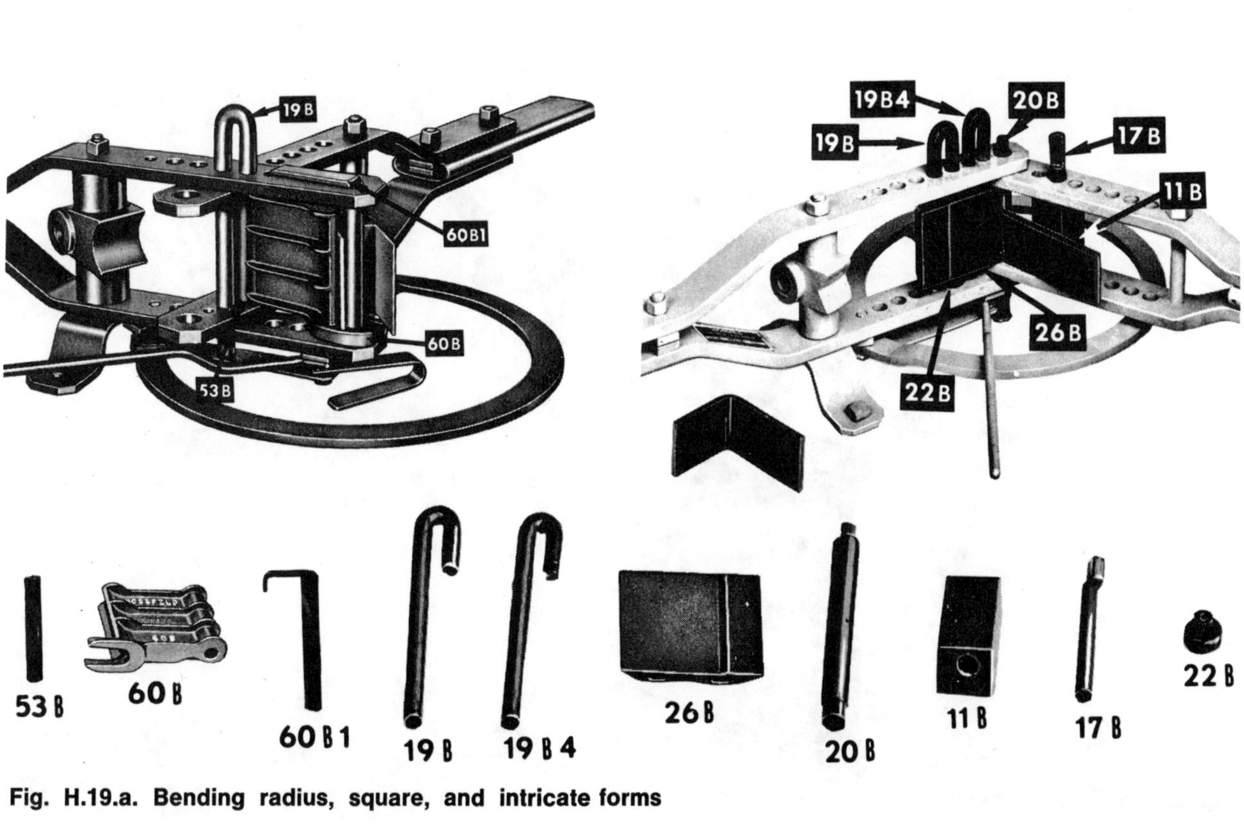

Fig. H.19.a. Bending radius, square, and intricate forms

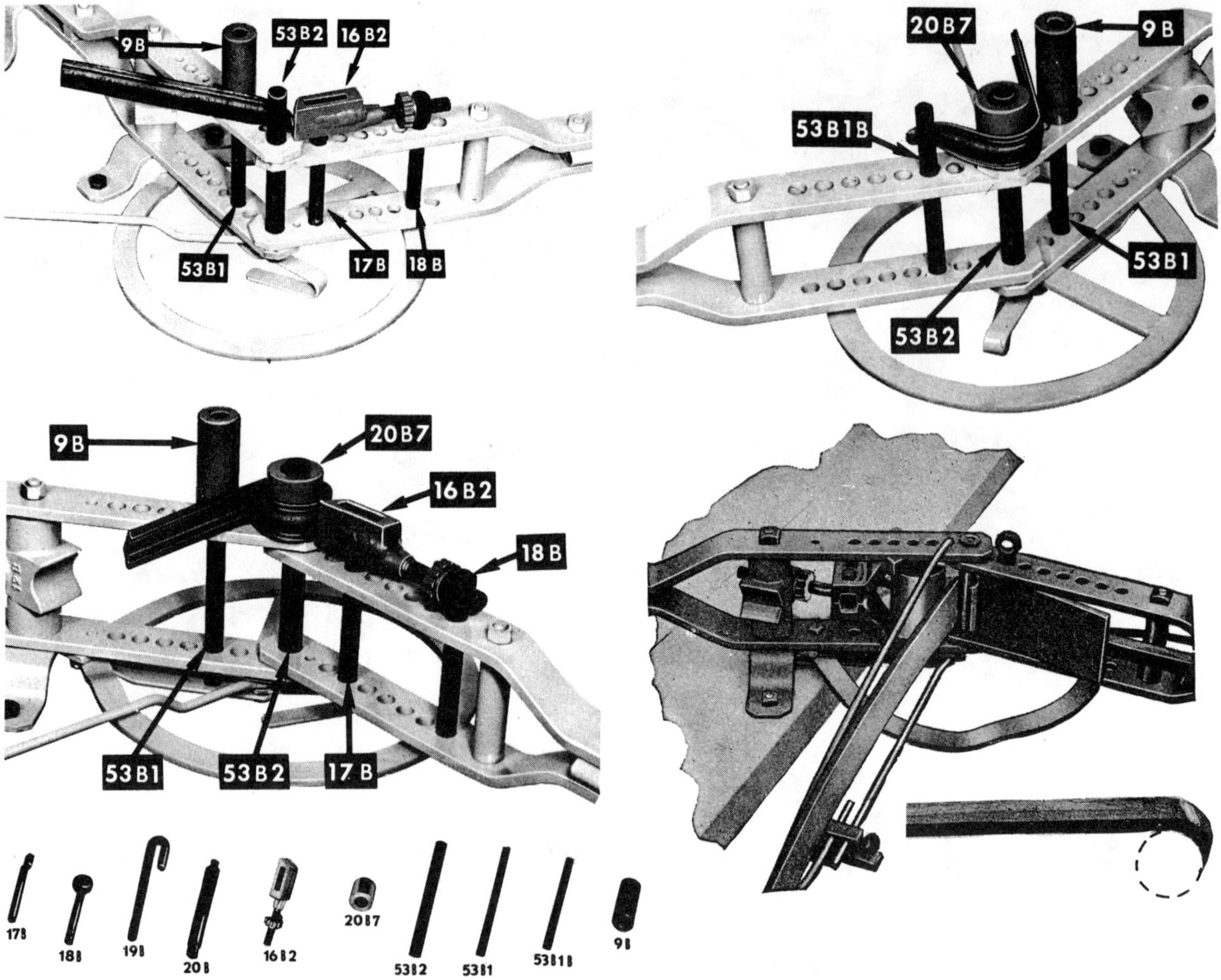

Fig. H.19.b. Bending lamb's tongue and volute and bending small diameter eyes

Fig. H.20.a. Forming bends

Fig. H.20.b. Forming curves or roll circles

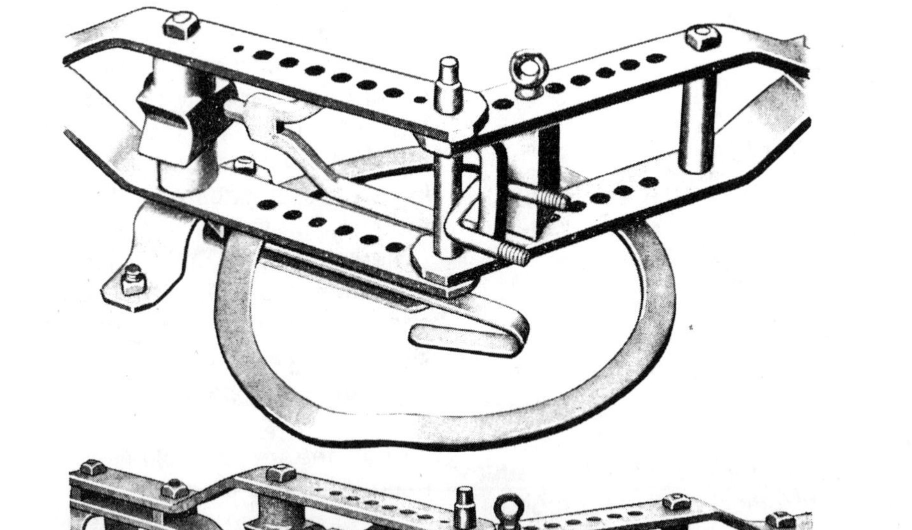

Fig. H.21. Bending square bends, "U's", and round or square eyes

BENDING ANGLE IRON AND CHANNEL

Because of their special shape, angle iron and channel cannot be bent sharply unless certain procedures are carried out first. If the angle iron or channel is to be bent with the flange(s) out, a cut must be made at the point of bending before bending is begun. See Fig. H.14. (p. 42.)

If the angle iron or channel is to be bent with the flange(s) in, a vee must be cut into the flange(s) or the metal will buckle and distort. See Fig. H.15. (p. 43.)

The size of vee removed will depend on the finished angle desired. Here again the hypotenuse theory will be useful. For example, to form a 90° bend on angle iron or channel, an amount equal to the width of angle iron or channel would be measured on each side of the point where the bend is to be made and the lines connected as shown in Fig. H.16. (p. 43.)
Note. It might help if a small hole is drilled at the apex of the vee, to prevent the edges from binding when the bend is formed.

BENDING SOLID METALS

The bending of solid shapes such as flat bar, square bar, and round bar present very little problems with these machines.

BENDING METALS MANUALLY

Pipe or tube is extremely difficult to bend by hand, producing a smooth, even bend, unless a hickey is used as shown in Fig. H.22. If a reasonably long piece of pipe or tube is used, the usual method is to place it on the ground, insert the hickey, stand on the pipe or tube and pull the hickey handle towards you as shown in Fig. H.22.

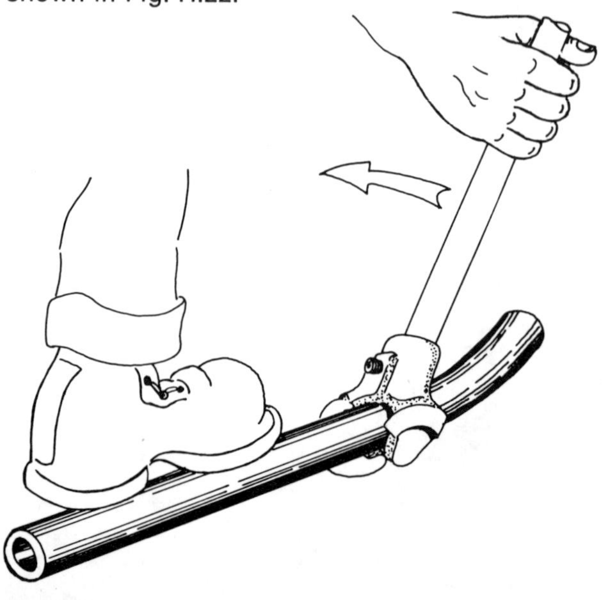

Fig. H.22.

The use of the bending hickey eliminates the possibility of the pipe or tube being squeezed out of shape or collapsing. The pipe or tube could be clamped to a solid object (like a bench), or clamped in a vise, but care must be taken not to squeeze the metal too hard and collapse it at the point of pressure. If the pipe or tube has to be bent in this manner, then it should be blocked inside at the pressure point and protected from clamp or vise marks. Leather, wood, or a soft metal like aluminum should be placed between the clamp or vise jaws and the outside wall of the pipe or tube. See Fig. H.23.

Finally, it is also possible to heat the pipe or tube at the area to be bent, then form it around a cylindrical shape of the correct dimensions. Here again there is the possibility of wrinkling or collapsing the metal.

Angle iron or channel can easily be bent by clamping it in a vise or to a solid object that has square corners, as long as the vee-cutting described previously has been carried out. See Fig. H.24. (p. 49.) If the angle iron is over ⅜ in. thick, then heat should be applied to the bending area. Flat, square, and round bar are bent in the same way. With the exception of round bar (where flat spots might occur), the metal may be hammered while still clamped to produce a sharp square corner, rather than a rounded one. See Fig. H.25. (p. 50.)

With metal bars, a length allowance must be made for bend; the length must be increased by half the metal thickness for each and every bend that is to be made. See Fig. H.26. (p. 50.)
Note. In order to produce a bend without twisting the metal, the line of bending must be parallel with the bending edge. See Fig. H.27. (p. 51.)

The allowance for bending must be above the jaws of the vise or bending area. See Fig. H.27.

If more than one bend is to be made on the same piece of metal, the allowance must be carefully calculated for each bend, and the sequence of bending must be planned ahead.

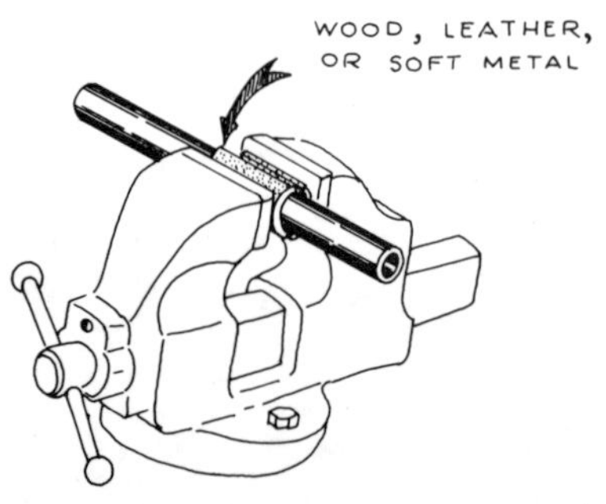

Fig. H.23.

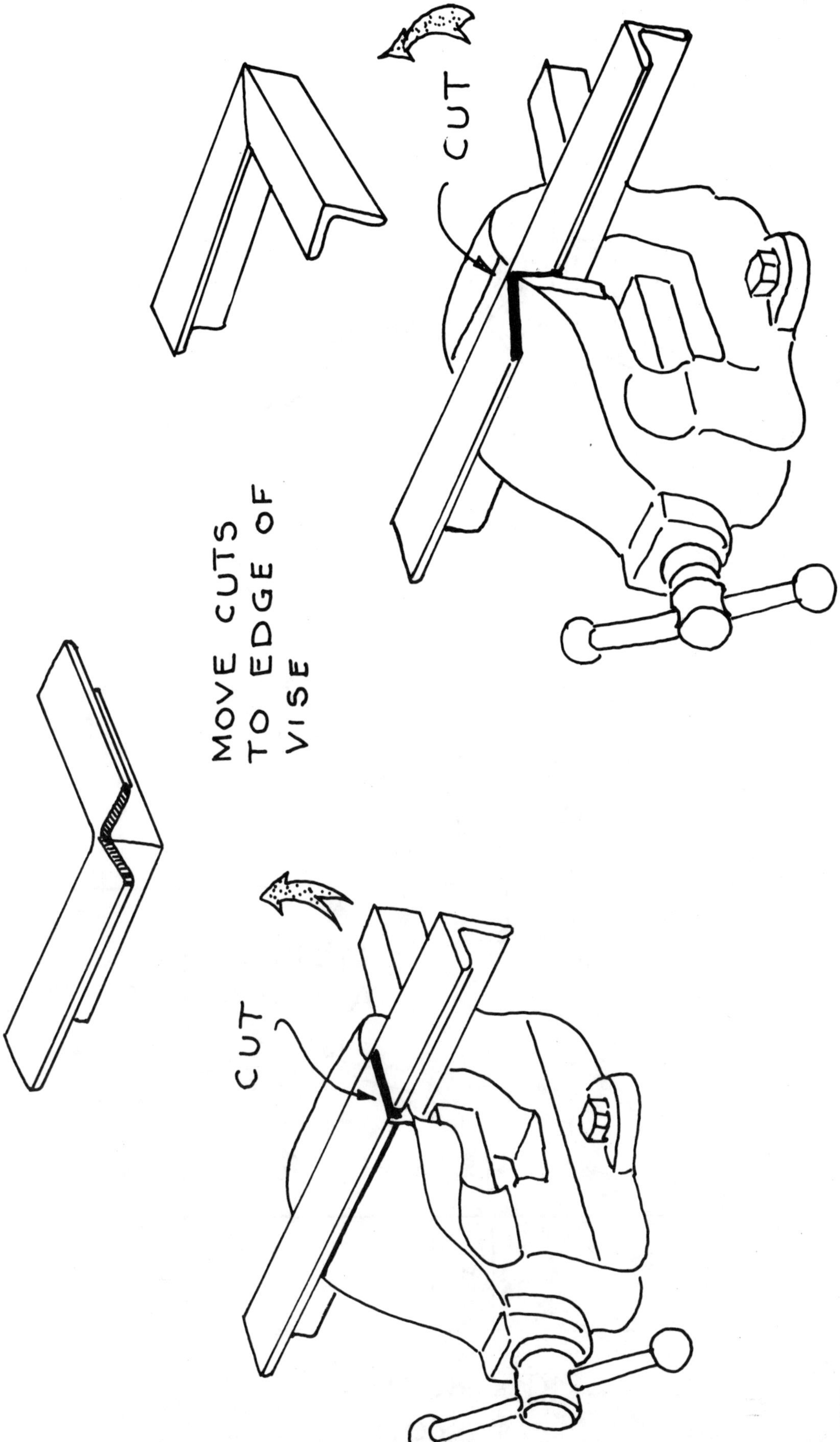

Fig. H.24.

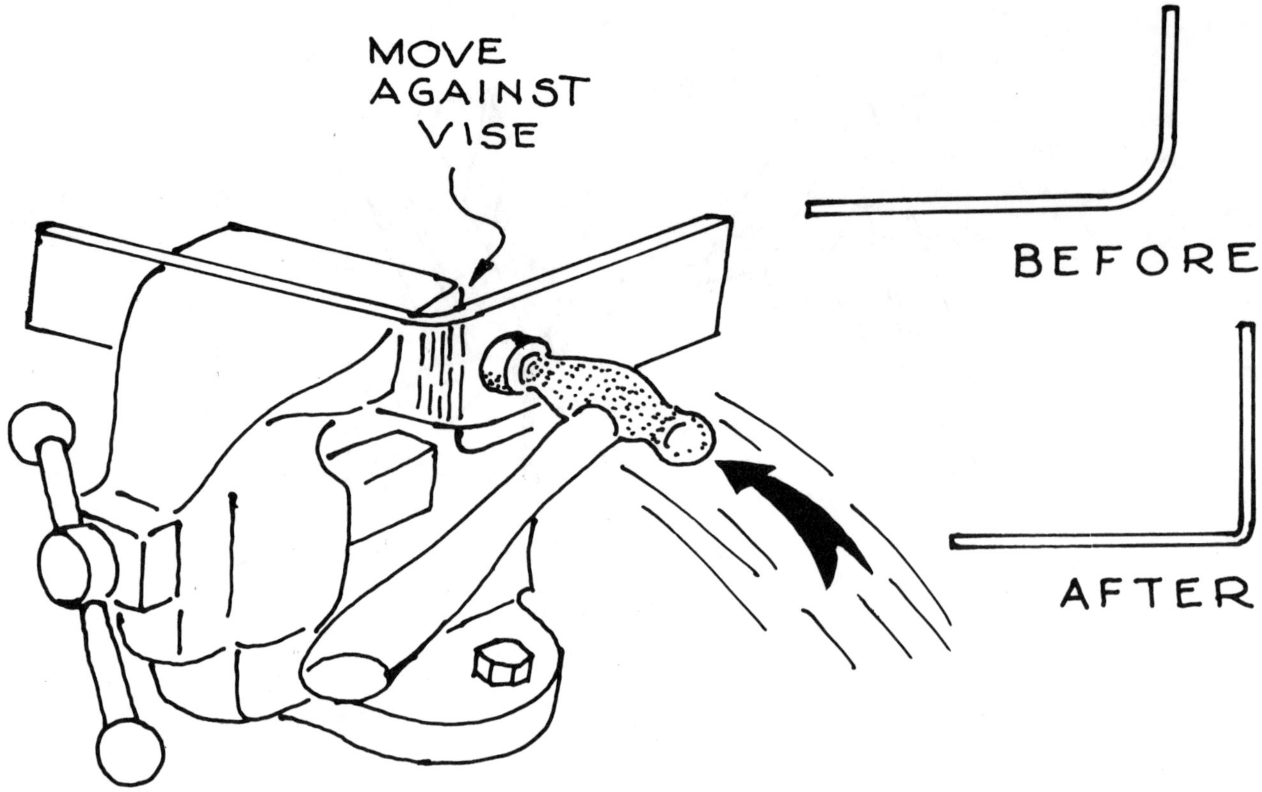

Fig. H.25.

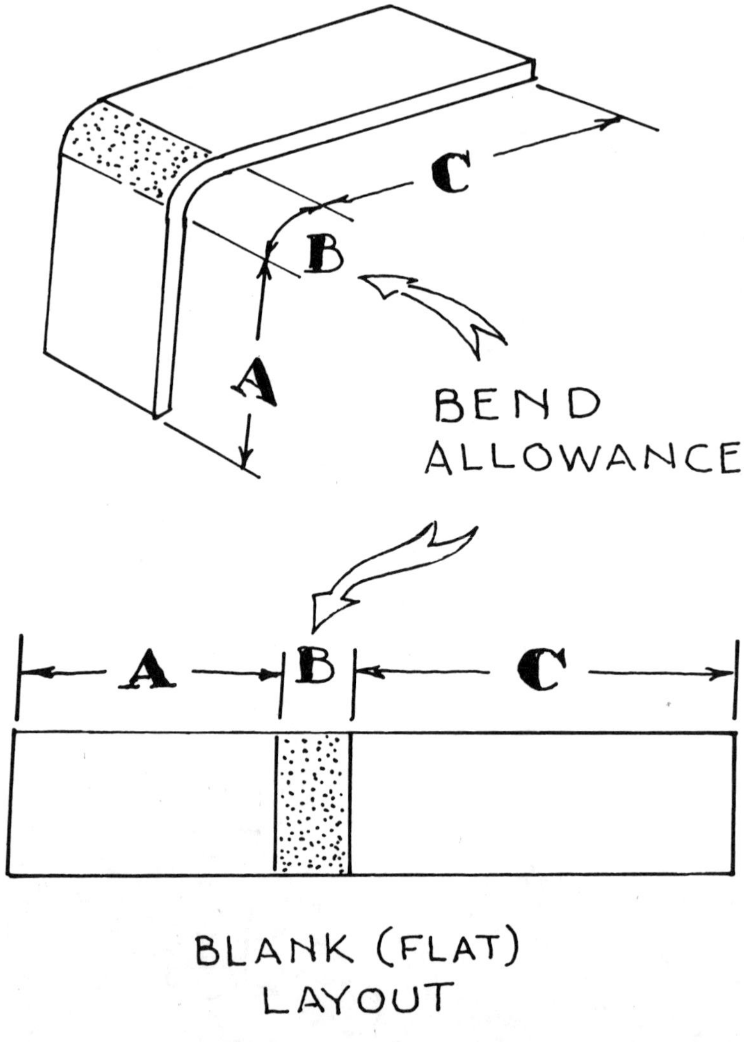

Fig. H.26.

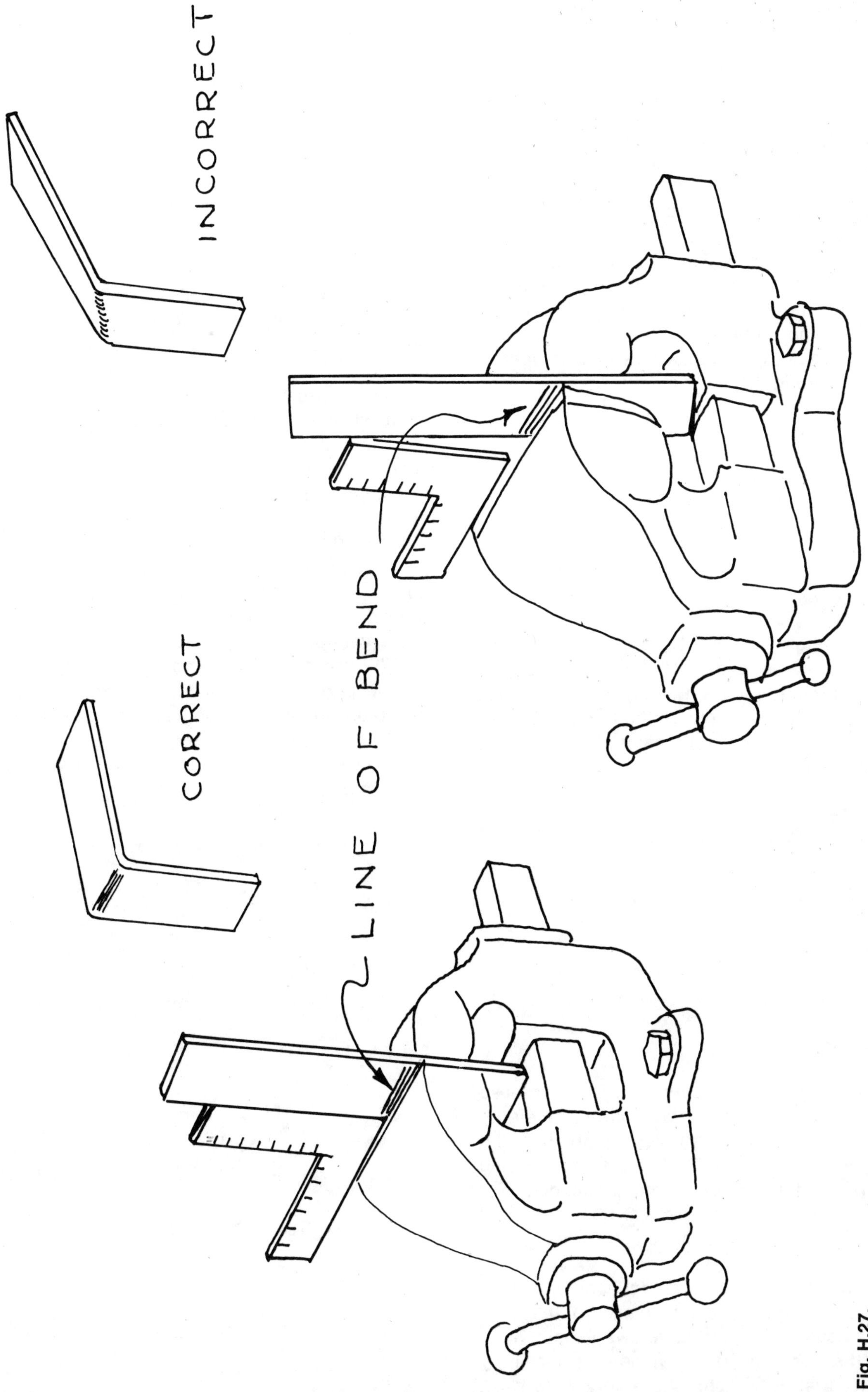

Fig. H.27.

TWISTING METAL

Metal bars are usually twisted to change their direction or for ornamental reasons. Flat and square H.R.S. (mild steel) are the most frequently used metals. Mild steel is very ductile; in flat or square bars it shows twists most effectively for decorative purposes.

Flat bars 1 in. wide by ⅛ in. thick are usually twisted cold. For thicker metal, (or if difficulty is encountered) heat may be used, but care should be taken to heat the metal uniformly. The metal has a tendency to "neck in", or become narrower, where excess heat is applied. See Fig. I.1.

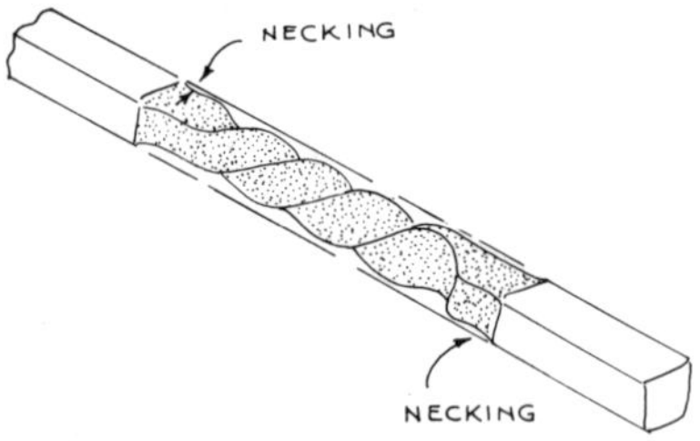

Fig. I.1. Necking

The metal may be twisted clockwise or counterclockwise and will of course produce left or right hand spirals. Usually this is a matter of personal choice, but once the twist is started, it must be continued in the same direction. See Fig. I.2.

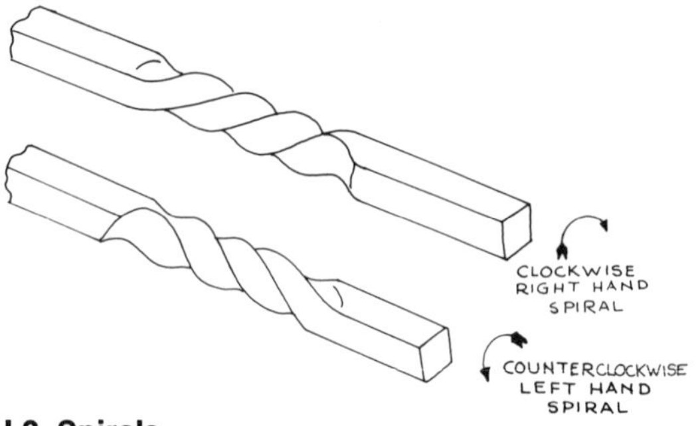

Fig. I.2. Spirals

Twisting metal may cause a slight decrease in the length of the bar (approximately an eighth of the width, for flat bar); for square bar, the decrease is negligible. Trial and error will help you arrive at an exact length, if needed. For example, twist a piece of metal 90°, 180°, and 360° and record the specific shrinkage in length.

The position of the twisting handle is very important. If the handle is close to the area to be twisted it will produce a very tight twist. If on the other hand the handle is far away from the area it will produce a very wide twist.

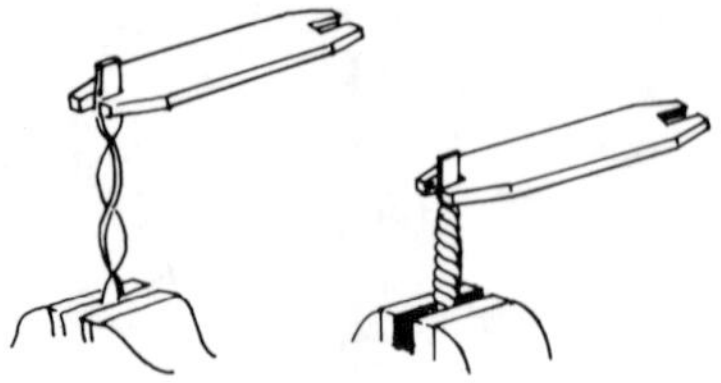

Fig. I.3. Wide and tight twists

Neither is wrong; again it is a matter of personal choice, but, as before, uniformity must be maintained to produce a uniform twist. See Fig. I.3.

Bars are usually clamped to prevent movement when twisting, the easiest method being to clamp the bars in a vise. However, a jig such as the one shown in Fig. I.4. would allow the metal to be twisted almost anywhere in the event the vises in the shop were required for other work. A different fixture could be made up for each size of square bar in stock, or the holes in the side bars could be slotted to allow the side bars to slide in and out for adjustment to the various sizes of bars.

The use of a "C" clamp here would serve a double purpose; it would clamp the fixture to any table or surface, and at the same time it would hold the bar in place and prevent it from jumping out of the fixture should pressure be applied in the wrong direction.

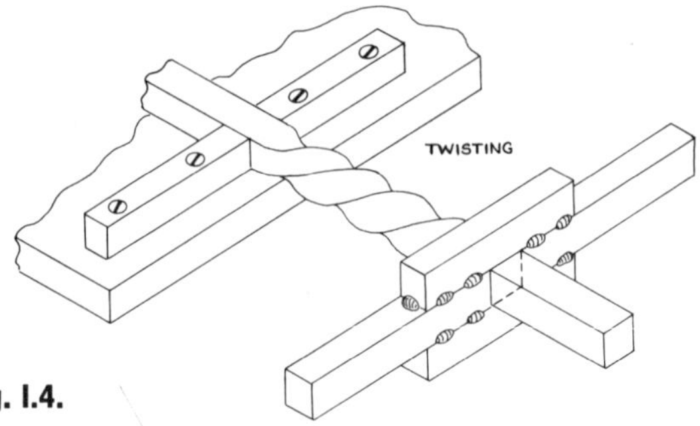

Fig. I.4.

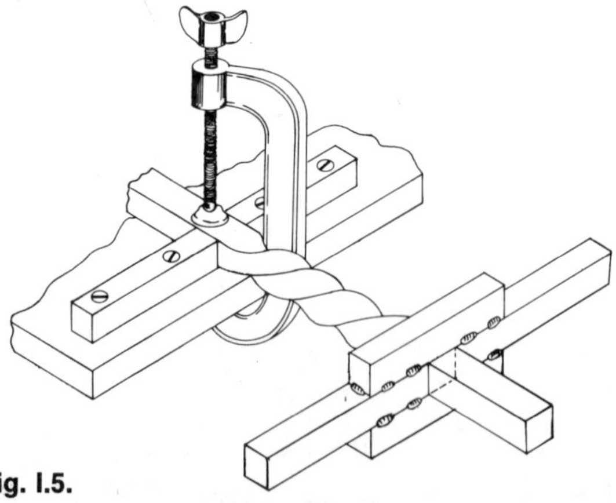

Fig. I.5.

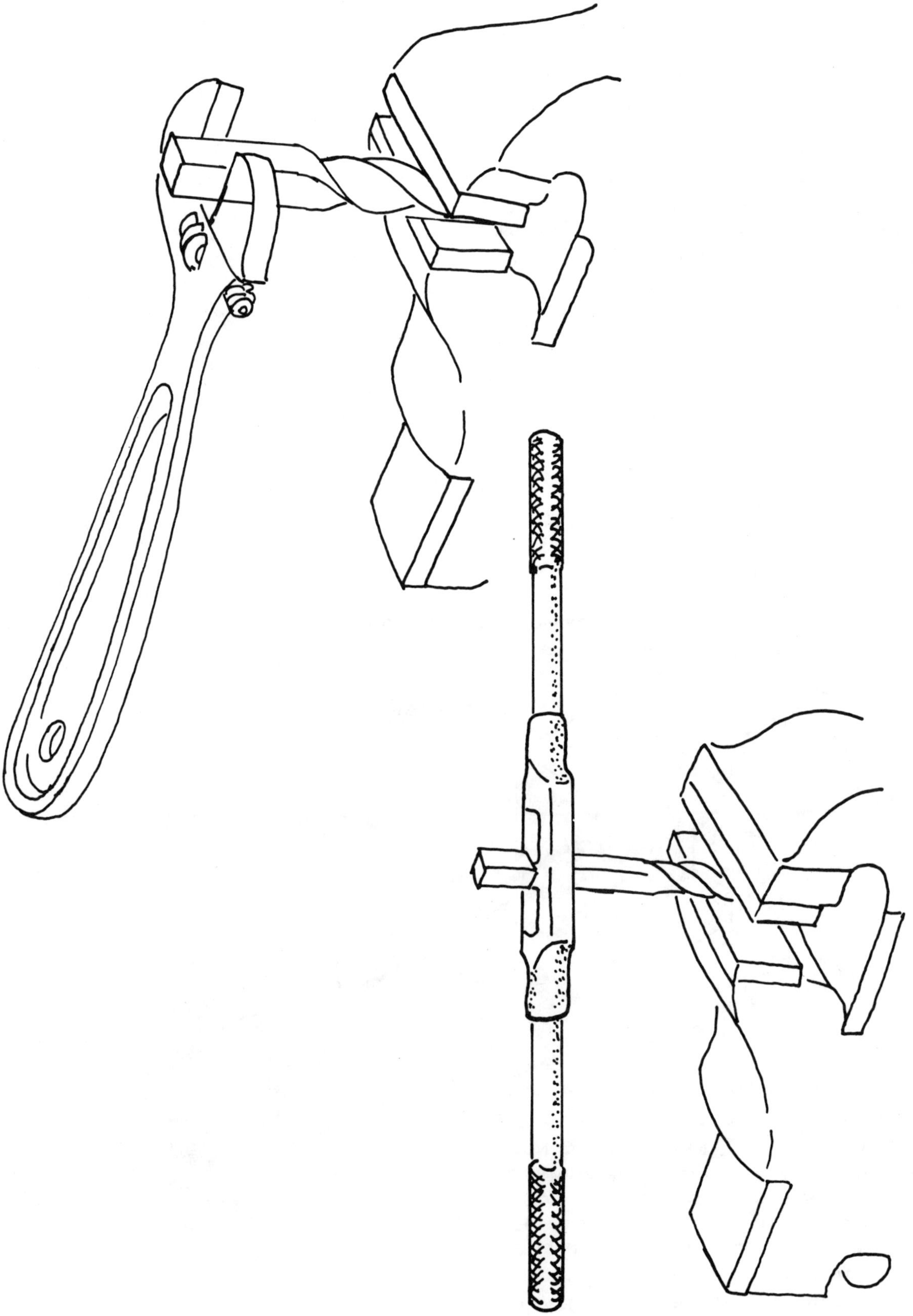

Fig. I.6. Adjustable wrench and machinist's tap handle

TWISTING TOOLS

Probably the simplest tool used for twisting metal is the adjustable wrench, or for small square bar, a machinist's tap handle as shown in Fig. I.6. (p. 53.)

For twisting heavier square bar, handles such as shown in Fig. I.7. can easily be made. For twisting flat bar the handle shown in Fig. I.8. is most effective.

Finally if many bars are to be twisted, a machine as shown in Fig. I.9. can be made quite inexpensively. It consists of an old manual transmission, a piece of channel bar, two pieces of angle iron and a piece of round bar stock.

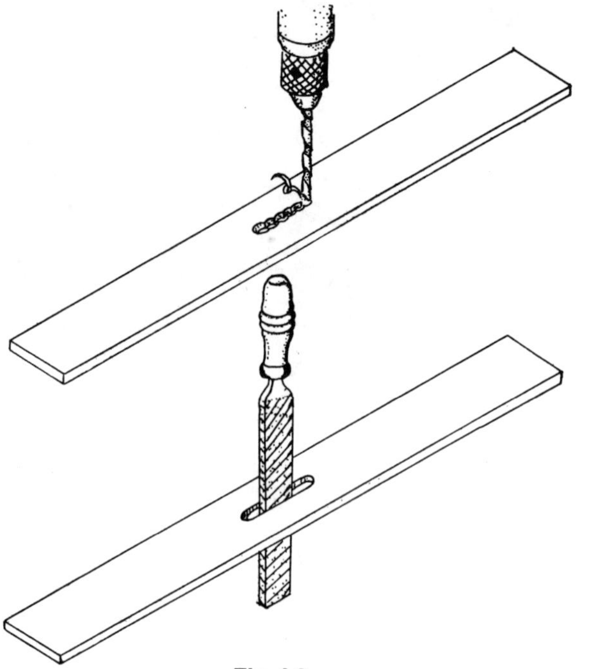

Fig. I.8.

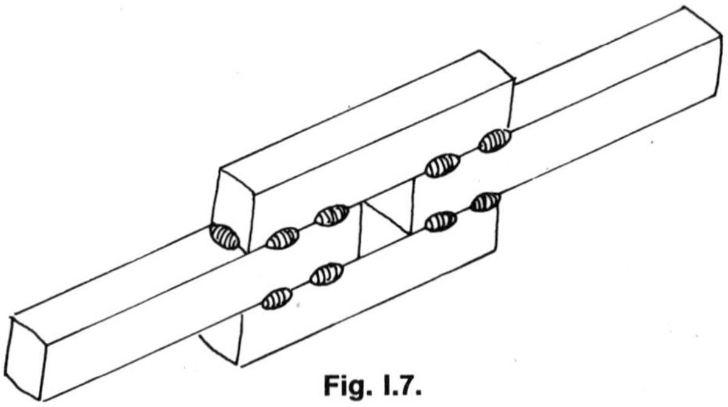

Fig. I.7.

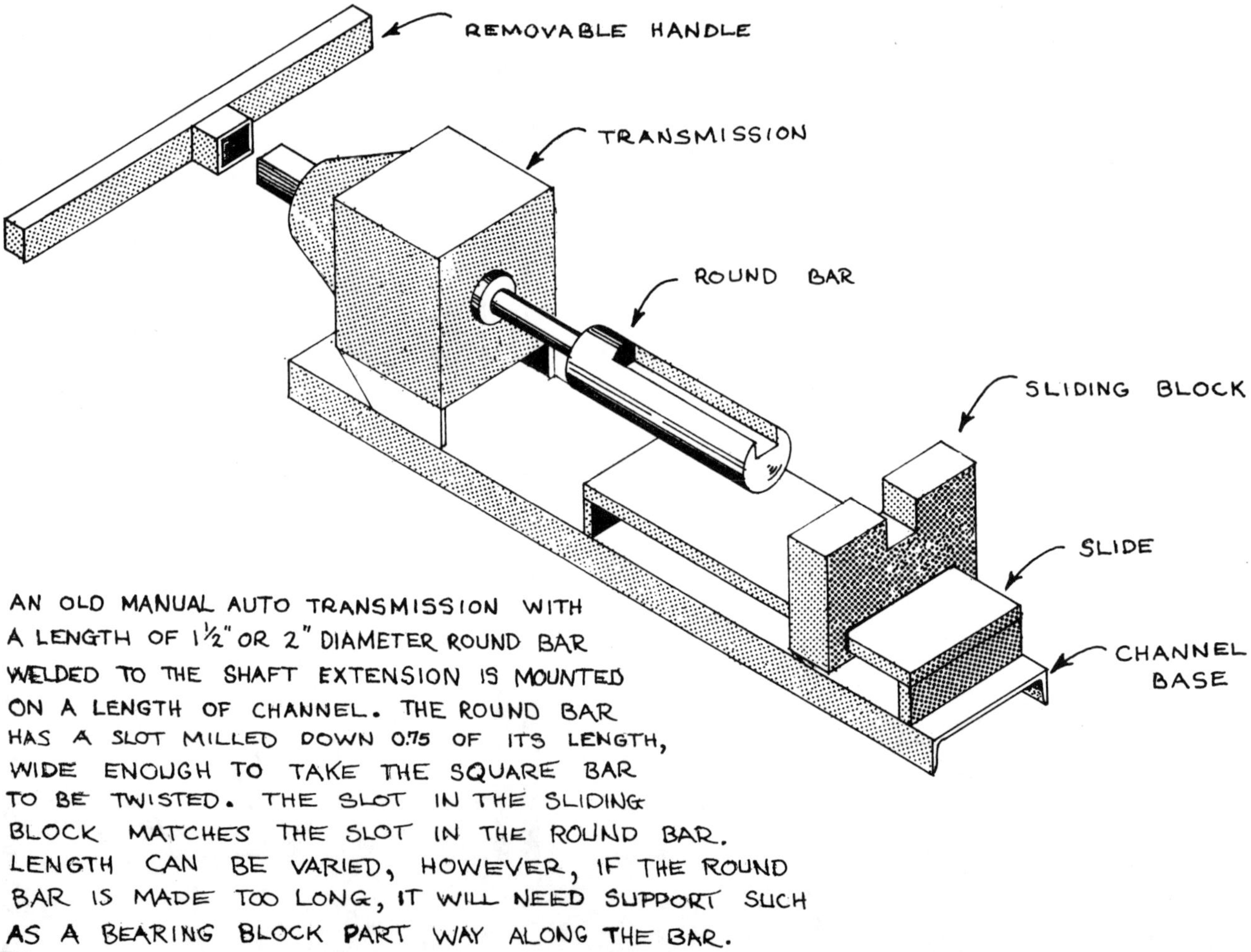

Fig. I.9.

POSITIONING THE WORK

Short pieces of flat or square bar are usually twisted when they are in an upright position in a vise. When larger pieces are being twisted, they would be too high to work at comfortably if placed vertically in a vise. Therefore, it is more advantageous if they are placed in a horizontal position and twisted. See Fig. I.10.

When long pieces are being twisted, the metal has a tendency to bend as it is twisted. This distortion or bending can be hammered out by using a rubber mallet and a heavy hardwood block. This will prevent the high points of the twist from being damaged, such as might happen if a steel hammer and anvil were used. See Fig. I.11.

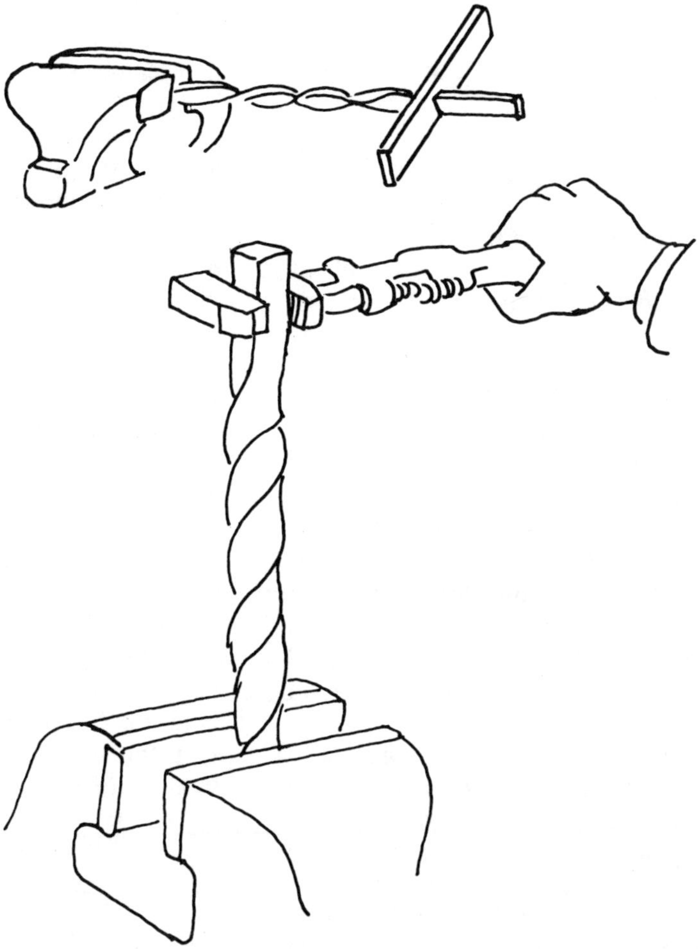

Fig. I.10. Twisting in a vise

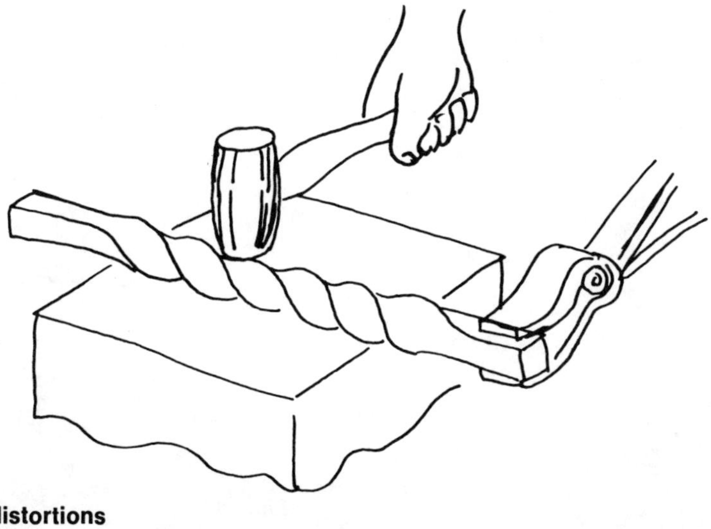

Fig. I.11. Hammering out distortions

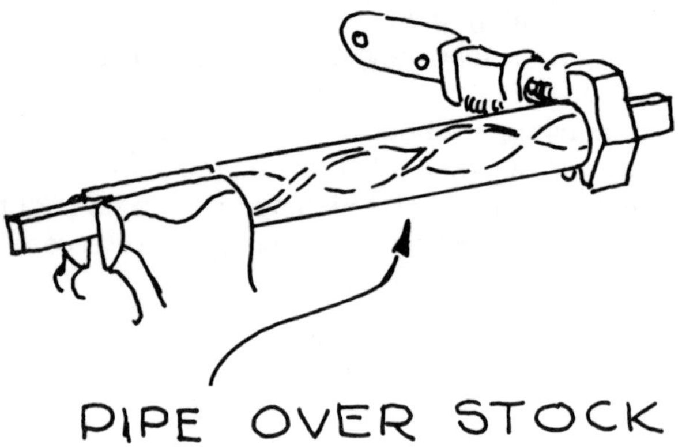

Fig. I.12.

Rather than twist and then hammer the bars, it would be better if a pipe the exact length of the twist and large enough in diameter to just slip over the bar were used to keep the metal straight. See Fig. I.12.

Start the twist with the twisting tool level or parallel with the floor. Turn the twisting tool clockwise until it is in a vertical position (90°), change hand positions and continue pulling the twisting tool clockwise until it is horizontal once again (through 180°). With the same motion, continue until the twisting tool is at 270° (vertical) again. Change hands again and turn the twisting tool until it is once more parallel to the floor (360°). See Fig. I.13. for illustration of these positions.

If only a short section of a long piece is to be twisted, mark the section to be twisted and clamp the bar in the vise or fixture at one of the marks and place the twisting tool on the other mark. For example, if a 150 mm twist is to be formed in the centre of a 900 mm long bar, place the bar in a vise or jig at the 380 mm mark and the twisting tool at the 530 mm mark, as shown in Fig. I.14., to start the twist.

Where the twisting is to be carried out on a piece of metal which will also contain a single or double scroll, the twisting should be completed before the scroll or scrolls are formed. However, occasionally the length of metal will be so short that the twisted portion will interfere with the die on the scroll making machine. If the distance between the scroll and the twist is very small, then the twisting will have to be completed after the scroll or scrolls are formed.

The twisting should then be performed in a vise using one of the slotted wrenches shown in Fig. I.15. (p. 57.), and the following procedure should be followed. 1. Make the required scroll. 2. On the straight part of the scrolled metal, mark off the length to be twisted in steps of about 40 mm. 3. Secure the scroll in a vise so that the bottom mark is level with the top of the vise jaw. 4. Place the slot wrench at the first mark above the vise jaws and twist the bar through 180°. 5. Raise the wrench to the next mark and twist again through another 180°. This can be repeated until the required amount of twist is obtained. See Fig. I.16. (p. 57.)

When forming a twist between a double scroll (that is, a scroll at each end of the same piece of bar), the twist should be made first for the same reasons as mentioned under single scrolls, unless the distance between the scrolls is large. Where the distance between the scrolls is small, the twist will have to be made in a vise using one of the slotted wrenches.

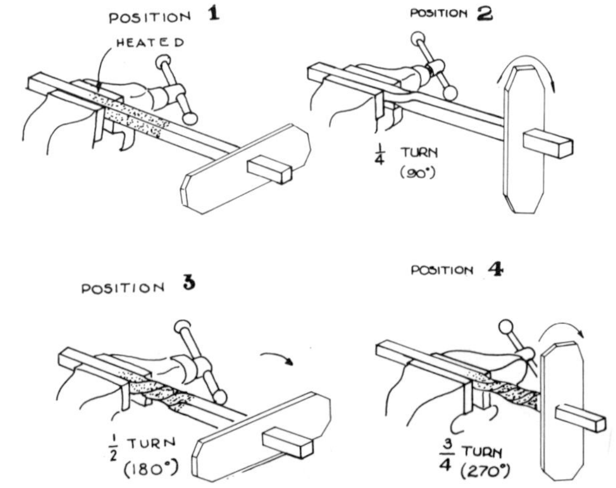

Fig. I.13.

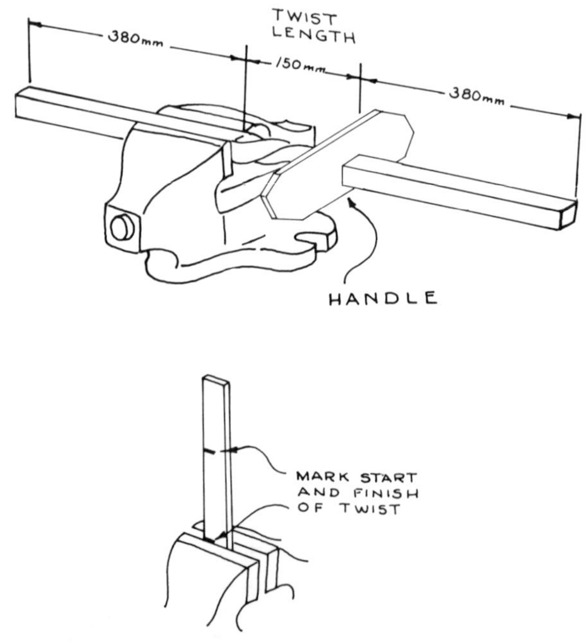

Fig. I.14.

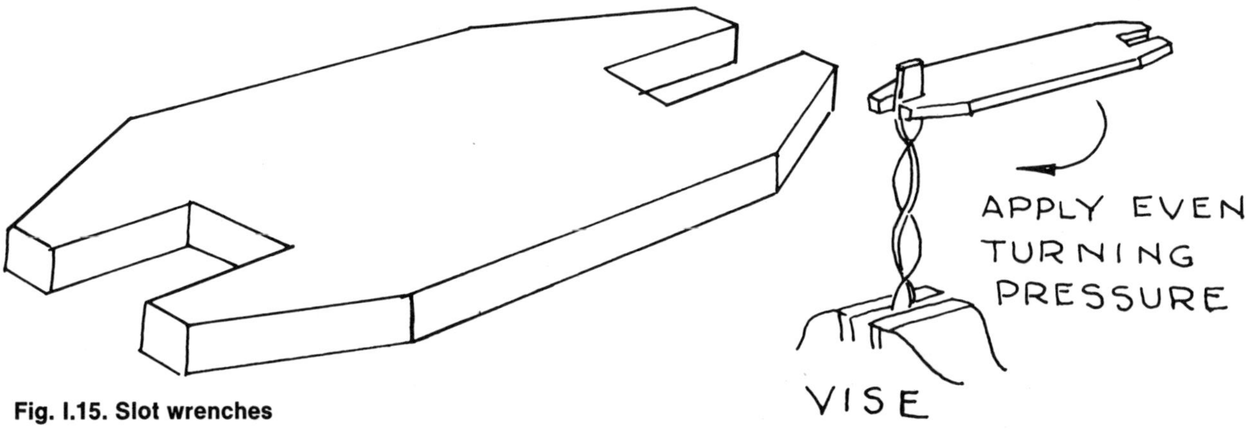

Fig. I.15. Slot wrenches

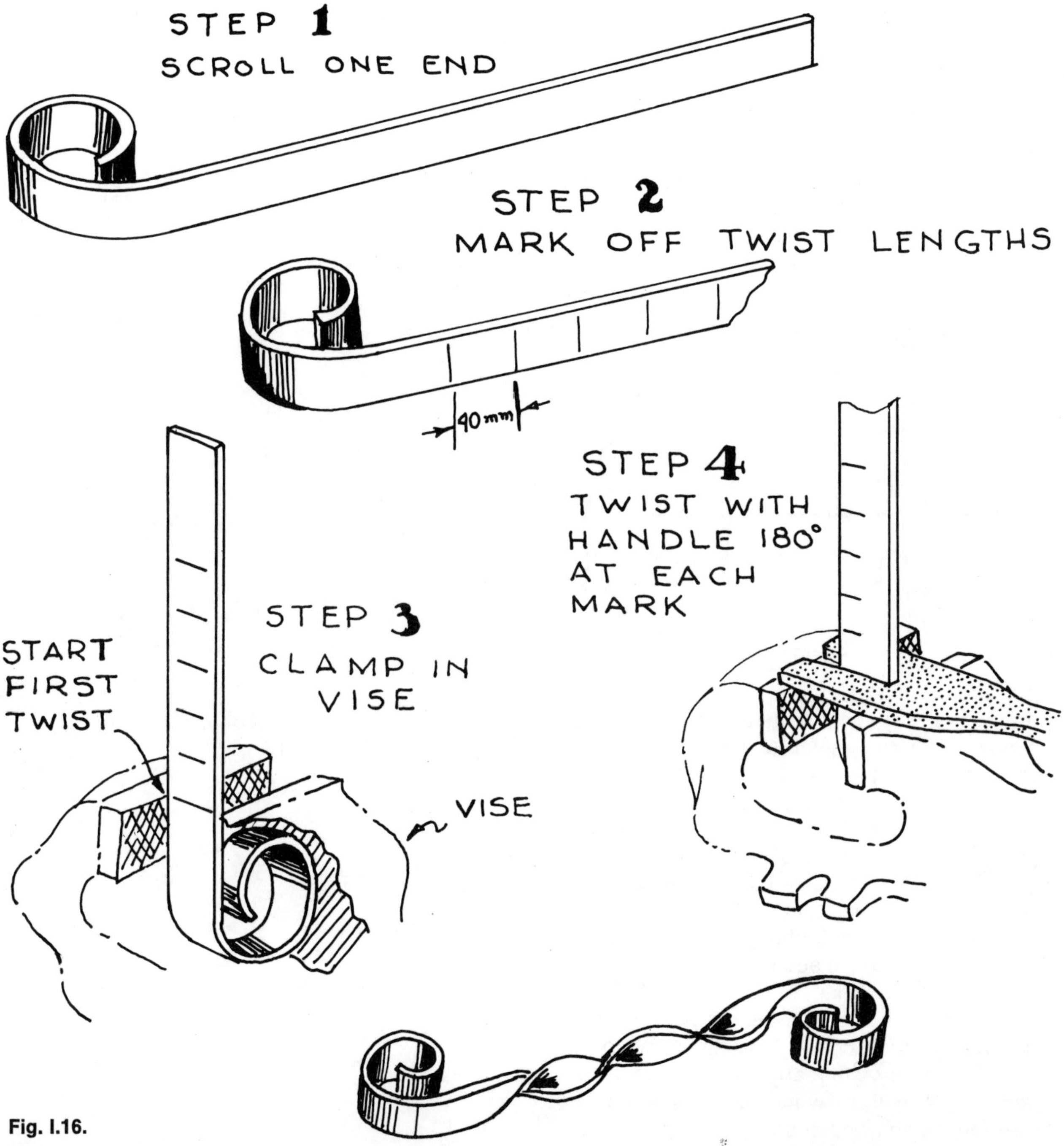

Fig. I.16.

FORGING METALS

Fig. J.1.

Forging is the shaping of metals by hammering, usually while the metal is hot. We normally associate forging with the idea of the old time blacksmith and his smithy as shown in Fig. J.1.

In today's society, forging is an industry in itself and is the subject of many books. In this section, we will only cover the very basic operations and techniques that might be carried out successfully in a small shop with the aid of the most simple pieces of equipment.

Forging derives its name from the forge, or heat source, used to heat the metal for shaping and joining. Most schools have small gas forges in their metalworking area. If a forge is not available, then the oxyacetylene flame can be used for heating and shaping. In fact, because it is portable as well as faster, cleaner, and more concentrated than the forge fire, it has become quite a popular source of heat in many schools.

Besides the necessary heating equipment, there are a few other pieces of equipment used for forging. The most common hammers used are blacksmith's, sledge, and ball peen. The anvil is another important piece of equipment. Certain names are given to parts of the anvil as listed below:

NAME	PURPOSE
1. Body.	Base or body of the anvil.
2. Cutting block.	For cutting operations.
3. Face.	For hammering upon.
4. Hardy hole.	For holding certain forging tools.
5. Horn.	For forming round shapes.
6. Pritchel hole.	For bending and punching holes.

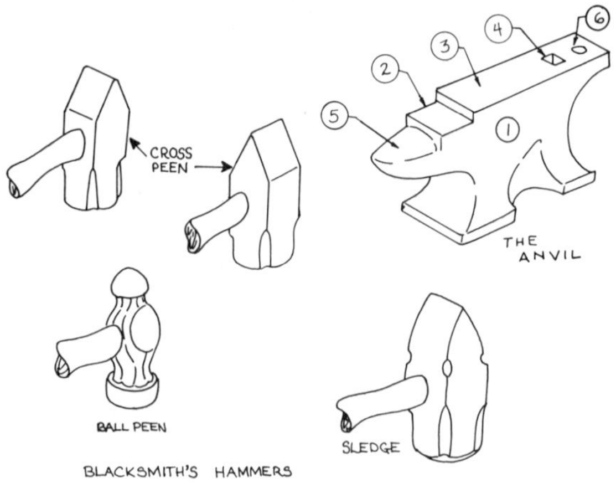

Fig. J.2.

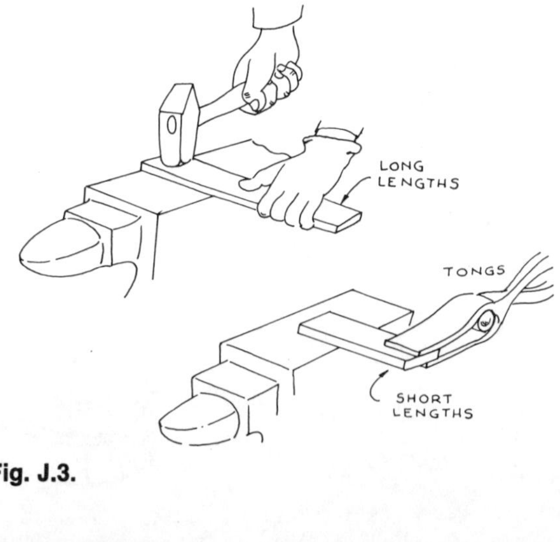

Fig. J.3.

Note. If one end of a long piece of metal is to be hammered, it is possible to hold the opposite end by hand, however when hammering short pieces of metal it is necessary to hold the metal with tongs as in Fig. J.3. (p. 58.) The common types of tongs used are shown in Fig. J.4.

NAME	PURPOSE
1. Curved lip tongs.	For holding round metal.
2. Pick-up or gad tongs.	For holding thick flat metal.
3. Straight lip tongs.	For holding thin metal.
4. Straight lip tongs with notch.	For holding square or round bar.

A small amount of anvil tools are necessary to complete the forging equipment. See Fig. J.5.

NAME	PURPOSE
1. Cutting tools.	For cutting metal on, usually placed in hardy hole.
2. Cutting tools.	For cutting through metal.
3. Flatter tools.	For flattening and smoothing metal.
4. Fuller or grooving tools.	For grooving metal.
5. Punching tools.	For punching square or round holes.
6. Swaging tools.	For shaping metal round.

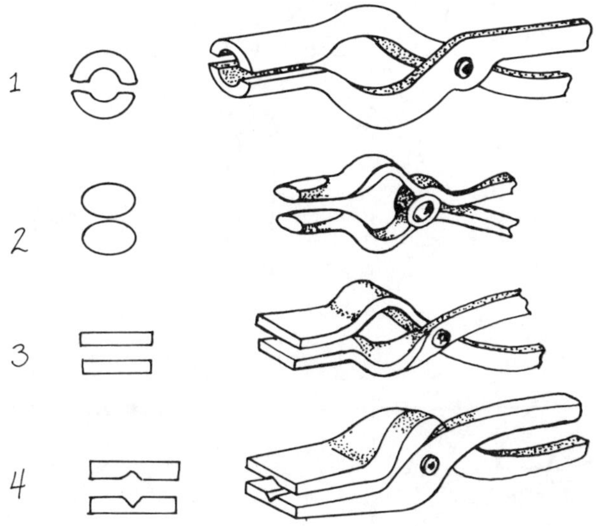

Fig. J.4.

Fig. J.5. Anvil tools

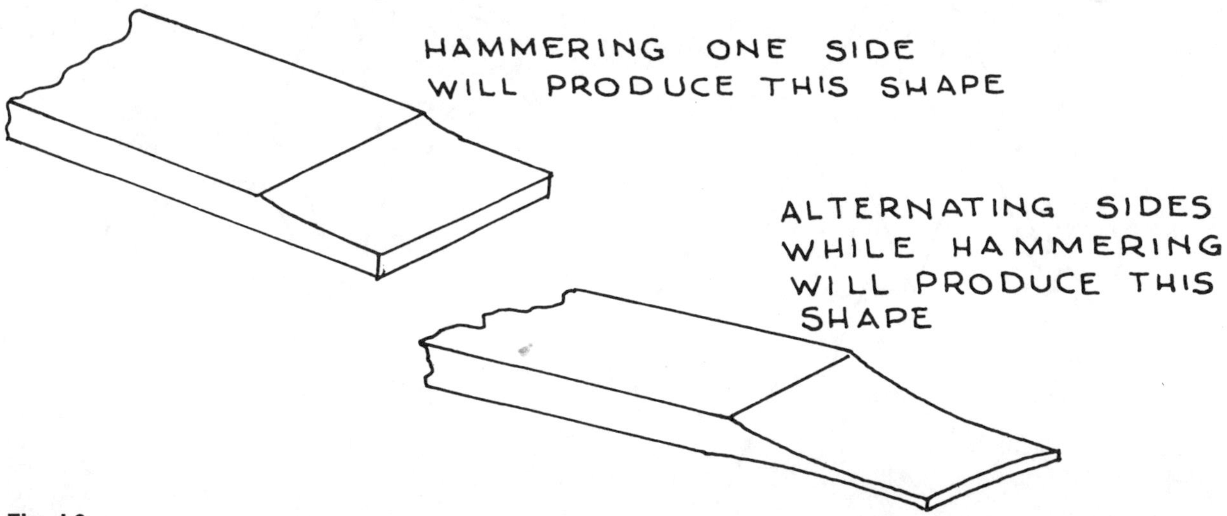

Fig. J.6.

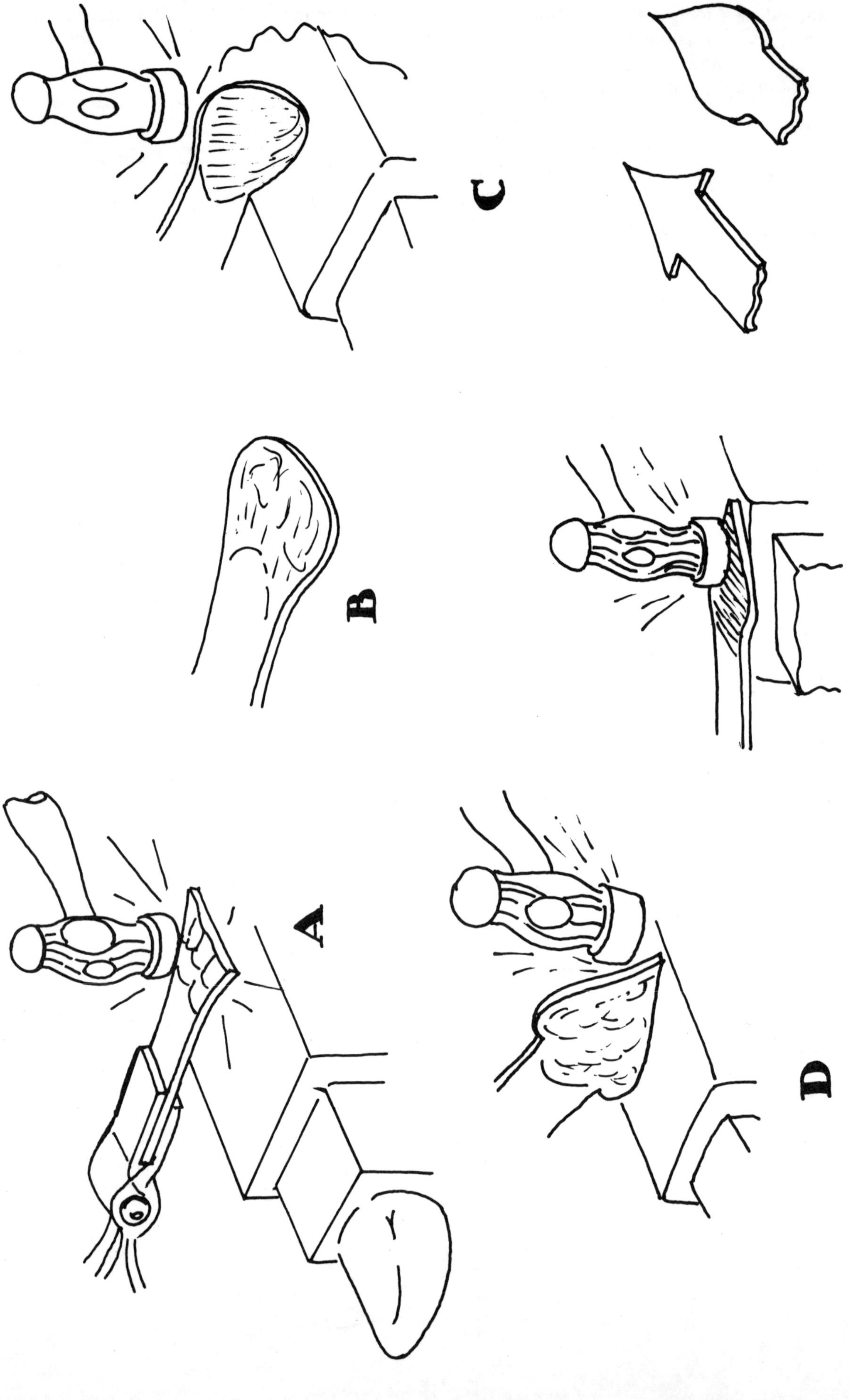

Fig. J.7. Forming a vee or arrow

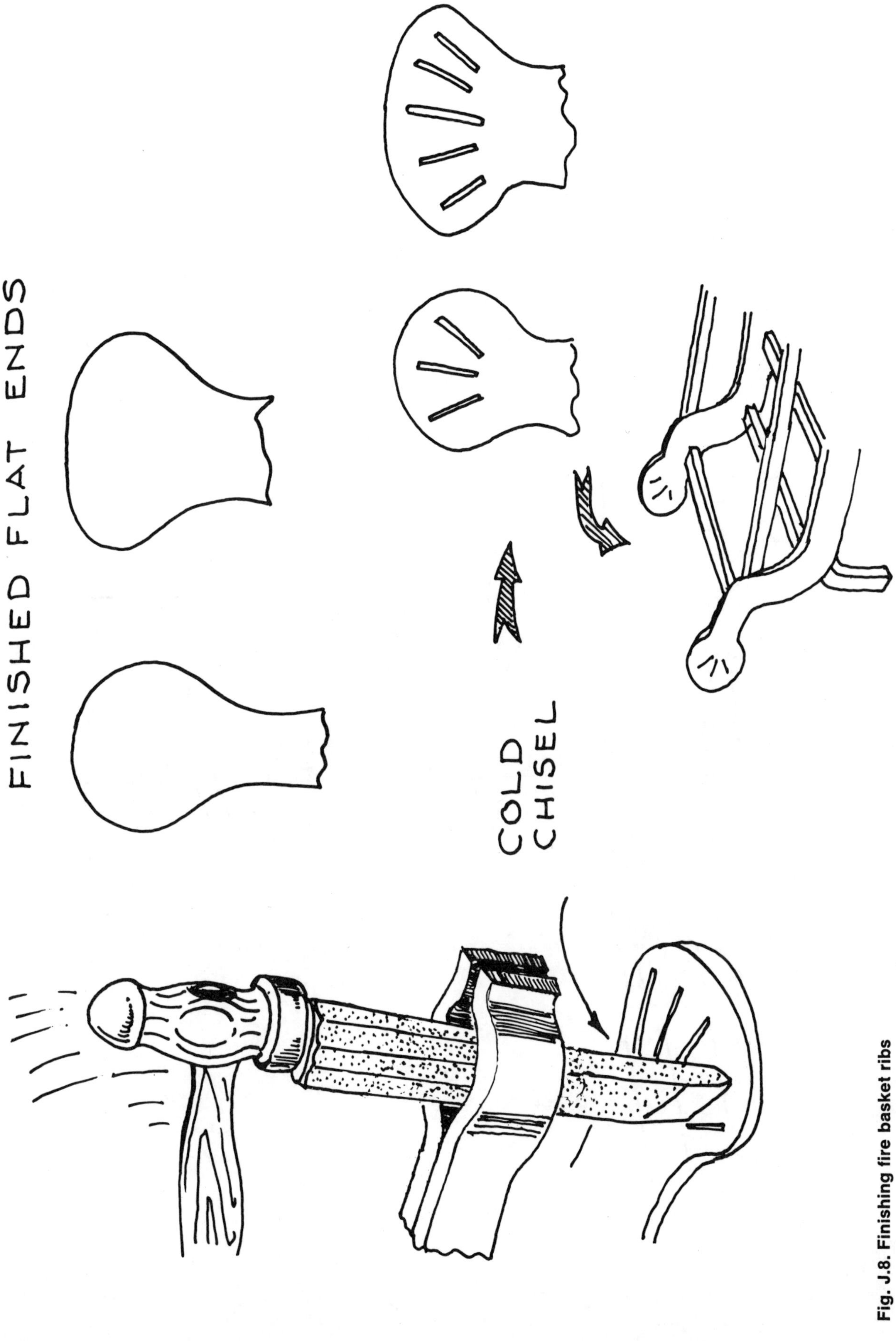

Fig. J.8. Finishing fire basket ribs

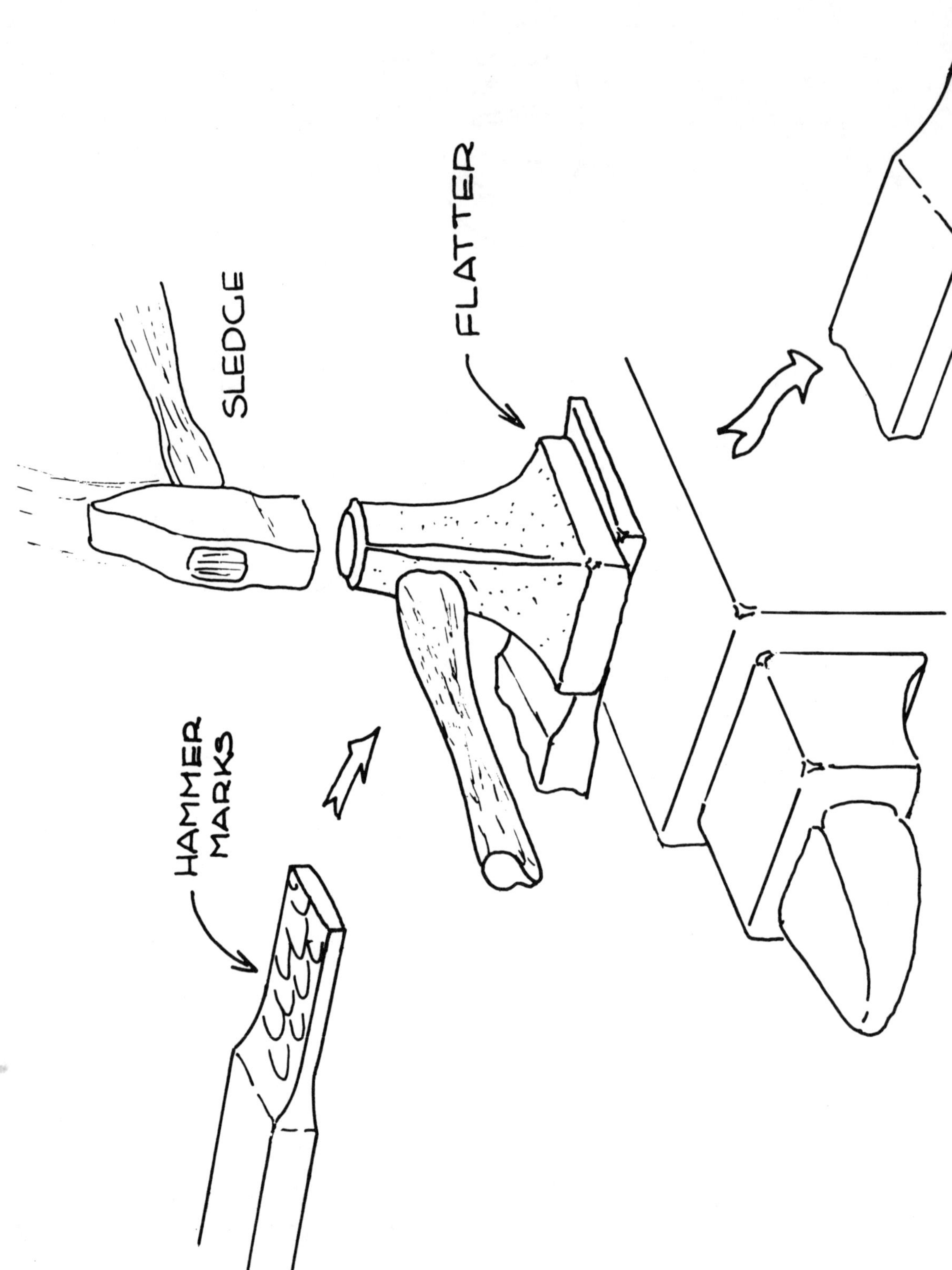

Fig. J.9. Smoothing hammer marks with a flatter

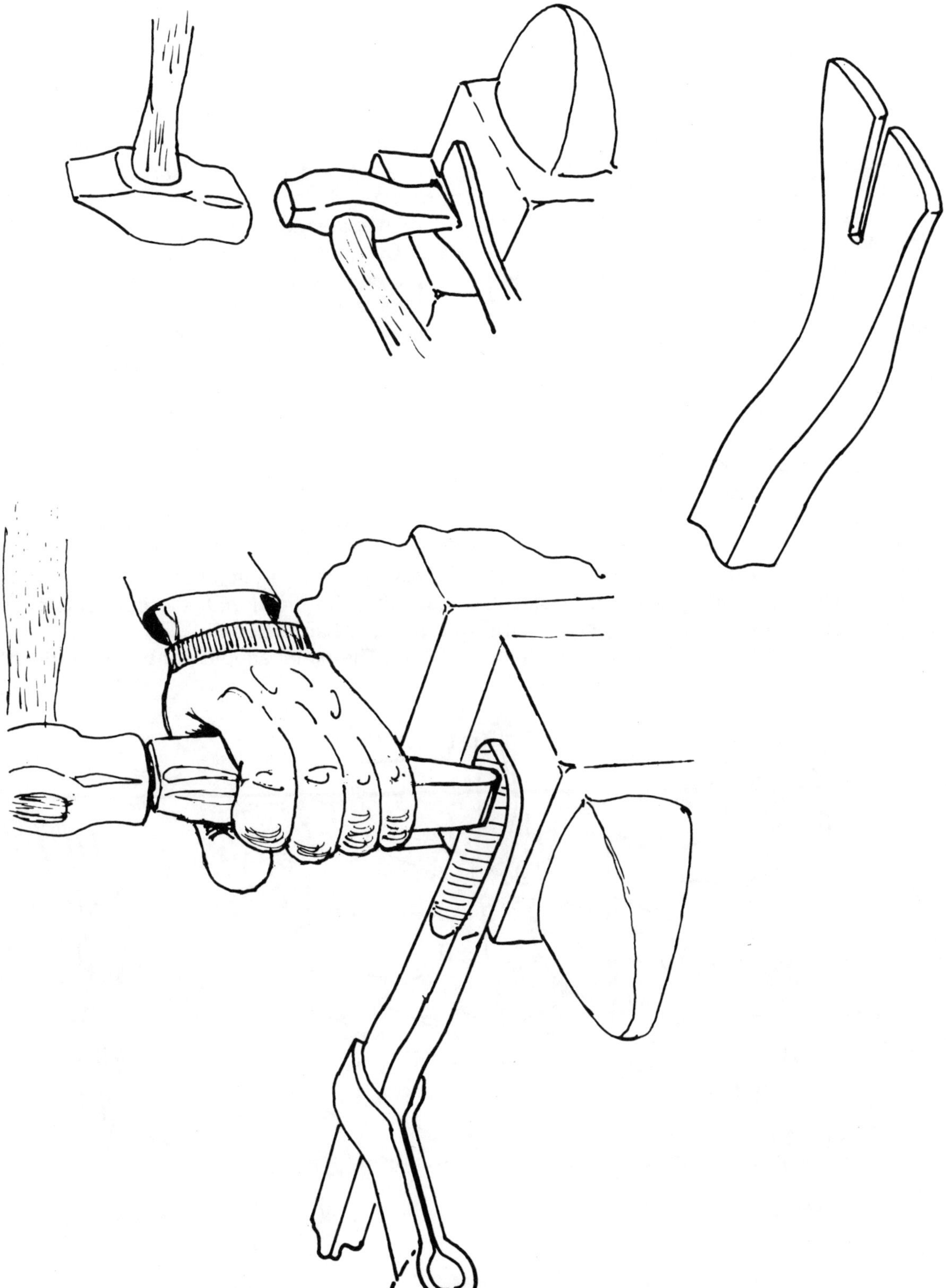

Fig. J.10. Splitting metal with a chisel

One common forging operation is the forming of a vee-shaped or arrow–shaped object such as might be used on ornamental railings, fire baskets, andirons, or the flukes on an anchor. Using a blacksmith's hammer or sledge, the metal should be drawn down to the approximate width desired. Before starting the first flattening operation, it must be decided if the shape is to be on one side or in the centre of the metal bar. See Fig. J.6. (p. 59.)

Heat the metal and keep it as hot as possible. Reheating should be carried out frequently. Turn the metal on its side and hammer the metal at the point with a ball peen hammer, working slowly back from the point. The sides should be alternated frequently and occasionally the metal should be laid flat and any distortion hammered out. See Fig. J.7. (p. 60.)

It does not require much practice to form varieties of the above design. The two styles shown in Fig. J.8. (p. 61.) are very popular for finishing the ribs of a fire basket. The grooves or marks are accomplished by striking the hot metal with a cold chisel. The harder the blow, the deeper the groove will be.

When forging, the metal should be as hot as possible. The hotter the metal, the easier it will be to shape and form. However, since care should be taken not to burn it by overheating, the metal should be heated to a white heat without excessive sparking.

Hammering metal will often leave undesirable marks on the surface, especially for those inexperienced in hammering. Much of this unsightly marking can be flattened or smoothed out by the use of the flatter as shown in Fig. J.9. (p. 62.)

Note. Many forging operations require the cooperation of two students, one to hold and turn the metal and one to strike the metal. The metal should never be hammered unless it is securely held because it may fly loose and strike someone.

Sometimes it is necessary to split a piece of metal to form a fork–shaped end; e.g., to make a fishtailed decoration at the end of a piece of metal as shown in Fig. J.10. (p. 63.) This is accomplished by placing the hot metal flat on the cutting block of the anvil and splitting it with a cold chisel or one of the blacksmith's cutting chisels. See Fig. J.10.

Although the oxyacetylene cutting torch provides an excellent, fast, neat, and efficient method of cutting metal, occasionally it may be necessary to cut the metal with forging tools. This is done in much the same manner as splitting the metal with the cold chisel, cutting chisel, or hardy. See Fig. J.11.

Note. The metal should not be cut through from

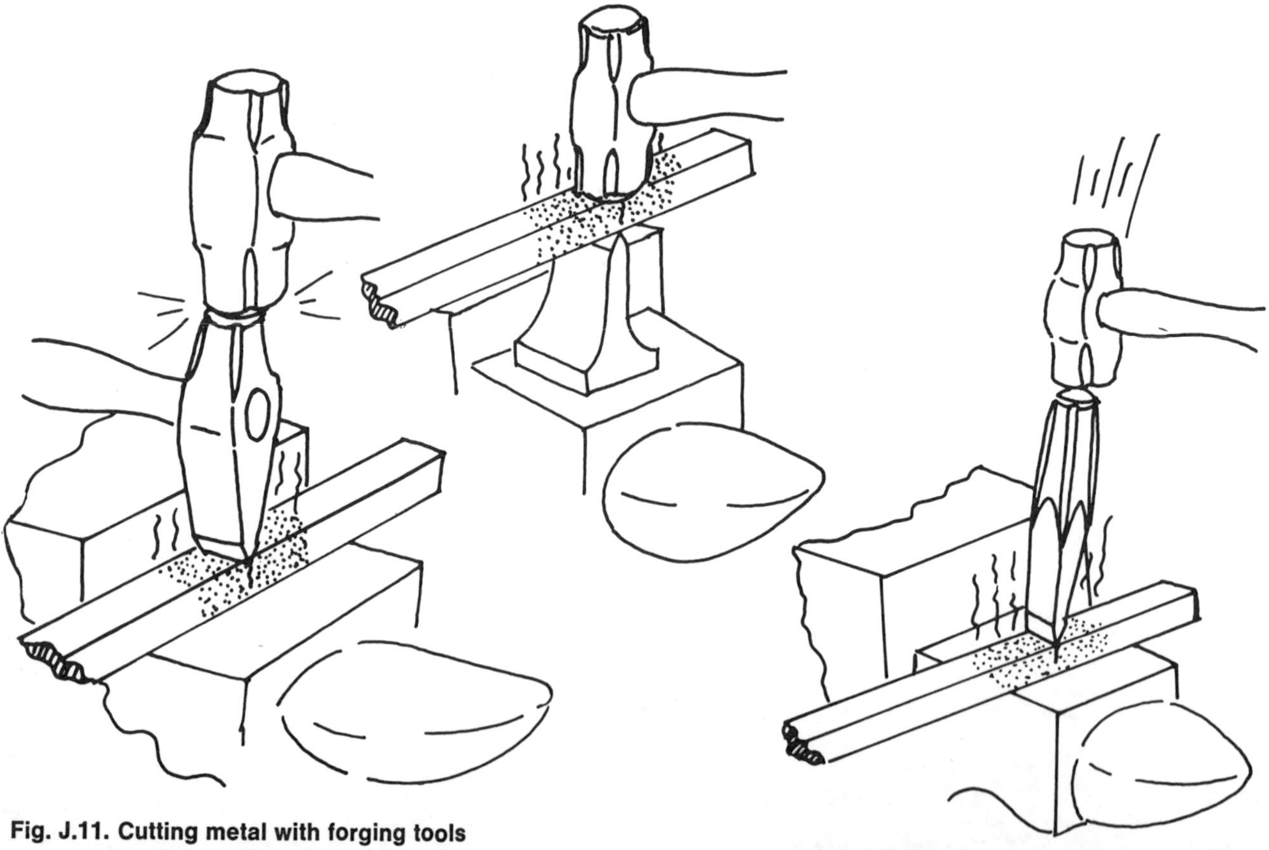

Fig. J.11. Cutting metal with forging tools

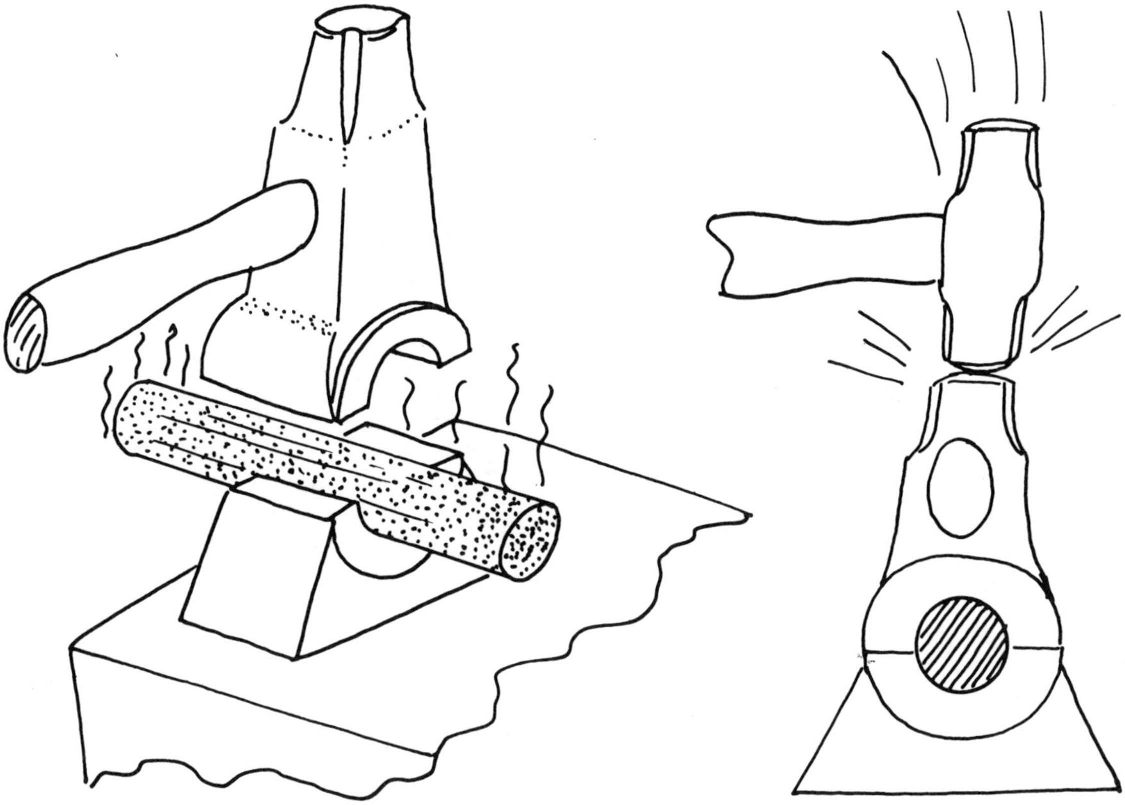

Fig. J.12. Swaging hot metal

the one side but should be rotated after each hit.

There are many other forging operations that may be required for certain other projects; some of these are shown below. It is advisable that the student practice on a piece of scrap metal of the same type to be used in the project.

To swage a hot piece of metal, it is placed between the swaging tools and the top swage is struck with a hammer. The metal is continually rotated after each strike to produce a smooth round surface. See Fig. J.12.

Grooving metal is similar to both cutting and swaging except that different tools (fullers) are used. The end result, of course, is a groove rather than a cut or a smooth round shape. See Fig. J.13. (p. 66.)

To punch holes in metal, simply heat the metal, place it over the hardy hole of the anvil, hold the desired punch in place and strike with a hammer. See Fig. J.14. (p. 66.)

Note. Punches may have square, round, or triangular ends and with the assistance of the school machine shop special punches could be designed for a specific shape. The metal should not be punched through from the one side only, because it will leave a burr on the underside. The metal should be reheated and turned over when the punch has penetrated halfway through it.

To form right angle bends on metal, small pieces may be heated and placed in the pritchel or hardy hole and bent manually. Larger pieces should be heated and bent over the edge of the anvil face by hammering. See Fig. J.15. (p. 66.)

Arcs and circles are formed by heating the metal as usual, placing it over the horn of the anvil, and striking it with a hammer. Care should be taken not to advance the metal too far each time it is struck with the hammer as this will cause flat spots in the arc or circle. See Fig. J.16. (p. 66.)

Drawing down the metal to form pointed or flattened and pointed ends (such as would be necessary for a fireside poker) is done as follows: if the metal is square bar, it is heated and placed on the anvil face, then struck with a hammer beginning at the point and working back up the metal for a distance of 50-75 mm. The bar is struck on alternate sides and is also held at a slight angle to the face of the anvil as it is struck. See Fig. J.17. (p. 66.) A flatter should be used here to finish the surface; if a flatter is not available, the metal could be finished with a hand grinder.

If the square bar is to be drawn down to a round tapered point, the hot metal is first struck on four alternate sides for first flattening, then on the four corners to form an octagonal shape, and finally all corners or edges are hammered to form a round shape. See Fig. J.18. (p. 67.)

To make the hook for the poker follow the same steps as above. Use a long piece of metal, bend it slightly and cut it off at the desired length. Then weld the hook to the main stem of the poker. See Fig. J.19. (p. 68.)

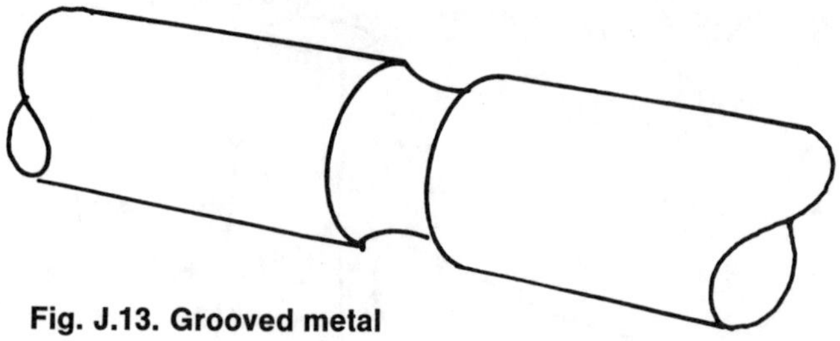

Fig. J.13. Grooved metal

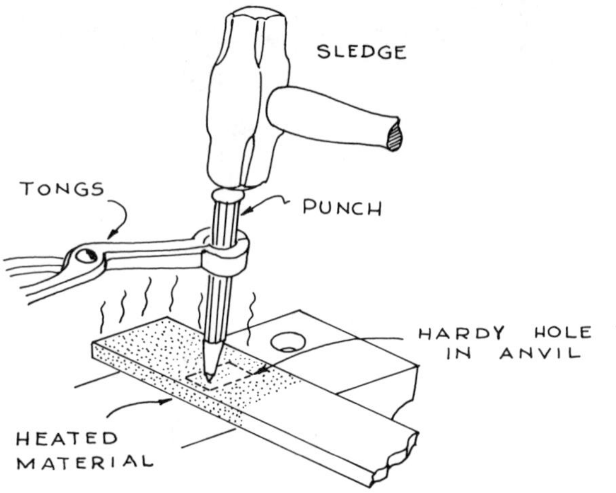

Fig. J.14. Punching holes

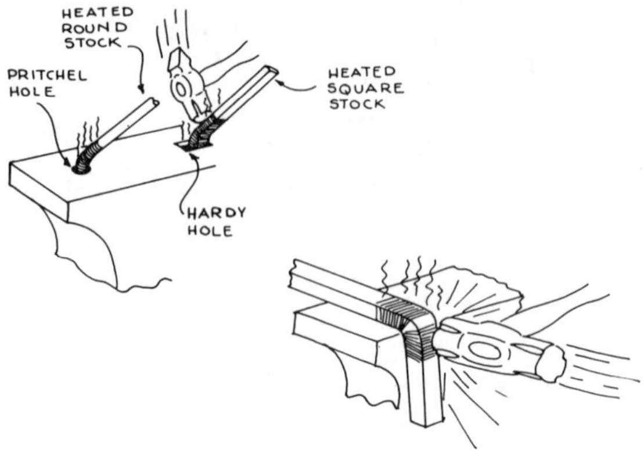

Fig. J.15. Forming right angles

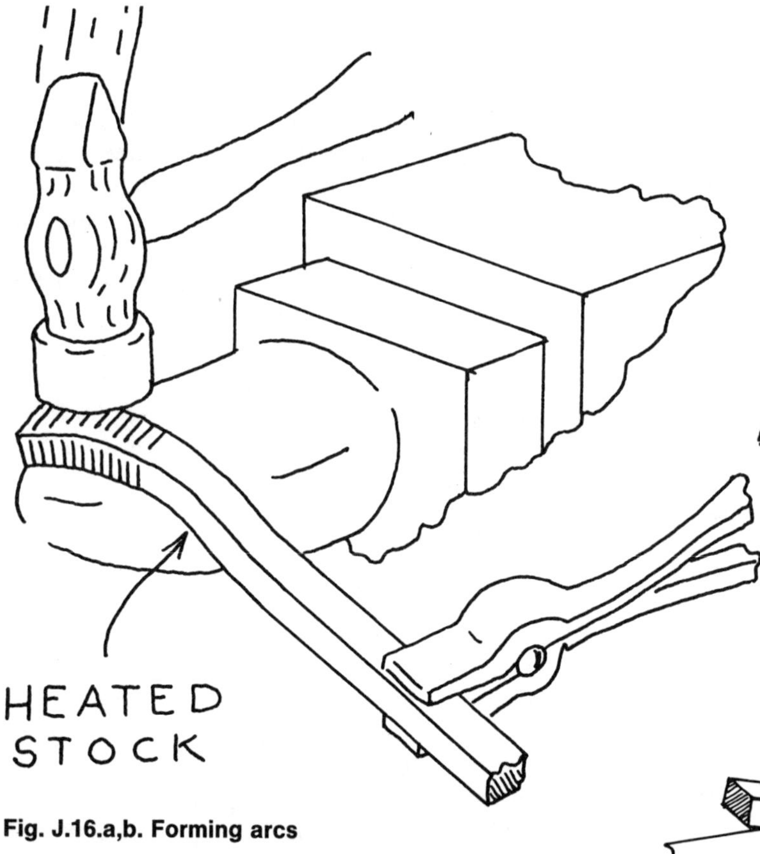

Fig. J.16.a,b. Forming arcs

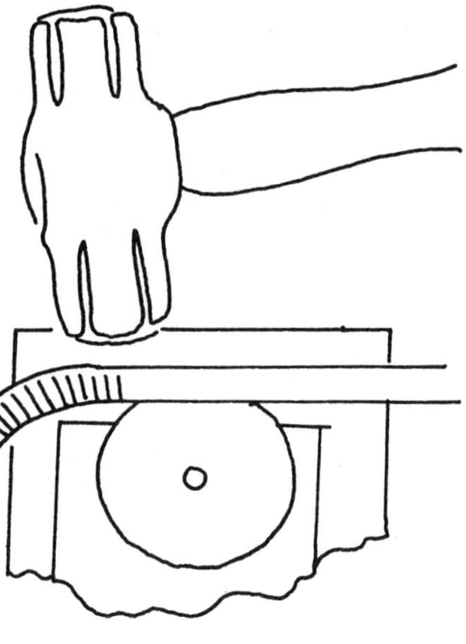

Fig. J.17. Forming points

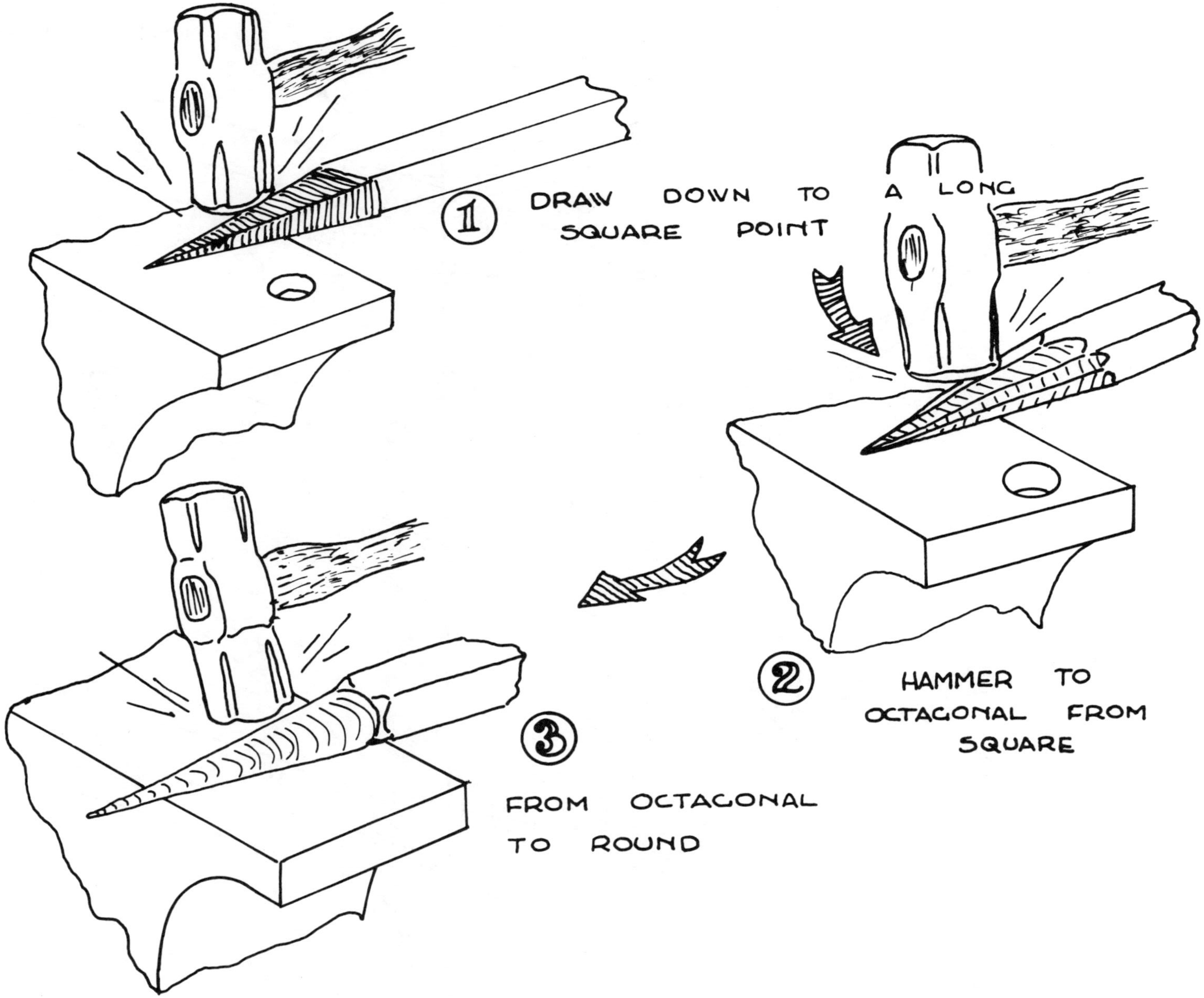

Fig. J.18. Rounding and tapering square bar

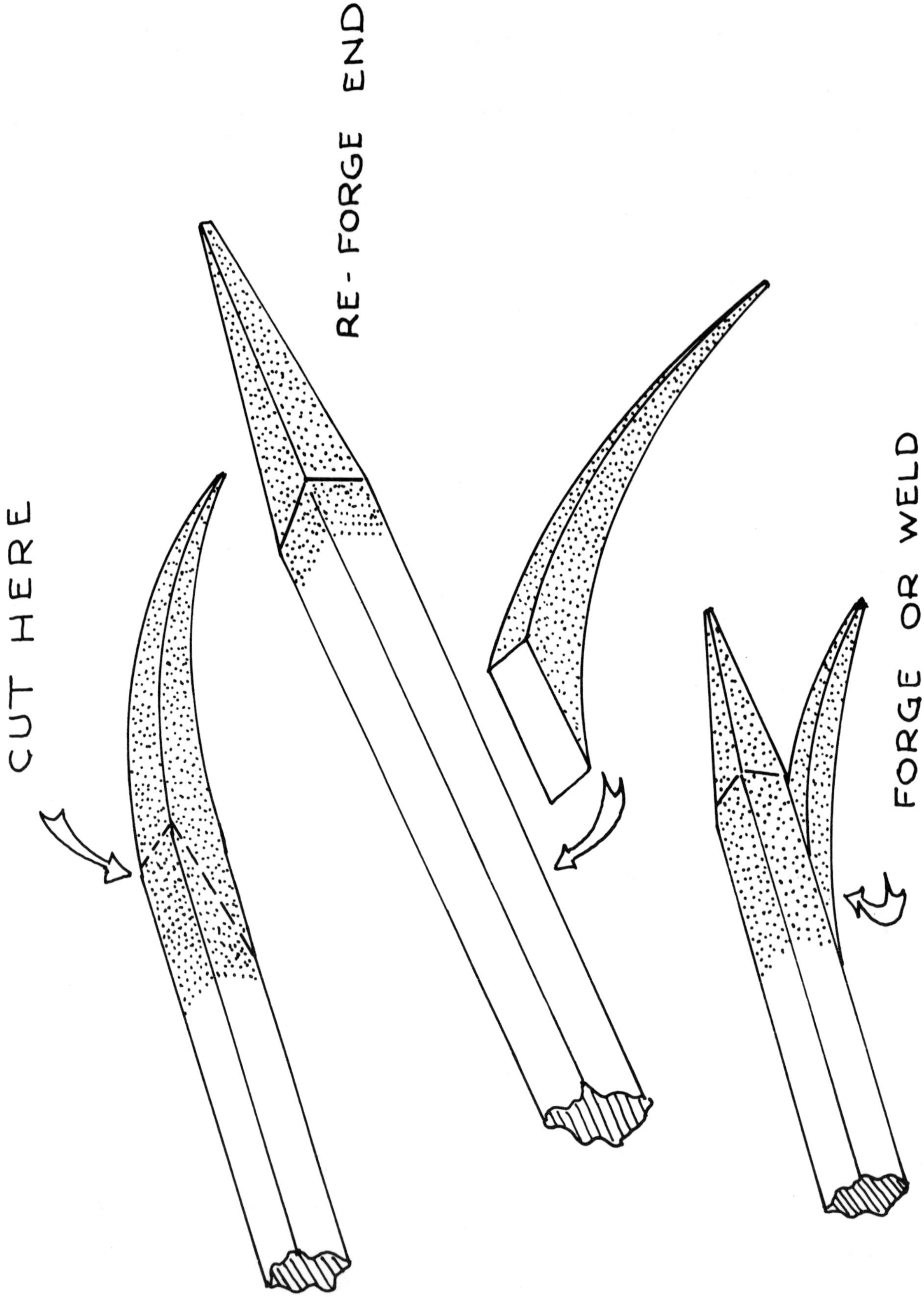

Fig. J.19. Making a hook for a poker

FINISHING, CLEANING, AND PAINTING METAL PROJECTS

SECTION K

If metal projects are left in their original conditions, they will discolour by dulling or rusting. Therefore, the student is advised to apply some type of protective coating to the metal. Not only will this coating protect the metal, in many instances it will improve the overall appearance of the project. Before the coating is applied however, the project must be thoroughly cleaned and all grease, oil, rust, scale, weld splatter, weld slag, and weld flux must be removed. See Fig. K.1.

This is also the time to inspect the project closely for unsightly lumps, nicks, grooves, and any other flaws. Protrusions (lumps) should be removed by filing or grinding. Dents, nicks, and grooves if shallow can also be removed by filing or grinding, or they could be filled up with welding or a commercial type body filler (commonly used in auto body repair work) and then smoothed as described.

Note. It is much better to repair *before* applying a protective coating *than after,* therefore, a little extra care at this point is advised.

Weld splatter, and deep grinding or filing marks will show up prominently after painting, so make sure all splatter is removed and when grinding or filing start with a coarse grit or file and finish with a smooth grit or file if possible. See Fig. K.2. (p. 70.) When grinding do not move the grinder in circles, as these marks will also show; instead move the grinder in straight lines.

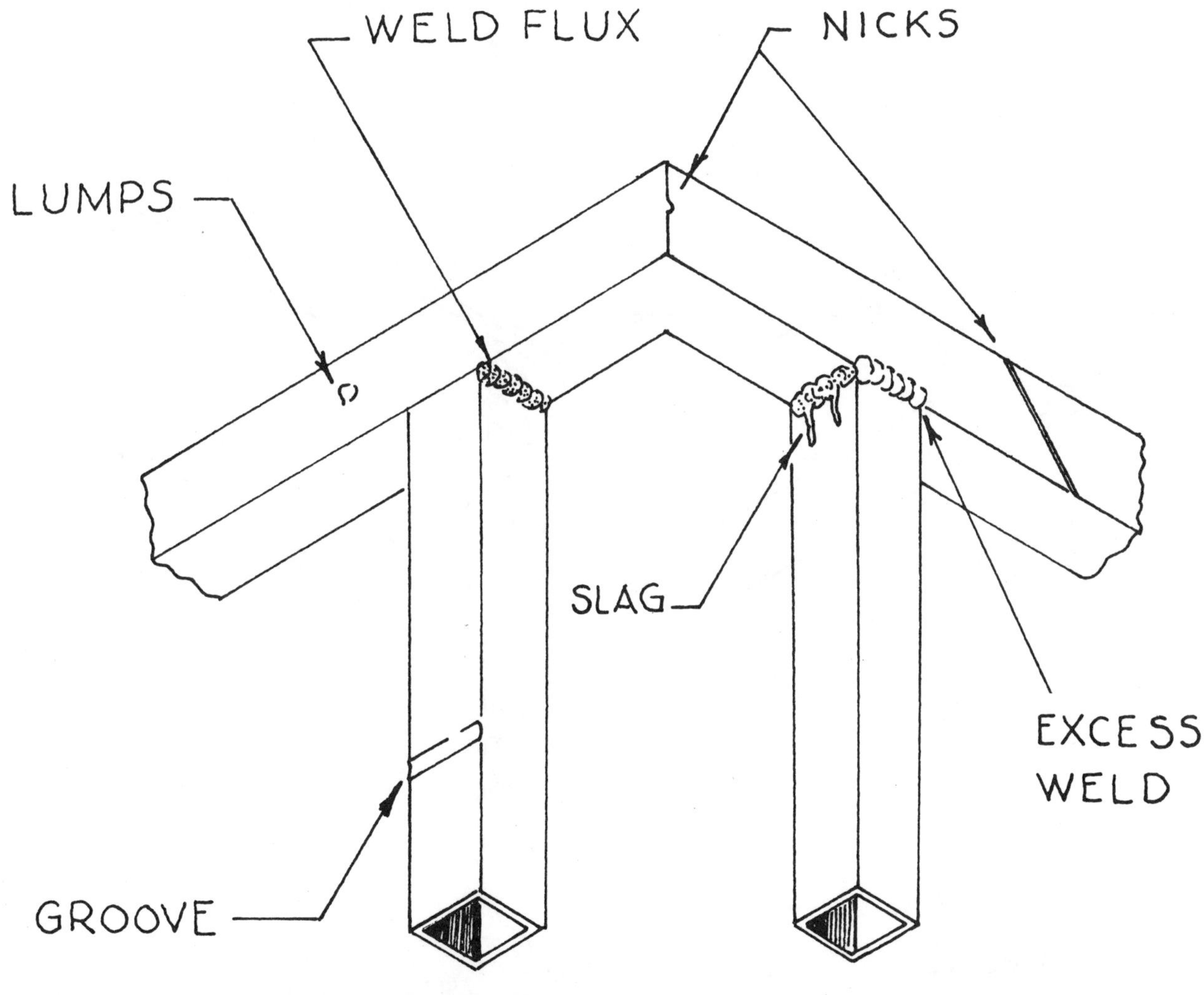

Fig. K.1. Clean the project before applying a finish

CLEANING METAL

The actual method of cleaning will be determined by the condition of the metal.

To remove.	Use.
Weld splatter. Heavy scale. Rust.	Power grinder/sander. Power wire brush. Flame & wire brush. Blasting (sand or shot).
Light scale. Light rust.	Hand wire brush. File. Abrasive cloths.
Weld slag. Weld splatter. Weld flux.	Chipping hammer. Hand wire brush. File.
Oil. Grease.	Kerosene. Lacquer thinners. Detergents. Commercial metal prep.

Note. Other agents such as a weak solution of muriatic acid or solutions of caustic soda can be used *but they must be handled very carefully and with extreme caution. They should only be used under supervised conditions.*

FINISHING METAL

After the metal has been cleaned and any defects repaired, the finish should be applied as soon as possible. If the metal is to be left in its natural state, all that is required is the application of a coat or coats of wax or clear lacquer.

There are a few variations that can be used to produce a different although natural finish. One variation that can be used is similar to the "blueing" techniques sometimes used in machine shops where the metal piece is polished with abrasives and steel wool to a high finish, heated, and placed in a tank of linseed oil. Similarly, if a metal project is polished, heating with the torch will make the metal change colour depending upon the amount of heat applied. The hot project can then be dipped in linseed oil or rubbed with a cloth soaked in linseed oil after it has cooled slightly *(extreme caution should be used here to see that the rag does not catch fire).* Finally a wipe with a damp clean rag will help to set the colour. The student may have observed this colour change in metal during previous heating and welding exercises. A test should be carried out on a

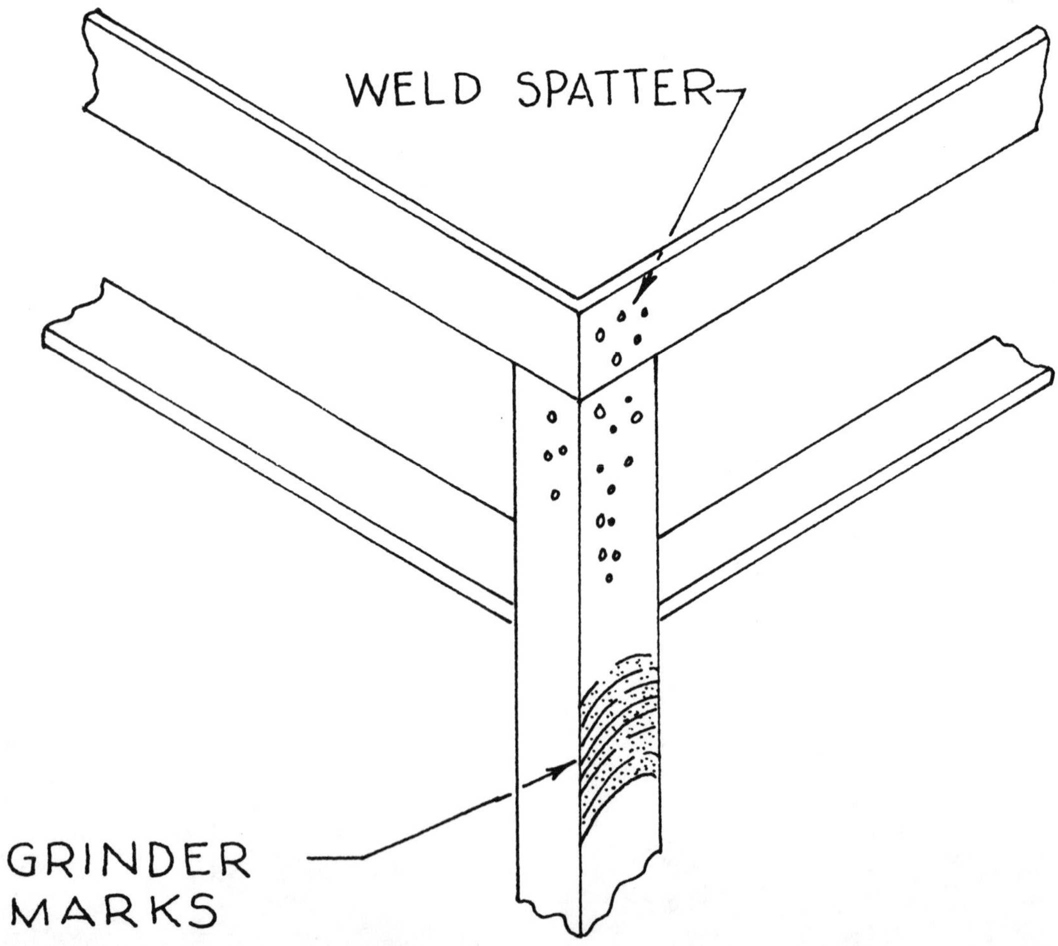

Fig. K.2. Remove splatter and grinder marks

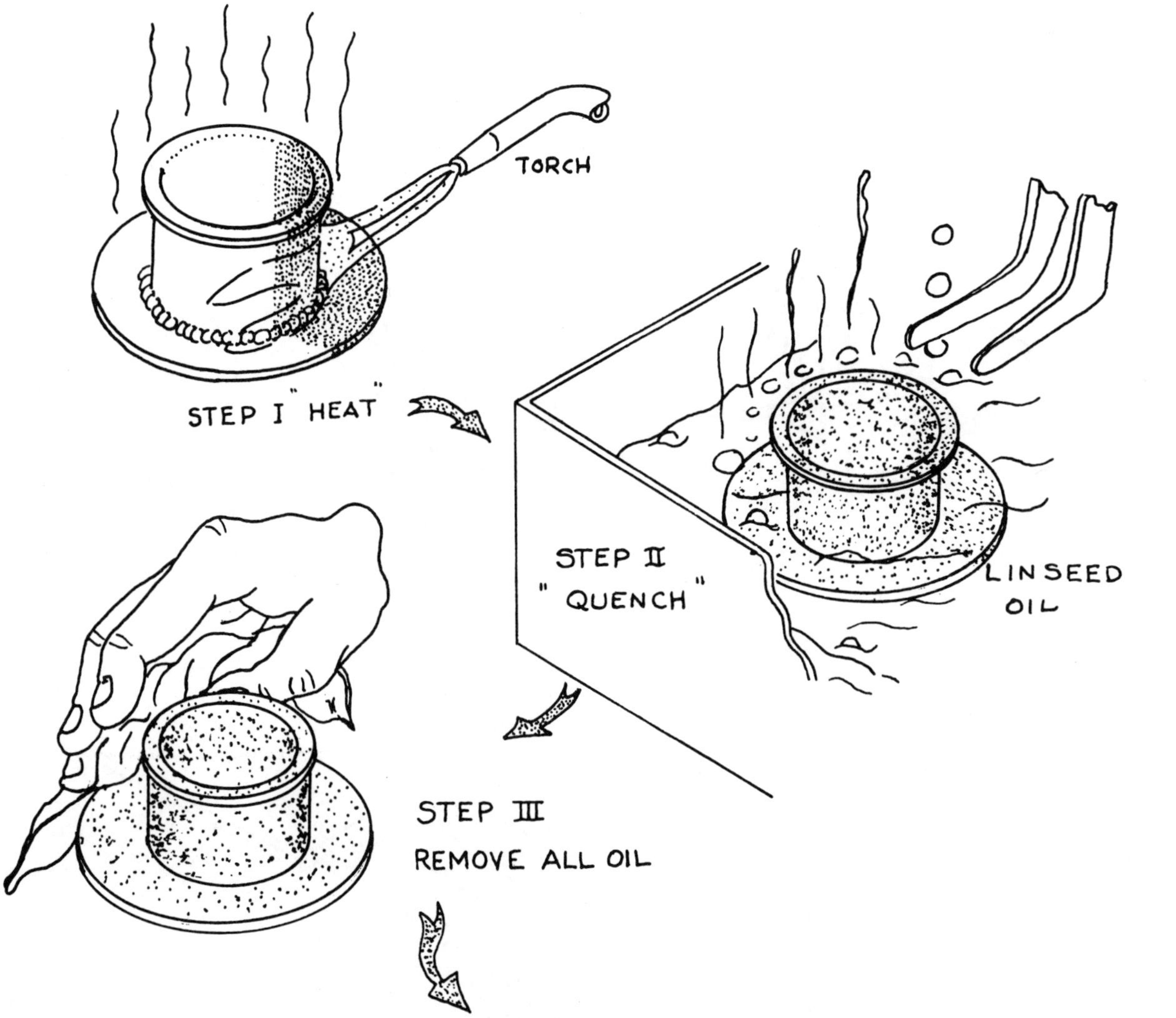

Fig. K.3. Finishing metal

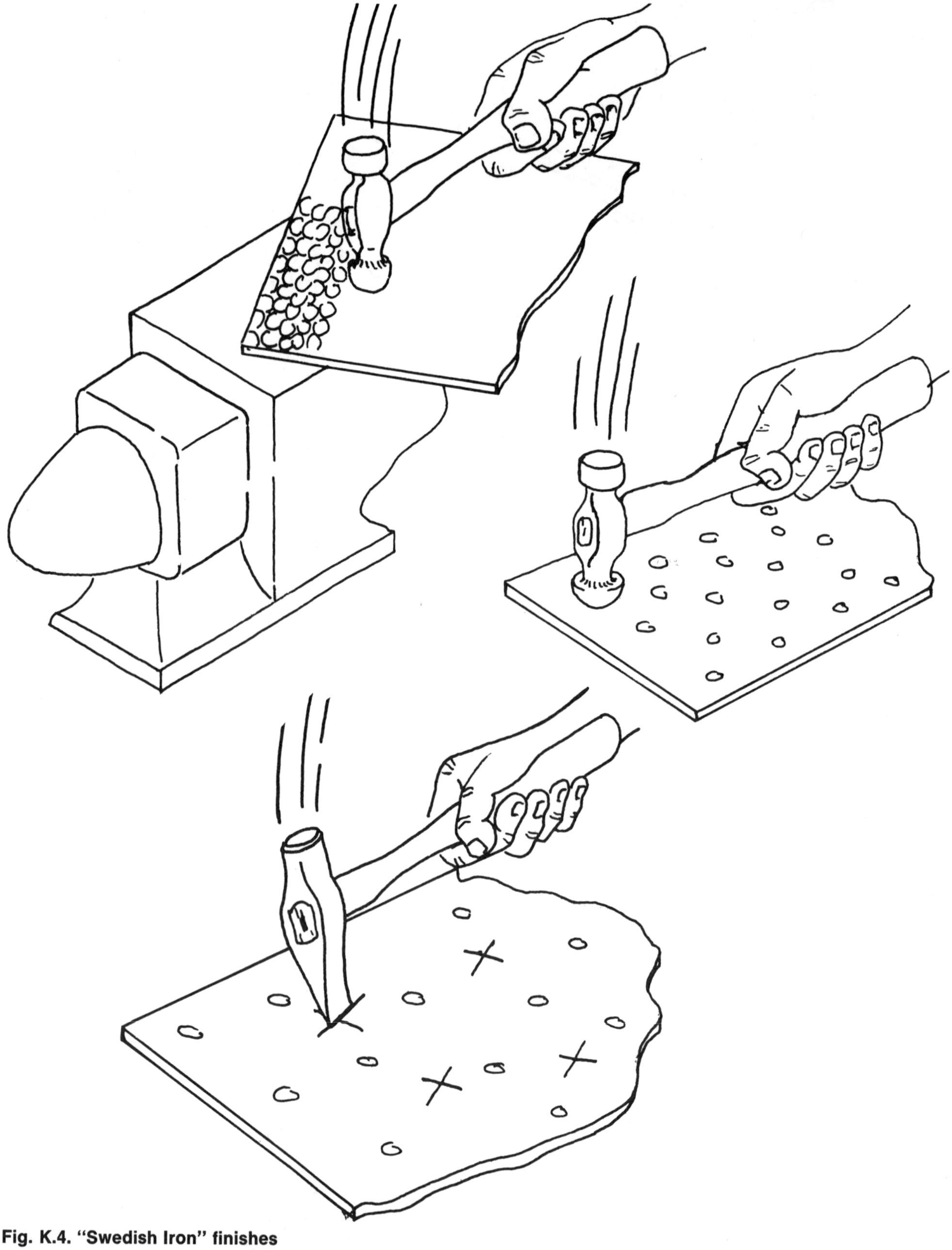

Fig. K.4. "Swedish Iron" finishes

small sample piece of metal to determine the amount of heat required and the method to be used.

Another variation is to hammer the metal to produce an antique or "Swedish Iron" finish. The piece of metal to be hammered is placed on a solid surface such as a blacksmith's anvil or a block of heavy steel clamped in a vise. The metal is then struck with a hammer, the heavier the blows the deeper the indentations. Use the same force for each blow. The design produced will depend on the type of hammer

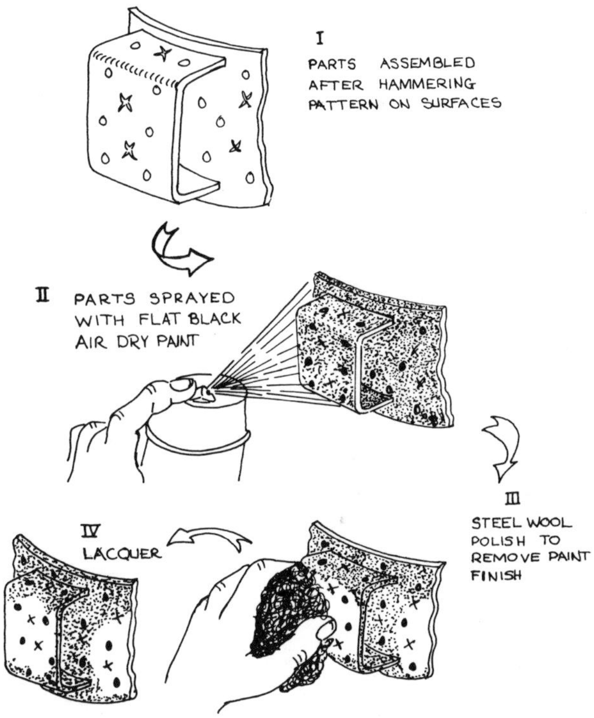

Fig. K.5.

used. The ball end of a ball peen hammer is most commonly used for this purpose.

Try to space the distance between the blows so that a uniform pattern is achieved. This pattern may be an overlapping design or a set space between each mark. Since the metal pieces may distort or bend with the hammering, it is better to hammer the pieces before joining. The distortion or bending may be corrected by turning the metal piece over and hammering. If one side only is to be marked, the straightening should be done with a rubber or wooden mallet to prevent marking. Usually however, both sides are marked. For this operation a piece of soft metal such as copper should be placed between the metal being hammered and the anvil or block to prevent damage to the first side.

Sometimes the hollows left by the hammer are accented by darkening. This is done by painting the project after it is assembled, then wiping it with a clean rag while the paint is wet or rubbing it with steel wool after the paint is dried. If the rag or steel wool is drawn evenly across the surface with little pressure, some paint will be left in the hollows. After allowing the project to dry, a coat of clear lacquer will protect the finish. See Fig. K.5.

PAINTING

Painting is another method of finishing the project. Many colours and types of finish can be achieved using paint. Paint is usually applied by brushing or spraying; small projects may even be dipped in paint. The types of paint most commonly used are primers, enamels, and lacquers.

Primers are paints that are applied before the finish coat. They protect the metal from rust, and help the finish coat to adhere better. Check manufacturers specifications to assure that the primer used matches the type of paint finish. After you apply primer, the project should be wiped with a clean rag or steel wool.

Enamels are used as a finish coat. They dry slowly and if applied by brush do not show brush marks as readily as lacquers. Lacquers usually dry quickly and the finished surface is more resistant to wear than the enamels. Better results will be obtained if lacquer is sprayed instead of brushed. If more than one coat of lacquer is to be applied by brushing, it will be found that the previous coat will soften and become tacky so that the brush will not move freely. Enamels and lacquers may be flat (dull) or glossy and can be obtained in a variety of colours.

Instead of painting the projects one colour, sometimes a very pleasing effect can be obtained by a combination of colours, e.g., black/gold or white/gold. The gold can be applied by spraying on top of the black from a distance further away than that required for actual painting. This is often called misting or overspray and the results will be small flecks or spots of gold on a black (or white) background. Another method is to spray the gold in small bands over the projects.

Spraying is preferable to brushing, as it will produce a more uniform coat. However, there may be a lot of waste and expense. Aerosol cans are most convenient, especially for small projects, but will prove to be the most expensive method of painting.

Note. The metal must be clean. All repairs should be done before painting. The metal should be primed (especially if it is to be used outdoors). The correct type of primer should be used. All safety precautions should be observed, i.e., fire hazards should be avoided. Ventilation, masks, and special containers for oily or paint–soaked rags should be used. Painting, particularly spraying, should not be carried out near welding or open flames. If possible, a separate paint room or booth should be used. All paints, solvents, and so forth, should be kept locked in a specially designated area away from all flames and clearly identified. All equipment should be thoroughly cleaned after use. A special container must be kept for oily rags or rags soaked in paint or solvents.

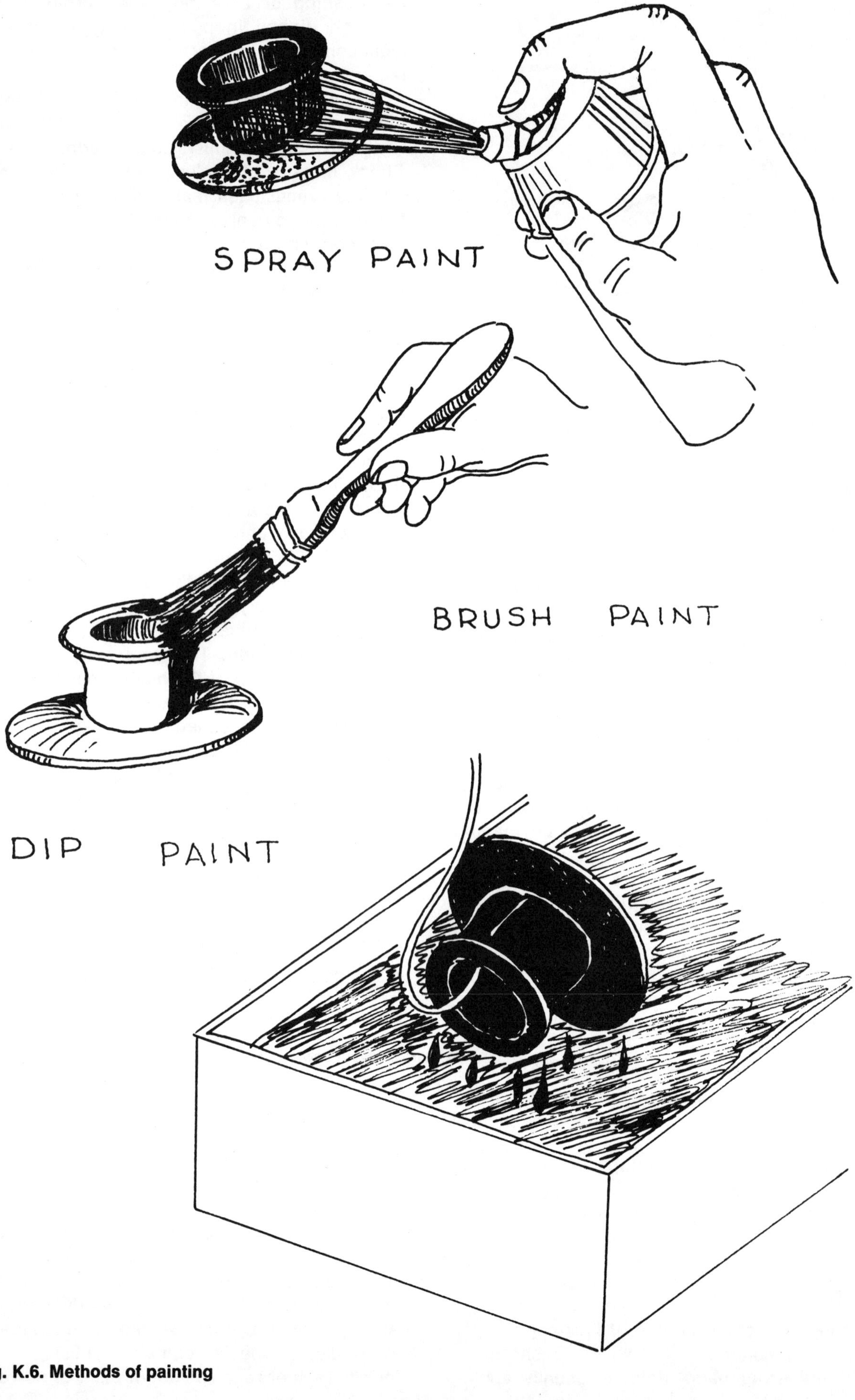

Fig. K.6. Methods of painting

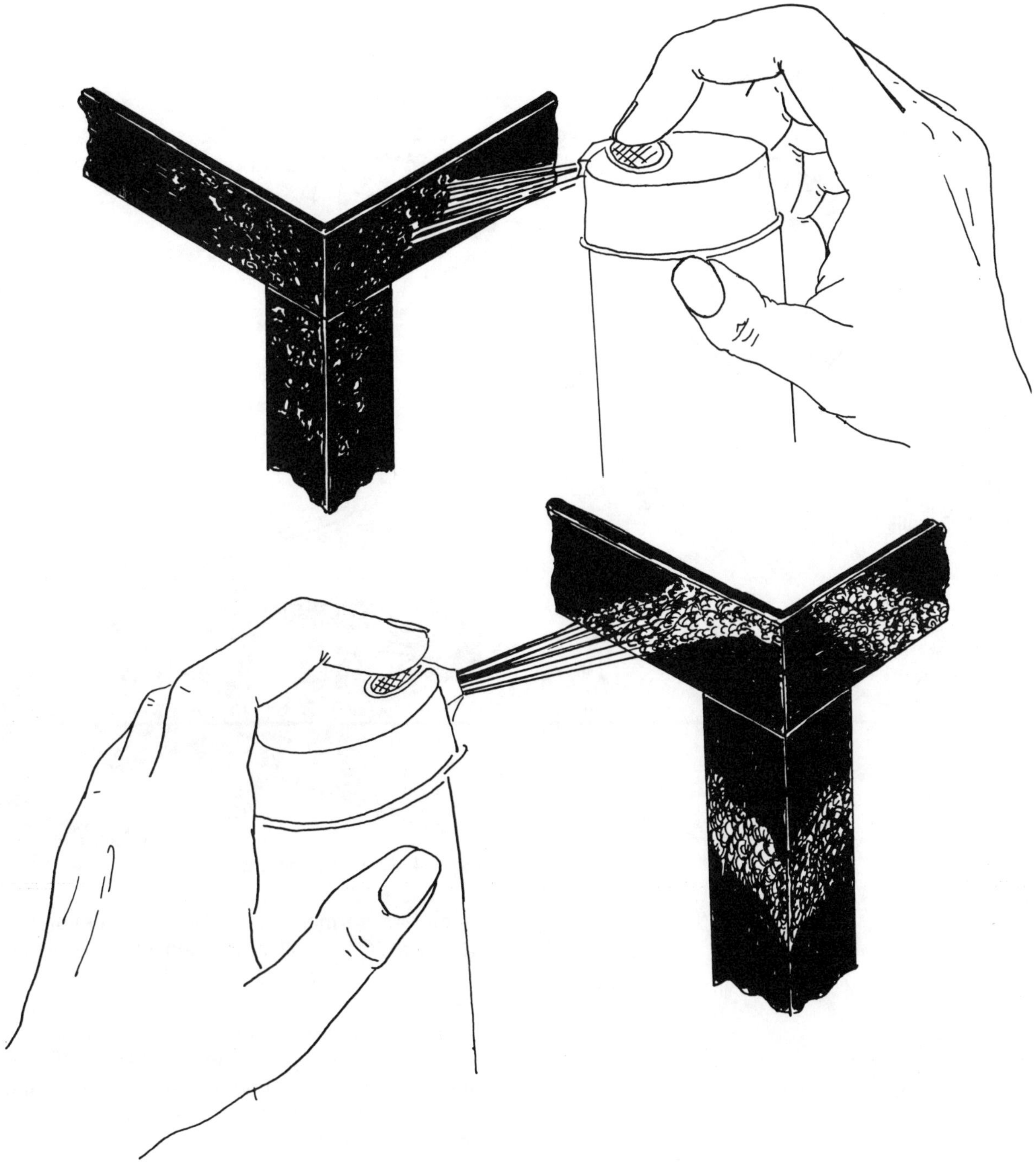

Fig. K.7. Misting and spraying bands

MAKING SCROLLS AND CIRCLES

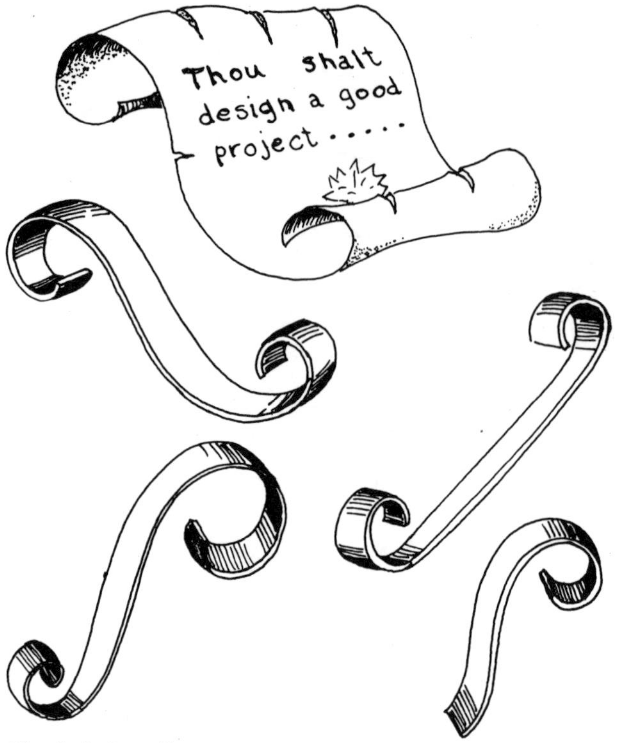

Fig. L.1. Scrolls

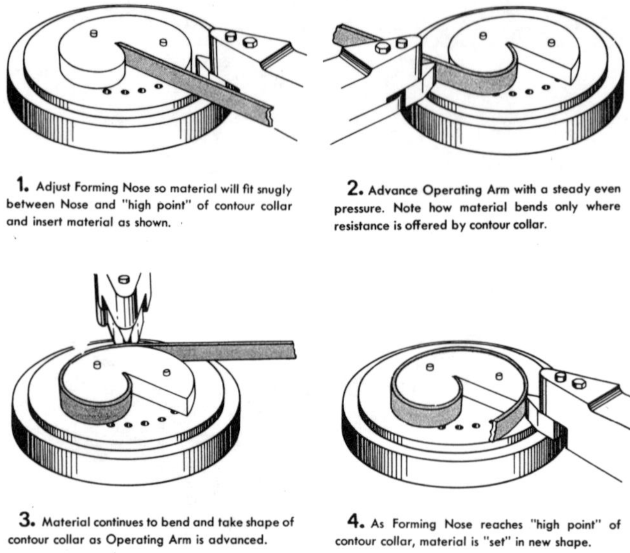

Fig. L.3.

Fig. L.2. "C" and "S" scrolls

Fig. L.4. Scroll-bending die

SCROLLS

"Scrolls" refers to ancient writings recorded on loosely rolled parchment or papers. In metalwork, "scrolls" refers to pieces of metal which are formed to resemble these scrolls.

Quite often a scroll is formed at both ends of a piece of metal. When both ends of the piece are formed in the same direction it is called a "C" scroll and when they are formed in opposite directions it is called an "S" scroll. See Fig. L.2.

These simple scrolls can also be made with a twist in the metal between the two scrolls. See the section on twisting metal. The twist must be made before the scrolls are formed at each end of the metal. This type of forming is accomplished with the Di-acro bending machine by using a collar having the same contour as the shape to be formed, as illustrated in Fig. L.3. Adjust the Forming Nose so it is located only the thickness of the material away from the "high point" of the contour collar. As the material will only bend where this contour collar offers resistance, the Forming Nose can lead the material around until it contacts the "high point" and exerts sufficient pressure to force the metal into the shape of the collar.

Sometimes it is desirable to produce different-sized scrolls on the same piece of metal. This can be done by using extension attachments similar to the one shown here being used on the Hossfeld bending machine. See Fig. L.4. (p. 76.)

Note. Flat bar is shown in the two machines opposite because it is the most popular metal used for scroll making. However, light square or round bar may also be used as long as the thickness of the metal matches the slot in the die or collar. Metal that is *slightly* thicker may be hammered on the end to fit the slot on the die. However, care must be taken not to exceed the manufacturer's specifications. Similarly thinner metal may be used if a shim or thin piece of metal is inserted in the slot to make up for the difference in thicknesses. For a single or a few scrolls, it is common practice to hammer the end of a welding rod to a chisel point and jam the thinner metal in in place of a proper shim. By changing the direction of the arm movement, a slight change can be made in the appearance of the end of the scroll. See Fig. L.5.

The Metalcraft Former consists of two knurled rollers at the centre of the former. The eccentric roller is rotated until contact is made with the metal strip to be formed. Pulling the metal strip secures it firmly between the rollers. The metal should be kept tight and flat as it is wrapped manually around the raised spiral of the plate to produce the desired scroll.

It is also possible sometimes to make a bending or forming jig in the shop. One idea is shown in Fig. L.7. Here the metal is placed between the pins and pressure applied, at the same time the metal is moved forward after each application of pressure.

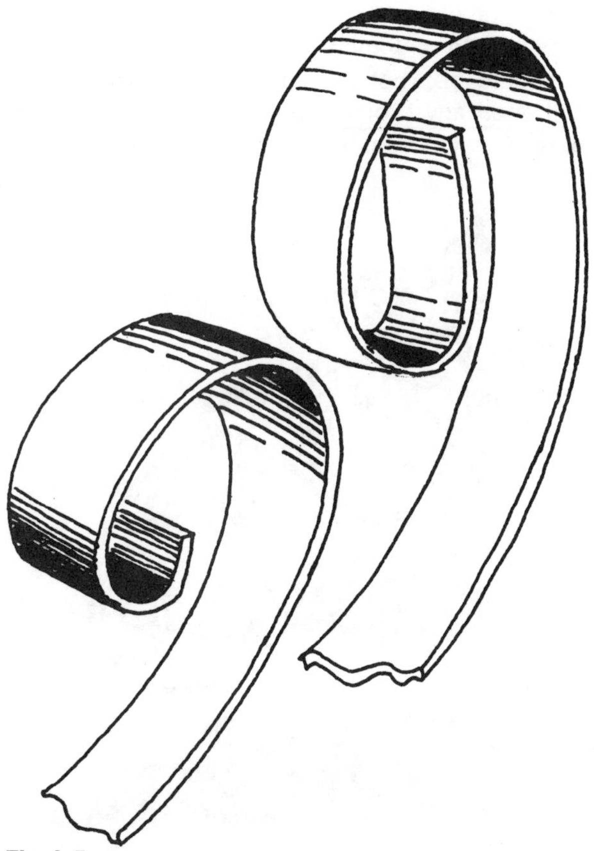

Fig. L.5.

Fig. L.6. Metalcraft former

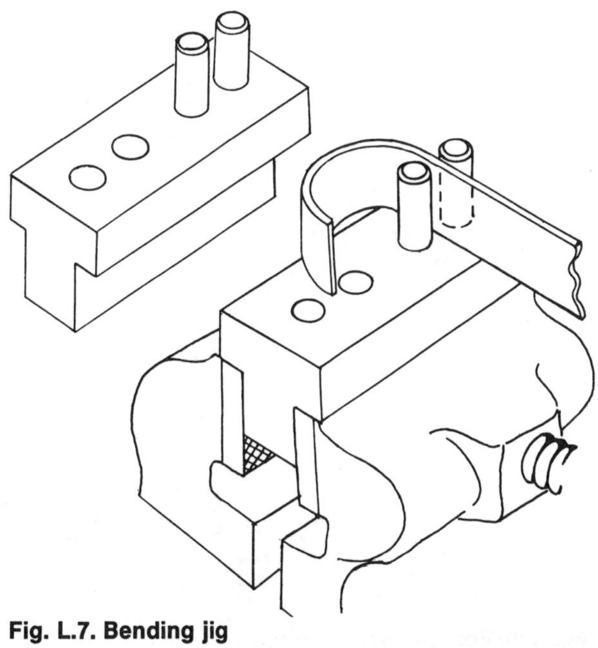

Fig. L.7. Bending jig

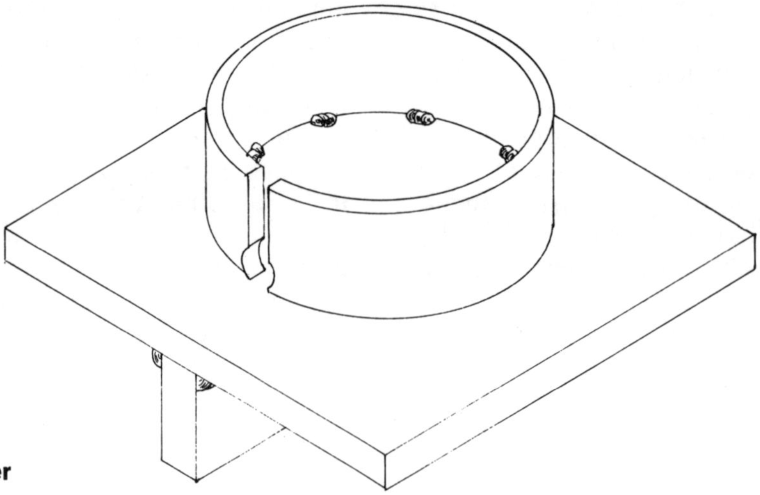

Fig. L.8. Scroll bender

A lot of patience and practice is required to form identical parts.

Another possibility is the fixture shown in Fig. L.8. Here the metal is inserted in the slot or hole and simply pulled around the pipe. It is of course necessary to have different–sized pipes to produce different–sized scrolls.

Finally it is possible to form scrolls with a hammer and an anvil much as the old-time blacksmith used to do. This again requires patience and practice to form identical pieces. When using this method, it is most important that the metal be moved across the horn of the anvil only a little each time. It is also important that the metal be struck at the point where it protrudes slightly beyond the horn. If these two hints are not observed, the scroll will not be uniform but will have flat areas. See Fig. L.9.

Note. When using any of the bending machines, to ensure uniformity in a large number of scrolls: form the first carefully; check it; mark the location on the die with chalk or pencil; and bend each scroll exactly to that mark. There are also available on the market many excellently designed ornamental scrolls like the ones shown in Fig. L.10. If properly applied, they would enhance the beauty of a project. It should be kept in mind that over-decoration will spoil any project and that each extra piece purchased increases the total cost of the project.

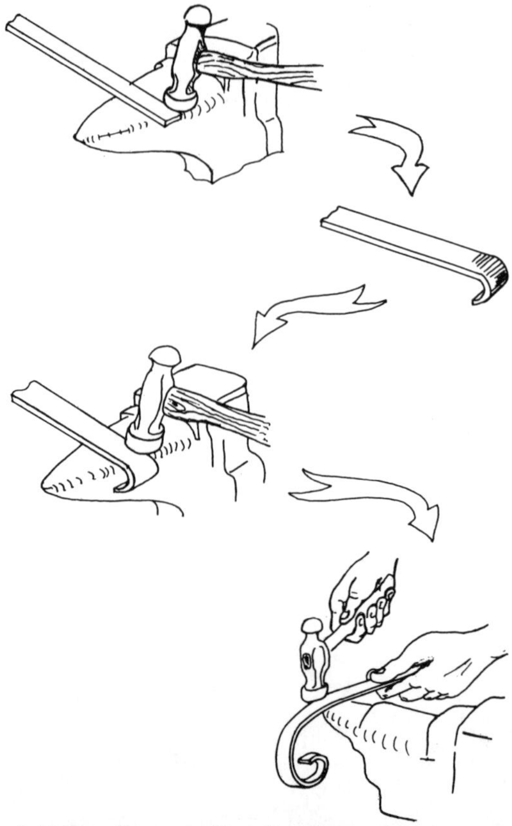

Fig. L.9. Forming scroll on anvil

Fig. L.10. Ornamental scrolls

CIRCLES

Quite often it is necessary to form complete circles when building projects. To find the amount of material required to form a circle, you measure the circumference (which is the distance around the circle). If the circle is to fit around a table top or some similar object, it is a simple matter to place a piece of string or soft solder around the outside of the top, straighten it, and measure.

However, the circumference can also be calculated mathematically in the following manner: circumference = diameter × 3.14; circumference = radius × 6.28. For example, find the amount or length of material required to form a 50 mm circle (inside diameter).

Diameter.	50 mm × 3.14 = 157 mm
Radius.	25 mm × 6.28 = 157 mm

Circles may be formed with any of the standard materials, i.e., flat bar, angle iron, square bar, round bar, pipe, and so forth. They may be large circles, small circles, eyes, loops, spirals, or springs. The following illustrations will show some of these operations.

With the Di-acro bending machine, a complete circle can be formed in one operation by clamping the material at one end and revolving the operating arm in a full circle (360°). However, this method produces long unformed lengths of material where the ends meet. To avoid this, bend the metal from the centre towards each end as shown in Fig. L.12.

Although each size Di-acro bending machine has a maximum width capacity determined by the height of the Forming Nose, it is possible to form much wider material by using certain attachments. With the use of these attachments, the actual width capacity is greatly increased.

Flat bar may also be bent edgewise to form a circle as shown in Fig. L.14. (p. 81.) However, since during this operation the outside edge is stretched and the inside edge is compressed, care must be taken to prevent metal distortion.

To make a small single ring or circle usually an "eye" is formed on a longer piece of metal and the excess metal is cut off as shown in Fig. L.15. (p. 81.) This is done because it is difficult to form small circles using the exact amount of metal needed in the circle. Also Fig. L.15. shows the two types of eyes, "the off–centre eye" and the "centre eye." For making a single ring, it is usually best to use the centre eye method because it will produce a neater finish at the ends of the ring when the extra metal is cut off. Small rings and eyes can easily be formed on the Di-acro or Hossfeld bending machine as shown in Fig. L.16. (p. 82.) and Fig. L.17. (p. 83.) They can of course be formed on flat, square, or round bar. Loops, spirals, springs, and coils are formed in a manner similar to the way the single ring is formed. See Fig. L.18. (p. 84.) If a number of rings are required for a project, the simplest method is to form a coil, then cut across the coil with a hacksaw as shown in Fig. L.19. (p. 84.) This will produce a number of identical rings.

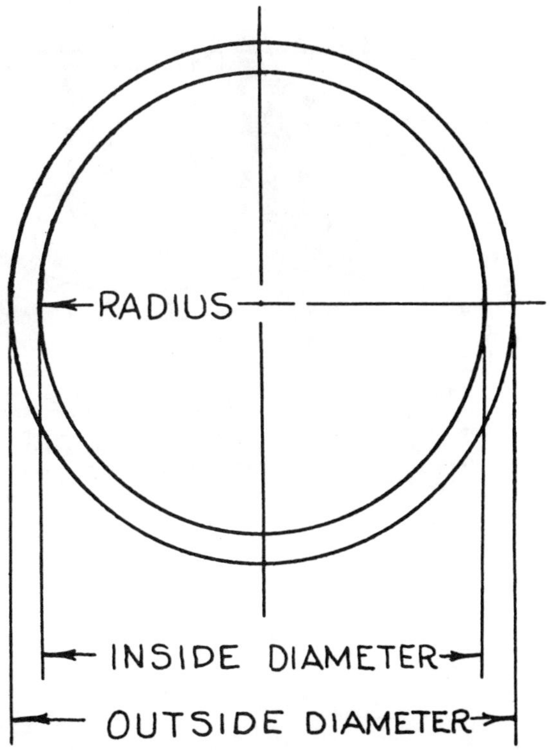

Fig. L.11.

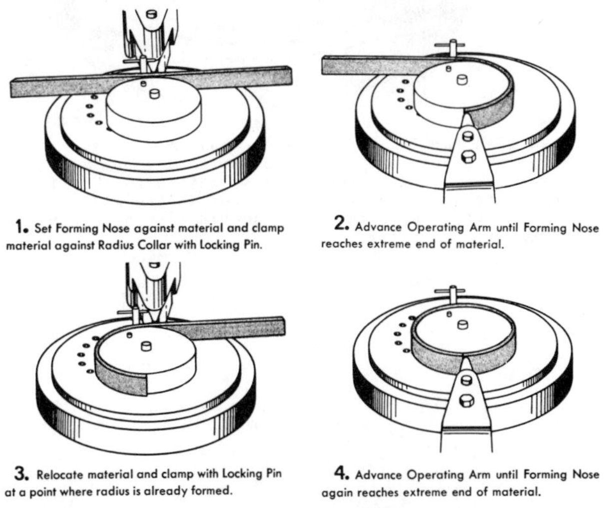

Fig. L.12.

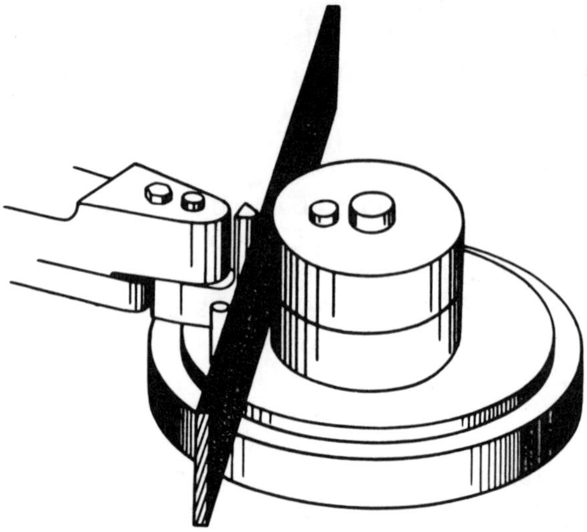

1. The Built-up Forming Nose illustrated above is available for all models of DI-ACRO Benders. Two or more standard Radius Collars of the same size can also be placed on top of each other to provide the required height.

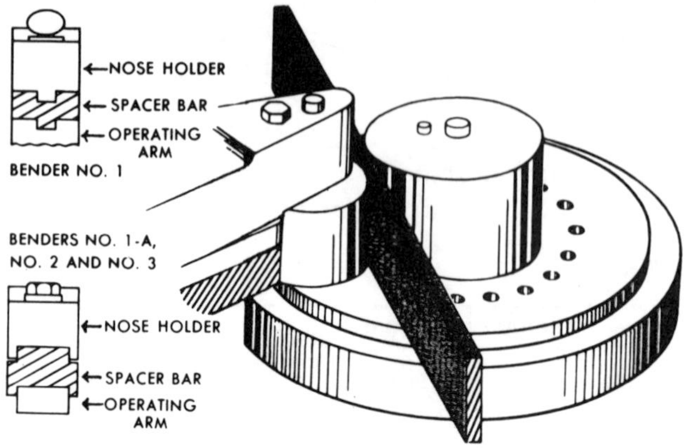

2. It is sometimes desirable to use a Forming Roller higher than the Nose Holder will accommodate. This can be accomplished by mounting a Spacer Bar of the desired dimension between the Operating Arm and the Nose Holder as illustrated.

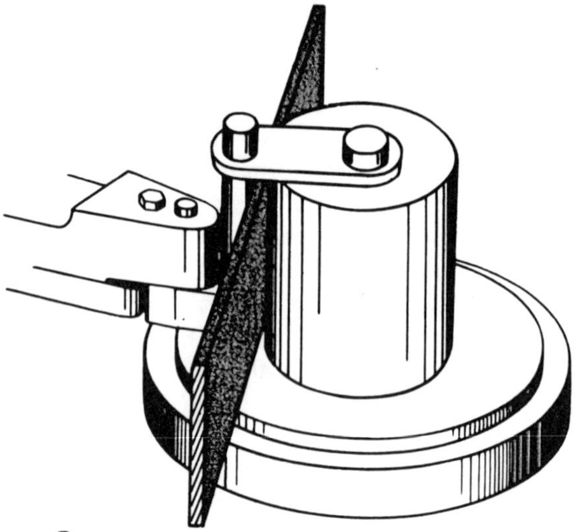

3. When forming very wide or heavy materials the Forming Nose and Radius Collar should be linked together as shown above to maintain rigidity and assure the same radius over the entire width of the part. The Locking Pin should also be linked to the Radius Collar.

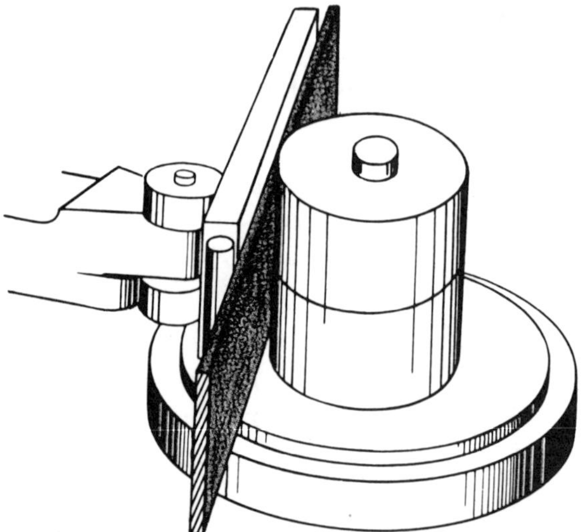

4. By using a longer Bearing Pin that extends above the top of the Nose Holder, a second Roller can be used to apply pressure against a wide Follow Block as illustrated. Heavy materials can often be formed by using double rollers without a Follow Block.

Fig. L.13.

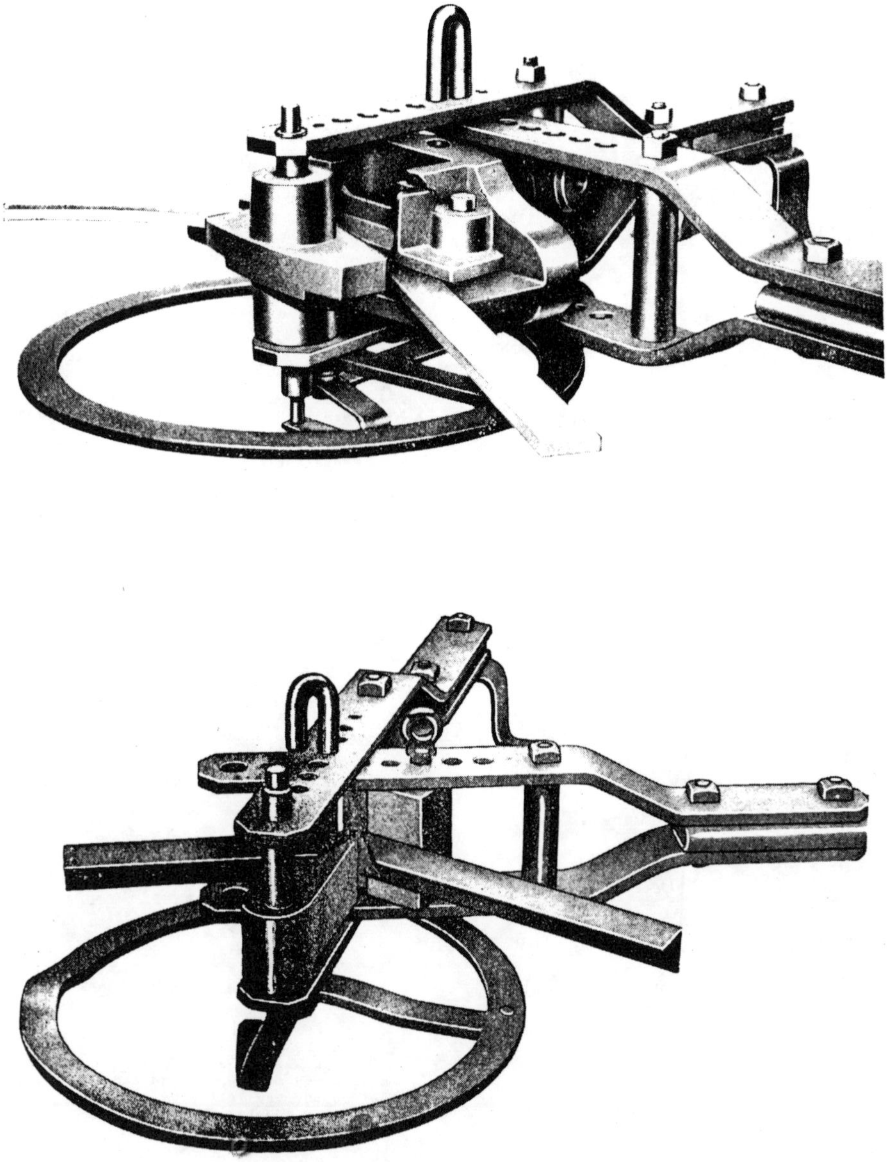

Fig. L.14. Edgewise bending and bending notched angle iron

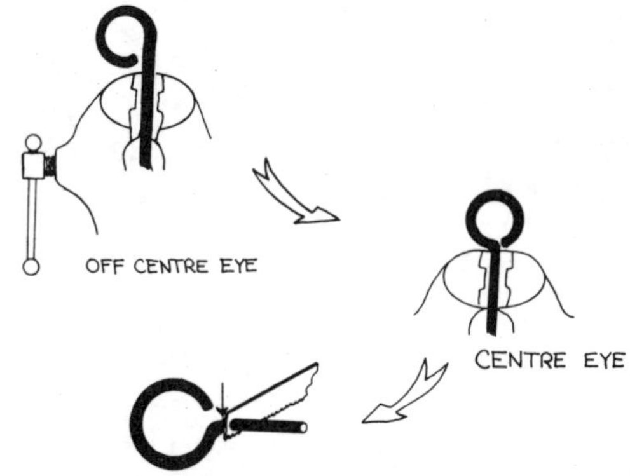

Fig. L.15. Making small single circle

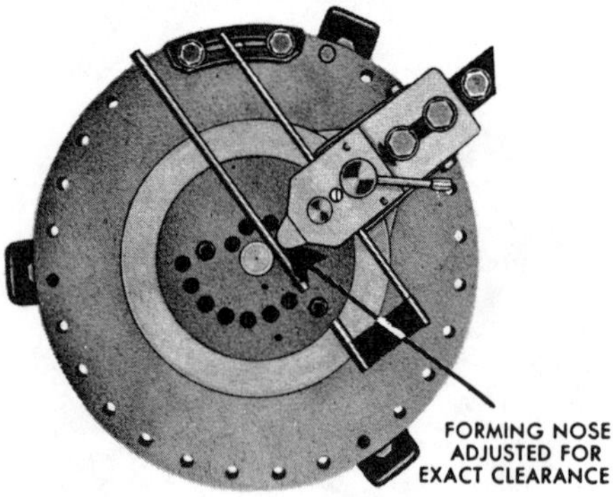

1. Mount Radius Pin or Radius Collar, of size required for desired bend, in center of Mounting Plate. Adjust automatic Forming Nose by moving Nose Holder so material will fit snugly between Nose and Radius Pin as illustrated, set Nose Holder Support against Nose Holder and tighten all bolts. Locate Locking Pin in the Mounting Plate hole nearest Radius Pin that will enable it to securely lock material.

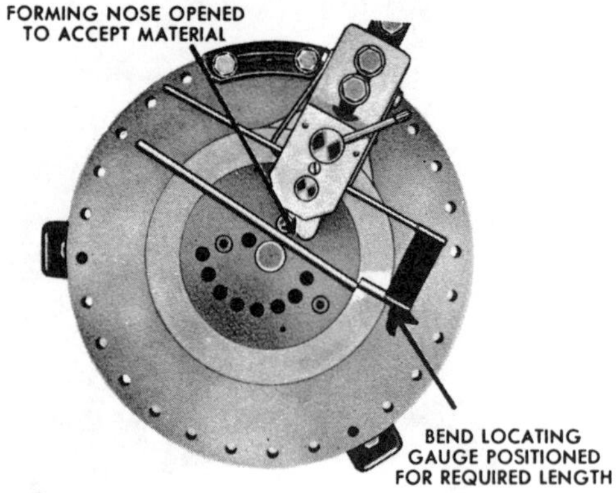

2. Adjust Angle Gauge for degree of bend and set Return Stop to limit opening of Forming Nose. Set Bend Locating Gauge to position Bend. Move Operating Arm in counter clockwise direction until Forming Nose opens, as illustrated. Then insert material between Locking Pin and Radius Pin and against Bend Locating Gauge. Lock material securely.

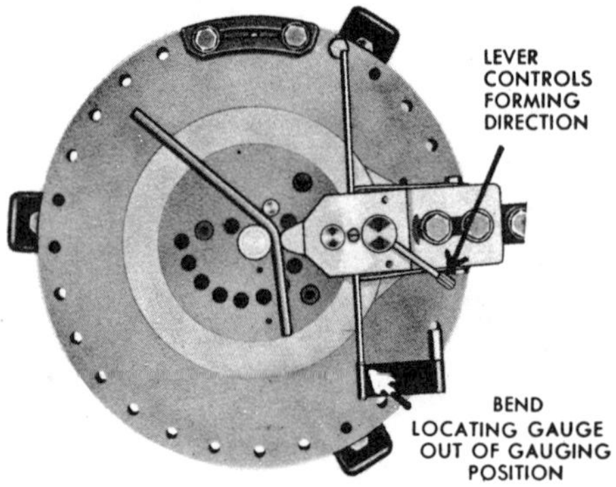

3. Move Bend Locating Gauge upward and begin bending operation by advancing Operating Arm in clockwise direction. When desirable to form in counter clockwise direction with No. 2, No. 3, or No. 4 DI-ACRO Bender, move Control Lever, shown above, to opposite side of Nose Holder.

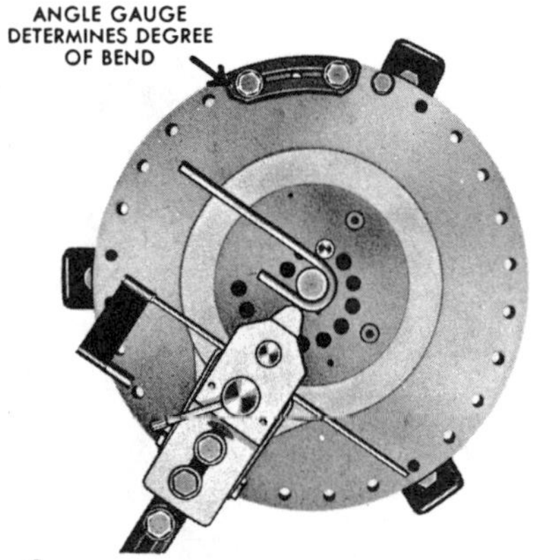

4. At this point a "U" bend has been completed. Any degree of bend can be accurately obtained and duplicated by adjustment of the Angle Gauge mounted in the outer rim of holes in the base casting.

5. A 360° bend has now been produced. In order to obtain a centered eye of a greater diameter or a bend of a larger radius, it is only necessary to use a Radius Pin or Collar of the proper diameter.

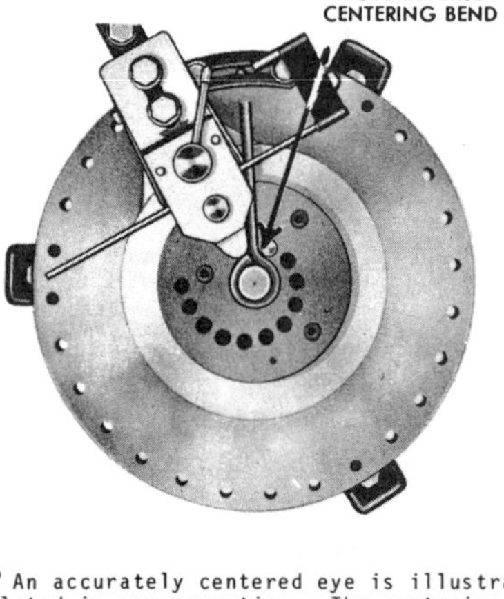

6. An accurately centered eye is illustrated, completed in one operation. The centering or reverse bend is obtained simultaneously as the eye is being closed by forcing the rod against the Locking Pin.

Fig. L.16. Centred eye bending

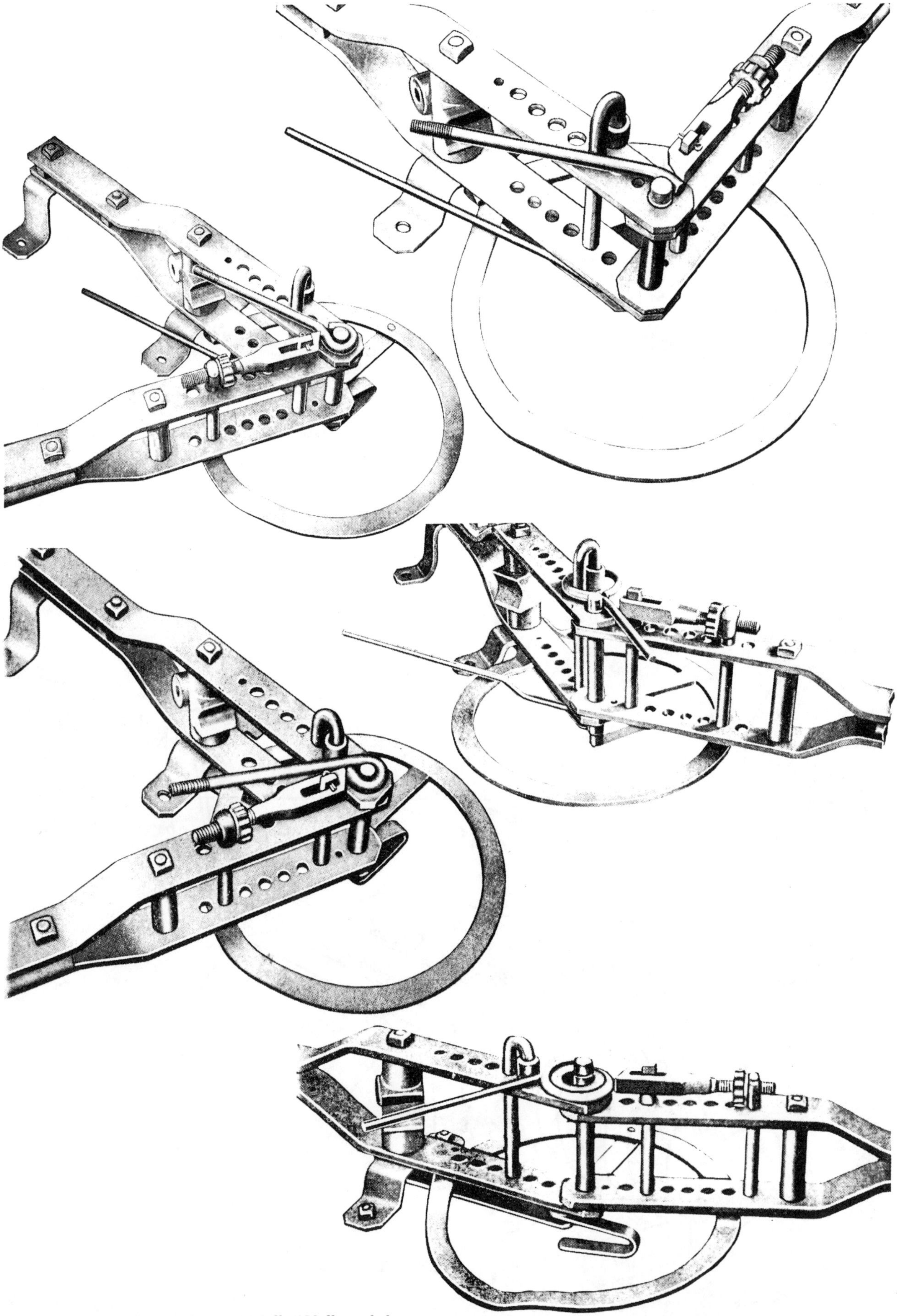

Fig. L.17. Bending eye bolts, "U's", "S's", and rings

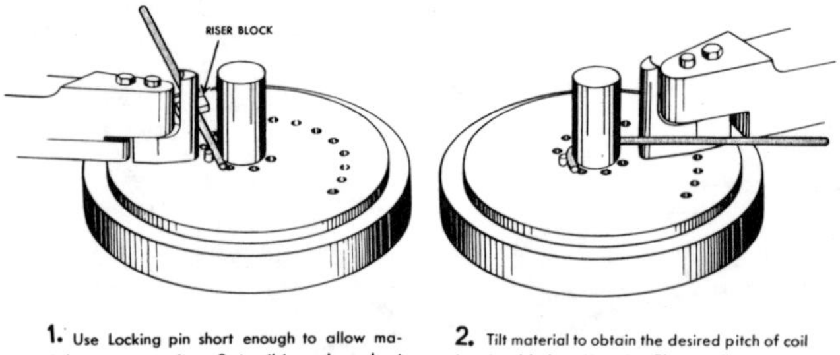

1. Use Locking pin short enough to allow material to pass over it as first coil is made and set Built-Up Forming Nose to clear end of material.

2. Tilt material to obtain the desired pitch of coil using riser block on Mounting Plate as illustrated to duplicate the pitch.

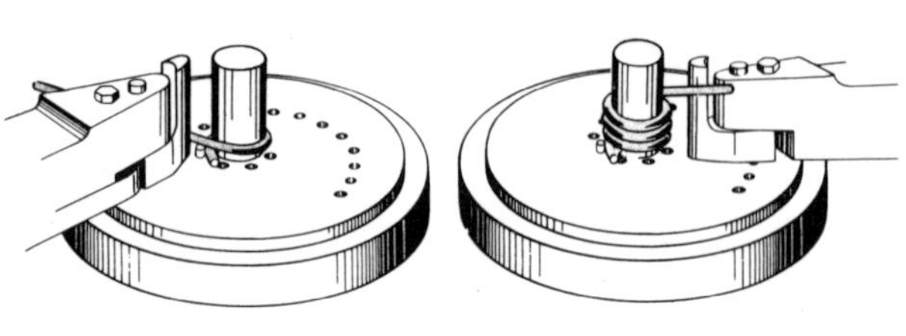

3. Note how Forming Nose clears end of the material and how material spirals as first revolution of operating arm is made.

4. The desired number of turns in the coil are readily and easily obtained by merely revolving the Operating Arm.

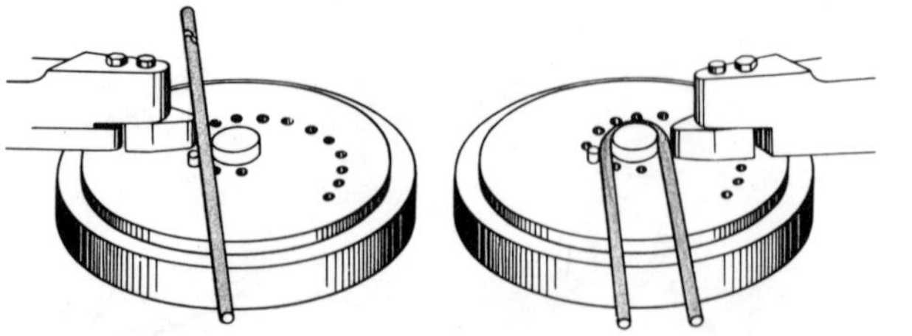

1. Use both a Holding Pin and Radius Pin short enough to allow material to pass over them as first loop is made. Set Forming Nose so material will fit snugly between nose and Radius Pin.

2. Set angle stop for a 180° bend. Revolve Operating Arm until it strikes angle stop thereby completing a U bend. This is the first step in loop and spiral bending.

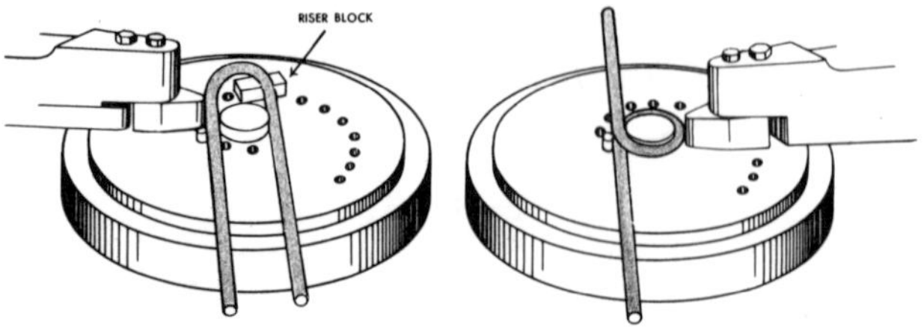

3. Advance material a distance equal to one half the circumference of Radius Pin and tilt material using Riser Block as shown on Mounting Plate to duplicate pitch.

4. Previously formed section will now spiral over center pin completing loop. By continuously advancing material as shown in Figure 3 and repeating cycle, any number of coils can be formed.

Fig. L.18. Bending coils and loops

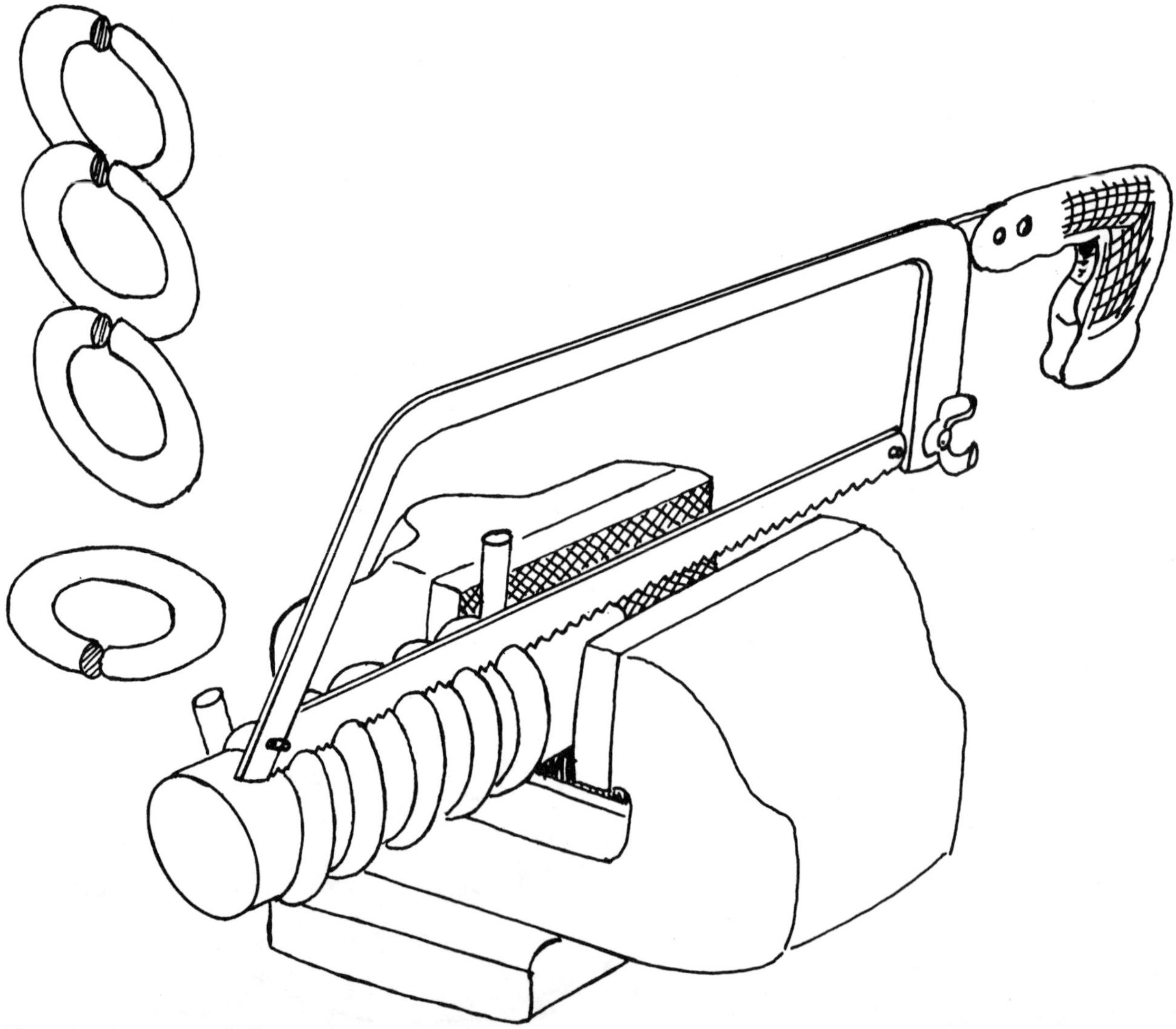

Fig. L.19. Producing identical rings

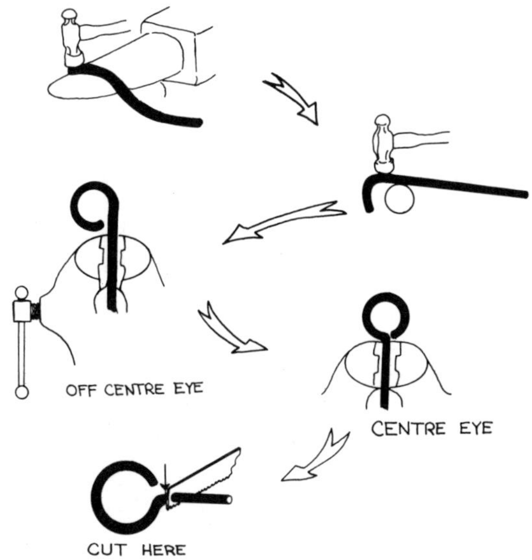

Fig. L.20. Forming circles by hand

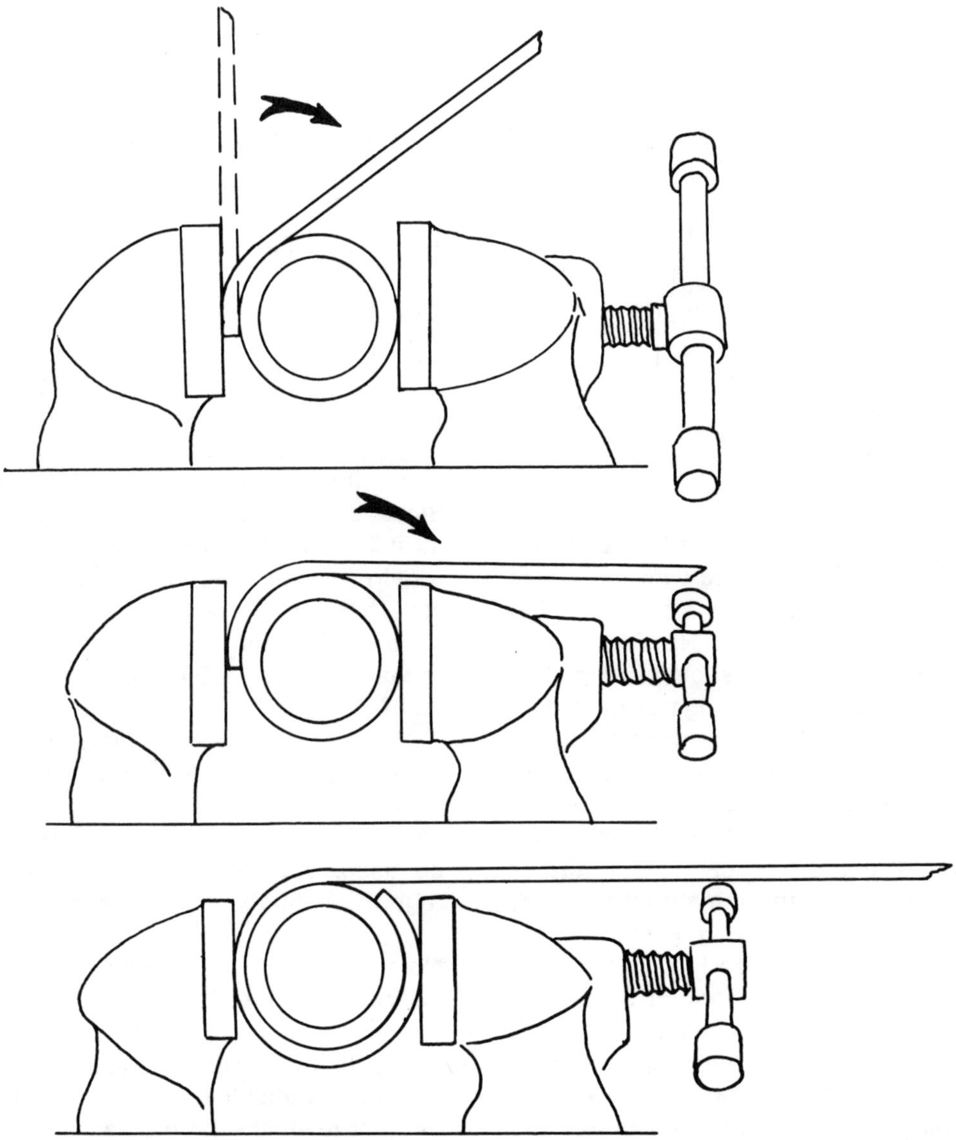

Fig. L.21. Forming circles with pipe and a vise

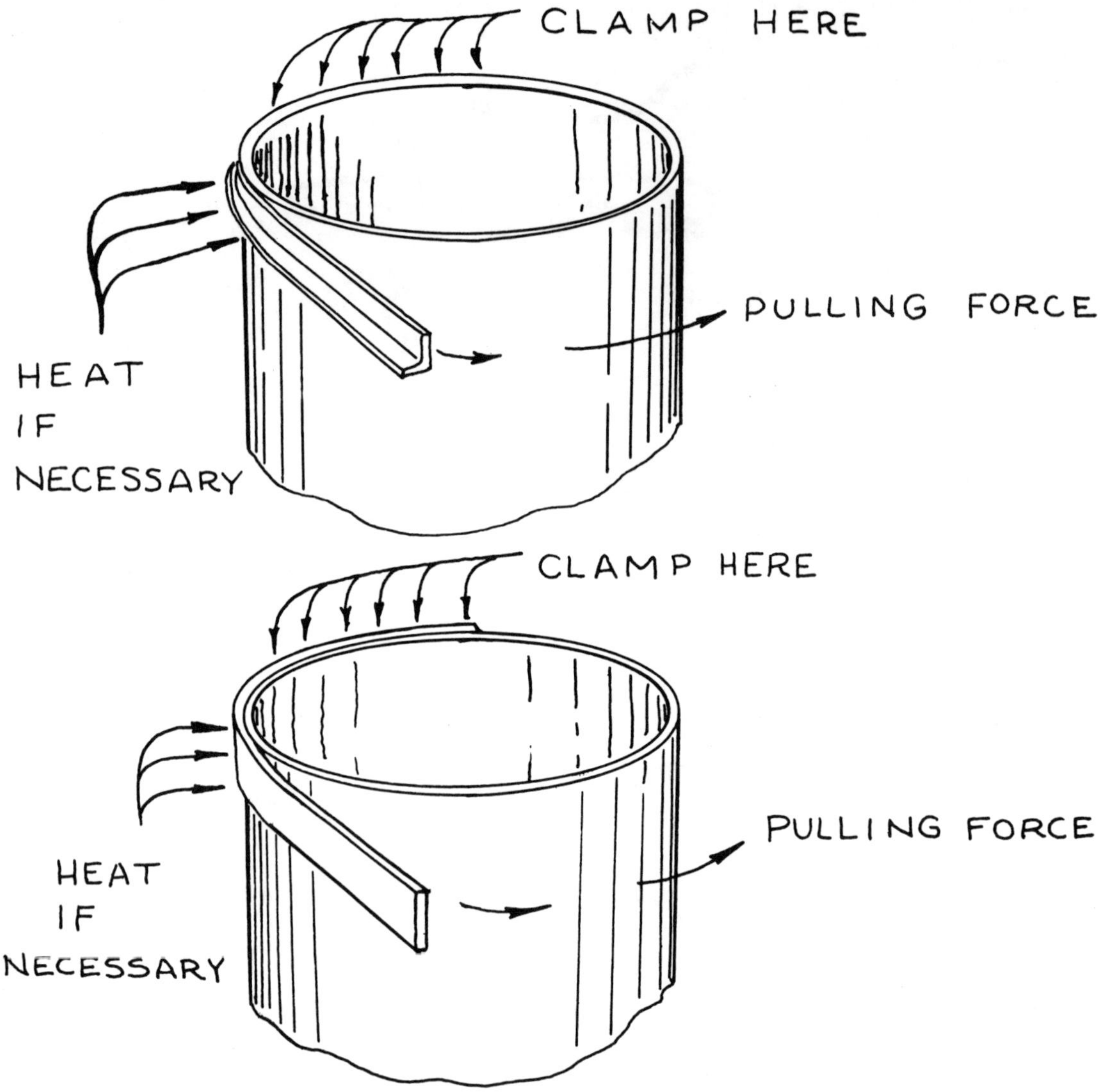

Fig. L.22.

As with the scroll making, if these bending machines are not available it is possible to form the circles by hand as shown in Fig. L.20. (p. 85.) Circles may also be formed with the aid of some cylindrical shape such as a piece of pipe and a vise as shown in Fig. L.21. (p. 85.) The jig or fixture for making scrolls shown in Fig. L.8. can also be used for making rings or circles.

For forming large rings or circles by hand, almost any cylindrical object may be used. The metal is clamped at one end and pulled or formed around the object; as the material is bent around the cylindrical shape further clamps are applied. If thick material is used, heating with a torch may be required to help the material flow around. See Fig. L.22.

Angle iron may also be formed in a complete circle, and may be formed or bent with the flange in or the flange out depending on the type of project being formed. However, this type of bending presents a problem in that the flanges may distort; therefore, extreme care must be taken. If a twist or distortion does take place it may be possible to hammer it out. Since it is necessary to compress the flange as it is formed inward, more pressure will be required than when bending with the flange out. Therefore the largest possible radius should be used to allow for compression of the material. See Fig. L.23. (p. 87.) and Fig. L.24. (p. 87.)

It is relatively easy to form circles by hand on angle iron with the flange out, but extremely difficult to form them with the flange in. The same technique as shown in Fig. L.22. would be used, but heat would have to be applied regardless of the thickness of the material.

Note. When using bending machines, if the horizontal flange is not as wide as the vertical flange (unequal leg-length angle iron) the tendency to twist will be less. It is possible but extremely difficult to form an angle iron circle by hand over the horn of an anvil in the same manner as forming scrolls or circles on other material.

Light channel may also be bent on the Di-acro bending machine providing the flanges are confined so that they do not bend or distort. They may also be bent flange in or flange out. It is recommended that the radius of the bend be 3 – 4 times the width of the flange to allow for stretching of the metal, although this ratio may be altered depending on the thickness and ductility of the material. See Fig. L.25.

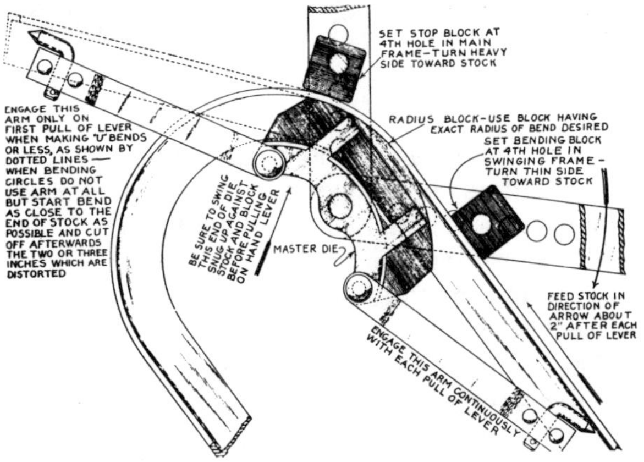

Bending Angle Iron Flange In.

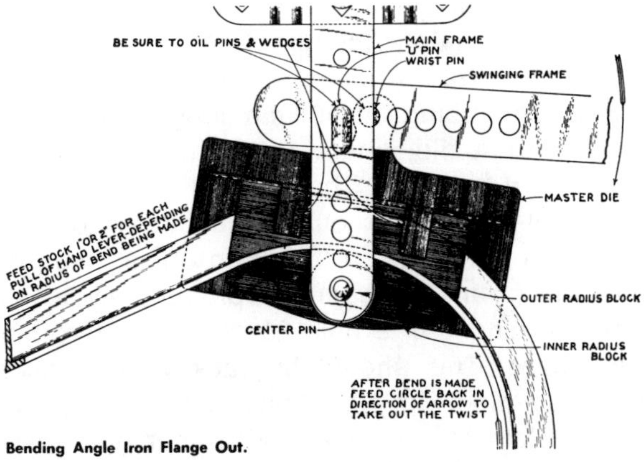

Bending Angle Iron Flange Out.

Fig. L.24. Bending angle iron

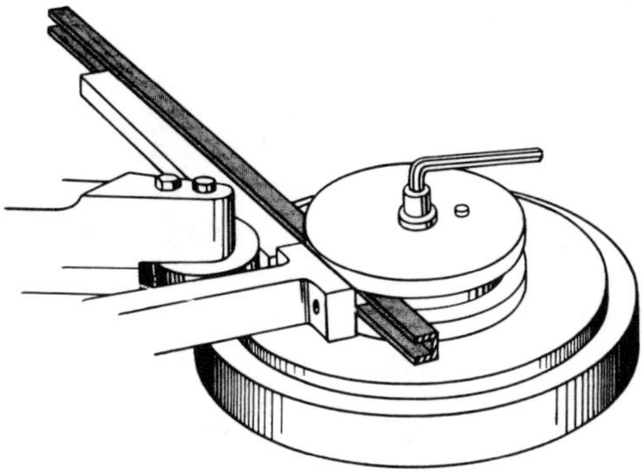

1. Position material in the Radius Collar. Insert Follow Block in the channel and tighten center bolt as illustrated in drawing.

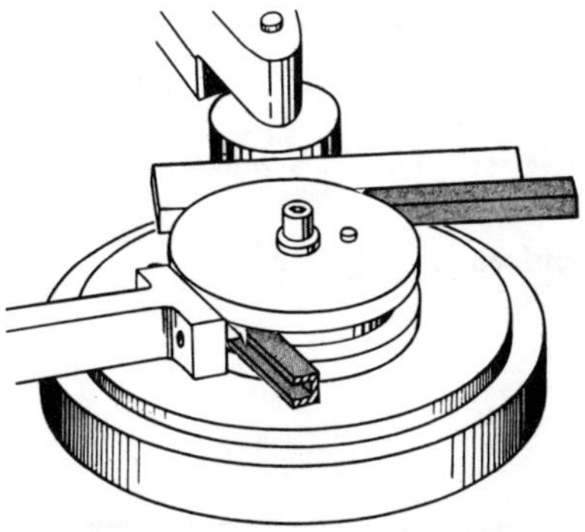

2. Clamp channel tightly. Advance Operating Arm with a steady, even pressure until it strikes stop pre-set for angle of bend.

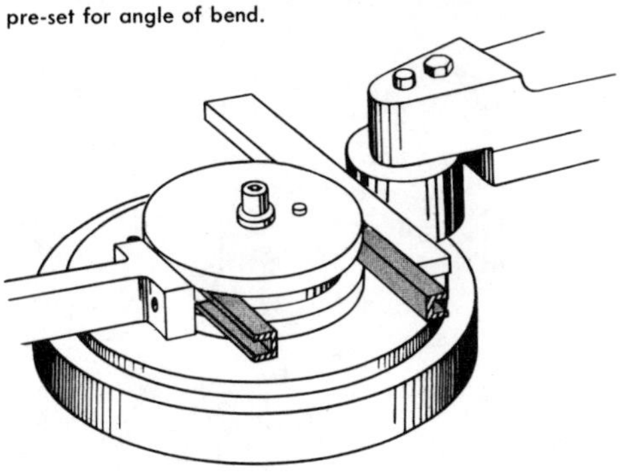

3. Loosen center bolt, release clamp, and slide material out of Radius Collar. Remove Follow Block.

Fig. L.23. Bending angle iron

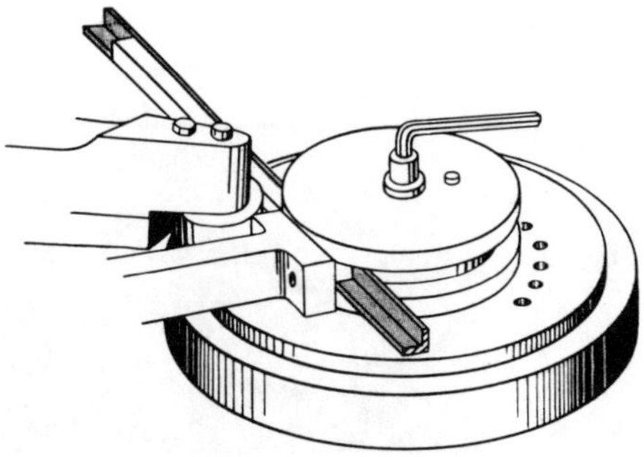

1. Position material in Radius Collar, insert Follow Block and tighten center bolt.

Fig. L.25. Bending channel

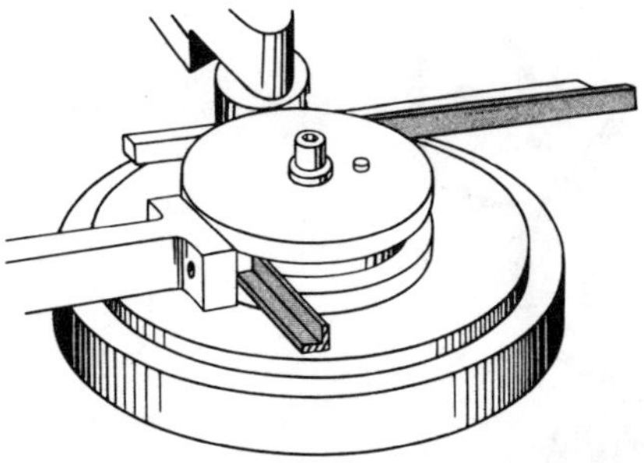

2. Clamp angle tightly. Advance Operating Arm with steady even pressure until it strikes Angle Stop.

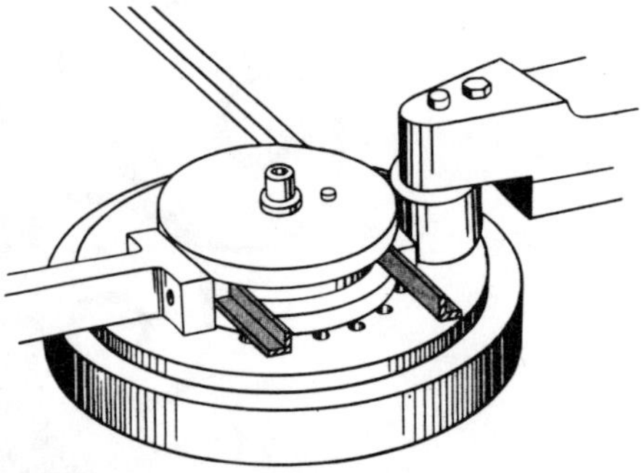

3. Loosen center bolt, release clamp, and slide material out of Radius Collar. Remove Follow Block.

HOLDERS AND TRAYS FOR CANDLES

There are many types and sizes of candlesticks, candleholders, and trays and it is a matter of personal taste and design to match them up. The holder that the candle sits on or in can range from a nail to a very decorative machine-pressed item.

The candleholders shown in Fig. M.1. are of the machine-pressed type and can be purchased through a supply house for a minimal charge. However, for the student who wishes to make his or her own the following suggestions are offered.

The simplest holder that can be made is a nail approximately 25 mm long welded to a flat base or tray. Another simple idea is to cut a piece of tubing 30–40 mm long, remove all burrs with a file, and weld or tack weld the tubing to a flat base. The inside diameter of the tube will equal the outside diameter of the candle. A variation on this is to cut a piece of solid bar (round or square) approximately 30–40 mm long and drill a hole 25 mm deep through the centre. This should be tack welded to a flat base. See Fig. M.2. Instead of leaving the tubing plain, it will look much nicer if a lip is formed on the end where the candle is to be inserted. This is done by flaring the end of the

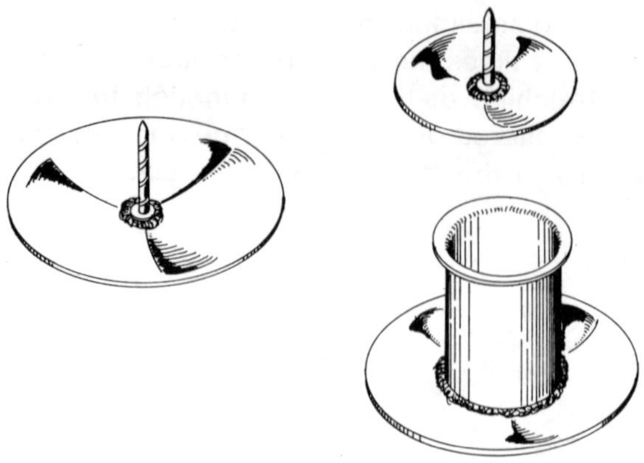

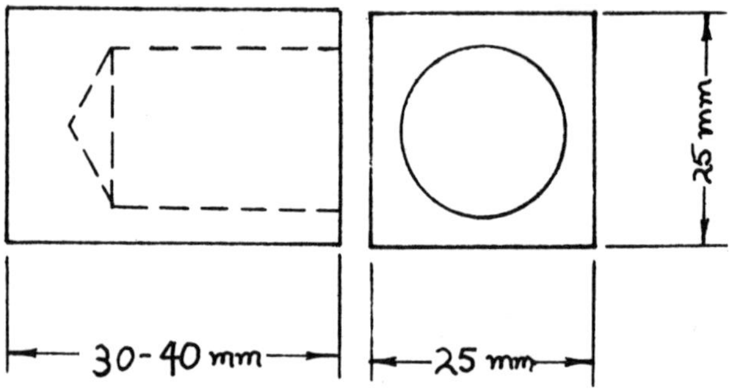

Fig. M.2. Simple holders

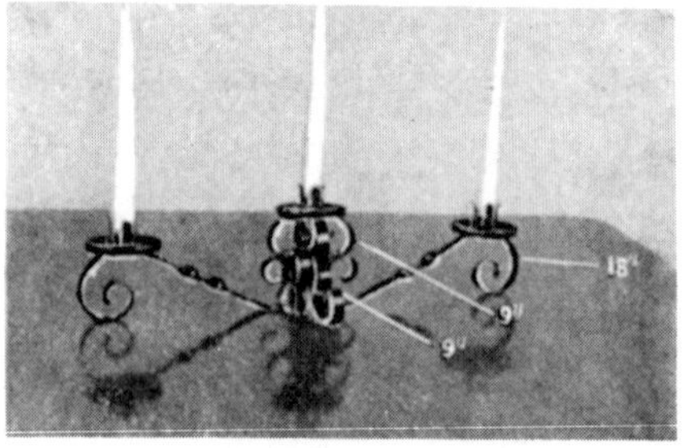

Fig. M.1. Machine-pressed holders

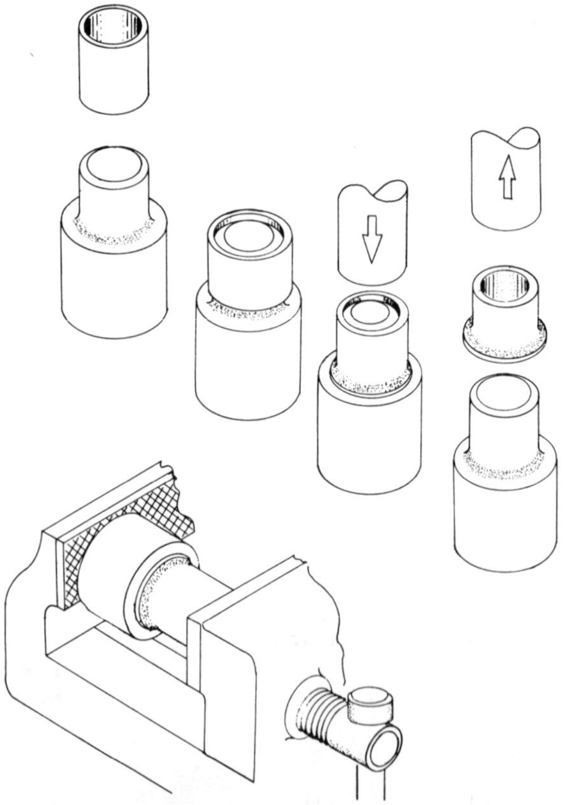

Fig. M.3. Dies for flaring holders

tubing. Place the ball end of one ball peen hammer in the tube and strike the face with the face of another ball peen hammer. There is always the possibility of chips flying off of one of the hammers. If this method is used, eye protection should be worn and other students should stand clear. A much safer method is to form a die as shown in Fig. M.3. (p. 88.), the die size to be determined by the size of tubing being used. If the shoulder of the die is correctly formed, when the piece of tubing is pressed or hammered down on the die a uniform lip will be produced.

Another method of forming a candleholder is to wind a piece of small diameter rod around a small piece of pipe or round bar which is clamped in a vise. See Fig. M.5. With a little artistic hammer work (forging), the spiral could be made into a snake.

For the more ambitious student, two variations on another type of candleholder are shown in Fig. M.4. The size of the circle will be determined by the size of candle that is to be used. The one shown in Fig. M.4. is for a 25 mm candle. Therefore, the inside circle will be 25 mm and the outside circle is 75 mm. This will form a 25 mm candle holder with 25 mm leaves which can be left straight or bent at the top as shown.

The trays shown in Fig. M.6. can be purchased at a supply house for a small charge or a blanking die as shown in Fig. M.7. can be made to form the trays. This die consists of a male and female part; again the sizes will be determined by the size of the candle. To prevent the molten candle wax from spilling over, the tray should be at least 12 mm bigger than the candle being used.

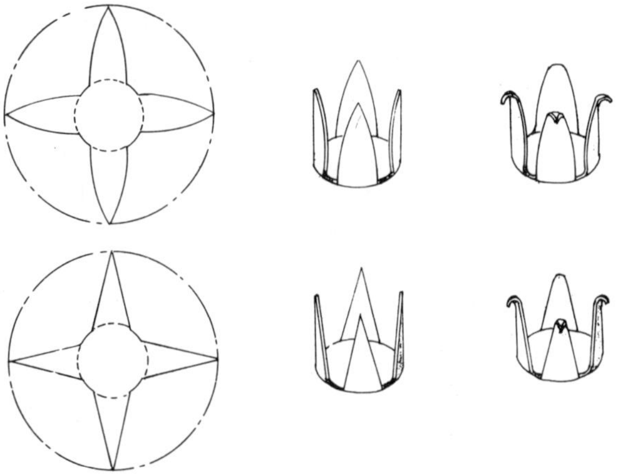

Fig. M.4.

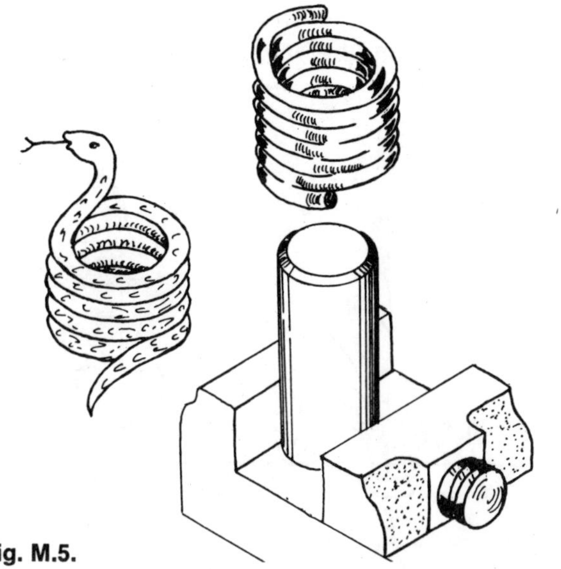

Fig. M.5.

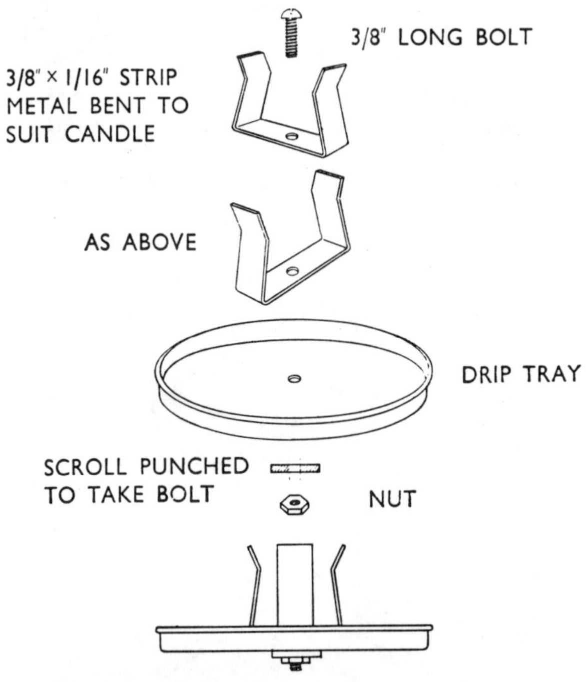

Fig. M.6. Exploded view of holder

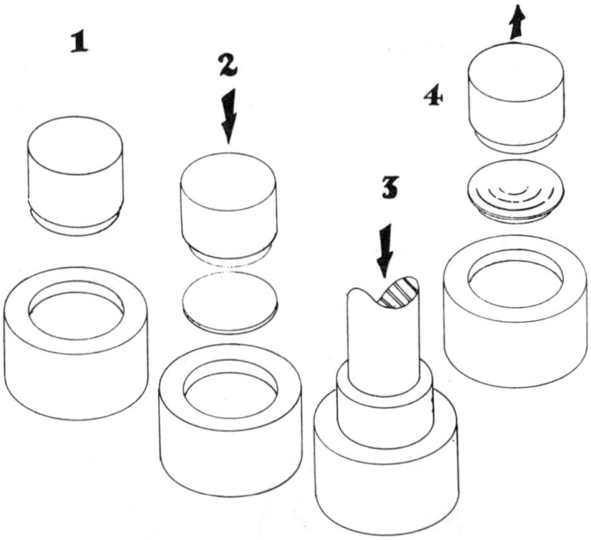

Fig. M.7. Dies for dishing drip cups

ORNAMENTS & ACCESSORIES

This section will cover some of the items that are used in project building but are not actually manufactured in the shop, although with the co-operation of other shops perhaps some of them could be made. Many of these items will enhance a project and in some instances can provide ways and means of cutting costs while still producing a well–finished attractive project. Hints will be provided as to use and methods of obtaining these items. Although we have specified mild steel hot rolled as the most popular metal for project building, some of these items can be purchased in other metals. For example, to provide a decorative finish to ornamental iron railings, andirons, or fireplace set stands, and other projects which have a protruding rod end showing, it is possible to purchase from the larger supply houses the following items.

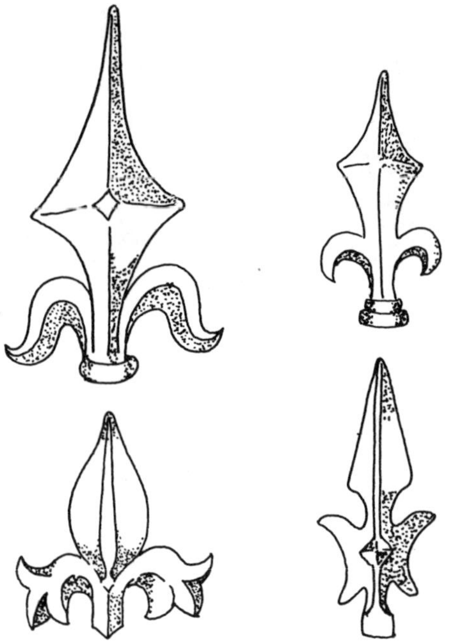

Fig. N.1. Pickets

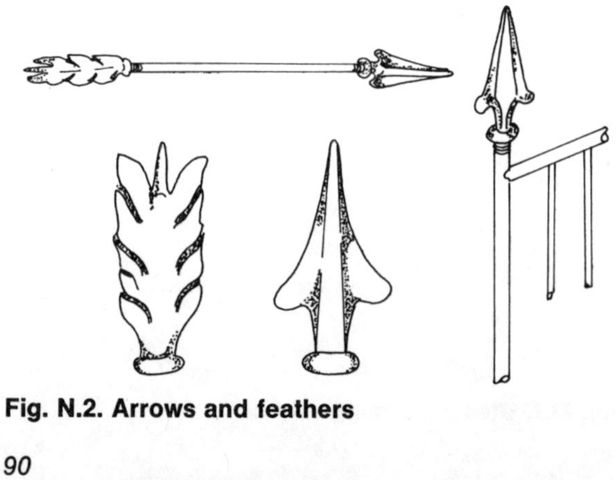

Fig. N.2. Arrows and feathers

Pickets have an approximate length of 100 mm — 150 mm and are made to fit ½ in. — ⅝ in. round or square bar.

Arrows and feathers.

Finials have an approximate length of 75 mm — 150 mm and width of 25 mm — 65 mm. They are usually complete with tapped hole in base, and would be excellent for handles on fireplace sets since there are no sharp edges as with the pickets. It is also possible to obtain the above items to fit steel posts for gates.

Plugs are available to seal off the ends of tubing for a neat finished appearance. See Fig. N.4.

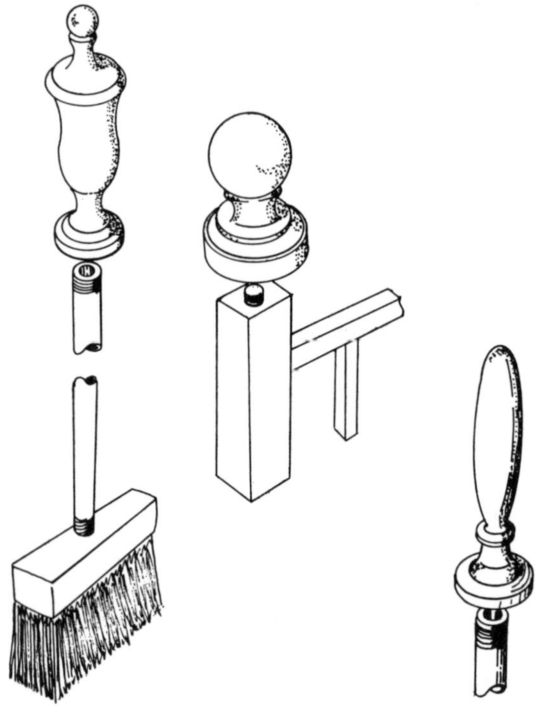

Fig. N.3. Finials

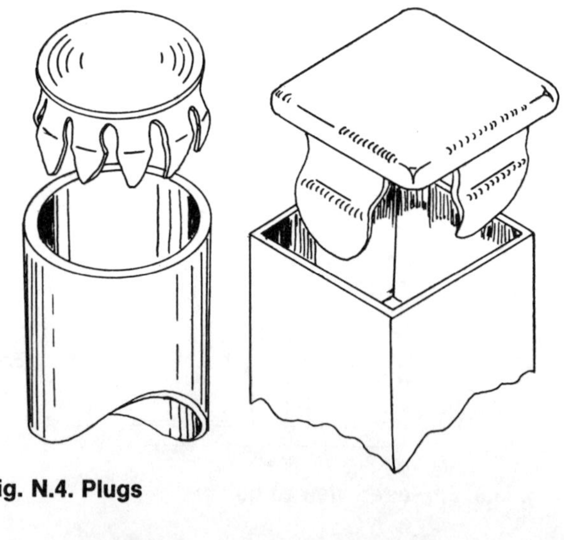

Fig. N.4. Plugs

Glides are designed to fit the ends of metal furniture to provide a finished appearance and to protect floors from sharp metal edges. There are two basic types, those which fit over and those which fit into the metal ends. Glides that fit over the metal ends (solid and tubular) are usually called crutches and are made of rubber or plastic material. They are relatively inexpensive and excellent floor savers. See Fig. N.5.

The type that fits inside tubular metal has many variations, from the simple type that can be hammered into a block of wood set in the end of the metal tube, to the more elaborate clip–in type which may be stationary or swivel. See Fig. N.6.

The glides shown in Fig. N.7. are fastened by using a wooden plug in the tubing as described above or by welding a nut or plate to the end of the tube. The nut then will receive the screw of the glide.

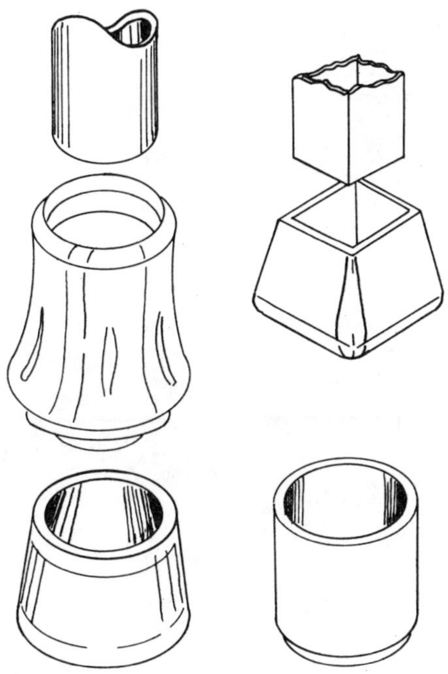

Fig. N.5. Glides

Fig. N.6. Glides for tubular metal

TABLE TOPS

Almost any material can be used for table tops, however there are many excellent sources of inexpensive attractive materials if the student or teacher is willing to do a little searching or telephoning.

One excellent source we have found is the manufacturer or installer of arborite counter tops, who cuts out a section to fit a kitchen sink or appliance. These cut–outs make excellent tops, and are usually given to schools free or for a nominal sum. They come in various sizes and shapes but a fixture like the one shown in Fig. N.8. can easily be made which will allow the sink top to be cut on a saw. It can also be used with an orbital sander for

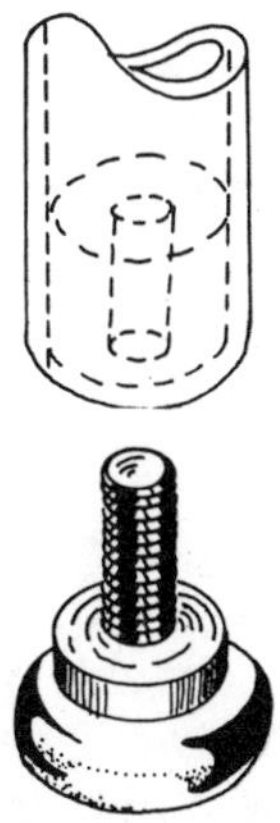

Fig. N.7. Glide

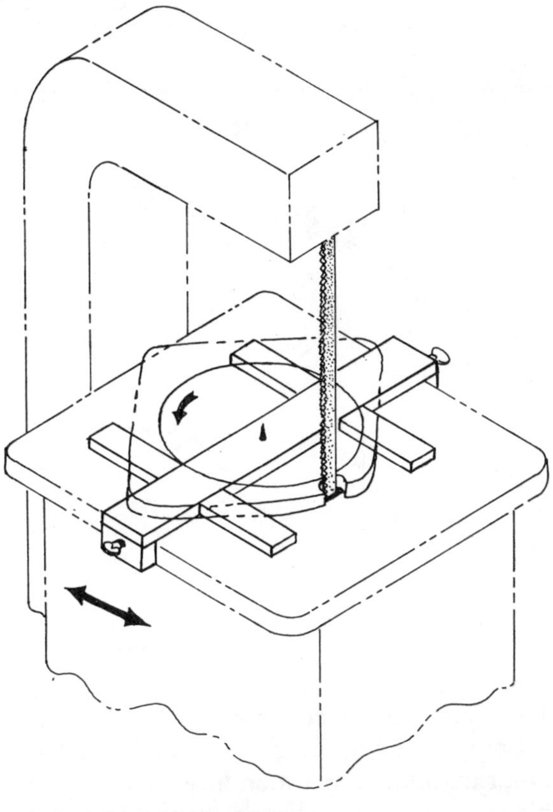

Fig. N.8. Cutting table tops

dressing edges. These tops look most attractive if a flat bar is formed around the edge and painted.

Glass tops can be rather expensive since the glass should be at least 6 mm thick and specially treated for breakage. We have found that companies who repair storefront windows will sometimes save and cut to size for a small fee large pieces of the original store window they are replacing. This will cost much less than glass purchased direct from a glass company. If the supply can be set up on a regular basis and the table frames are made to suit the size of the glass pieces that can be salvaged, window repair companies can be a dependable source of glass.

Another excellent source is the glass front from broken television sets if contact can be made with companies who take them for trade–ins. *Caution should be exercised here since some of these glasses become brittle with age and although made of safety glass may shatter at the slightest shock.*

STRIPPING

When putting glass tops inside angle iron frames, do not allow the glass to rest on the angle iron. There should be some type of cushion between the two. The simplest method is to purchase some stripping of the type used by automobile manufacturers. This stripping comes in a variety of widths and thicknesses. If it is self–sealing, remove the tape which protects the adhesive surface of the stripping. The stripping can be made of cork, neoprene, sponge rubber and so forth. See Fig. N.9.

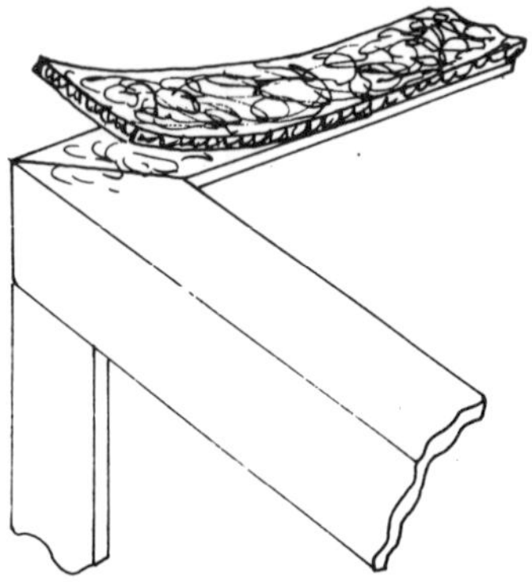

Fig. N.9. Stripping

It is also possible to use glass run channel which is used for automobile glass but this would leave a small ridge around the table which might be unacceptable. See Fig. N.10.

Sometimes the glass top has to be attached to four uprights rather than fitted inside a frame. This may be accomplished by the use of suction cups which can be screwed into the upright. Another method is to use a special interlocking tape fastener that can be applied to metals and glass with a special adhesive. See Fig. N.11.

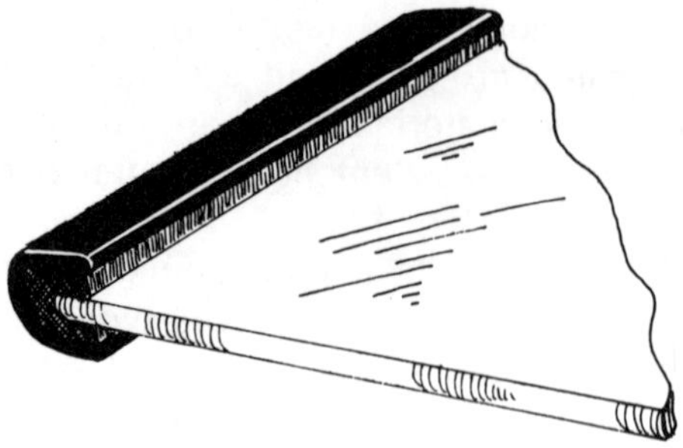

Fig. N.10. Glass run channel

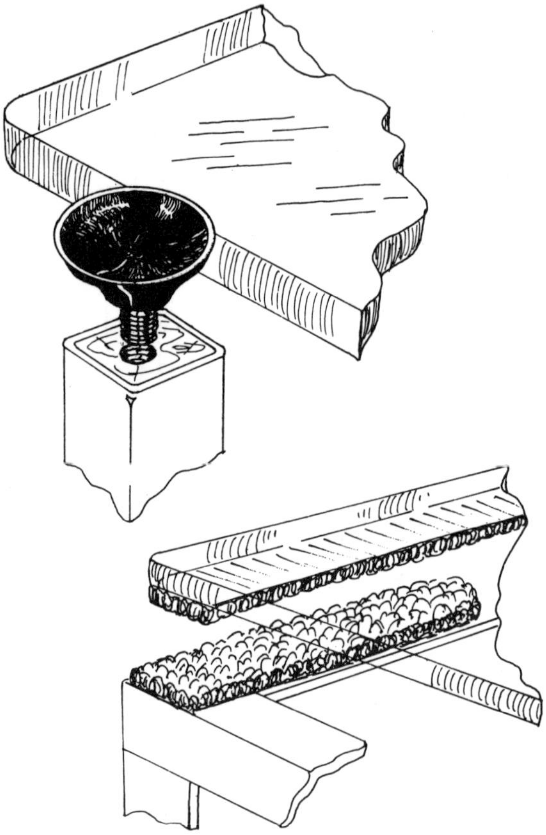

Fig. N.11. Suction cup and tape fastener

GROMMETS

When building lamps, quite often it will be necessary to pass the electric cord through a hole in the metal. If left like this, the metal edges of the hole will eventually fray the electric cord, with perhaps serious consequences. To prevent this a rubber grommet is used. Since there is a large variety of designs and sizes of grommets, it is necessary to specify the diameter of the hole and the thickness of the metal when ordering these items. See Fig. N.12. (p. 93.)

Note. If the local hardware store does not carry these items, contact an automobile supply house in the district.

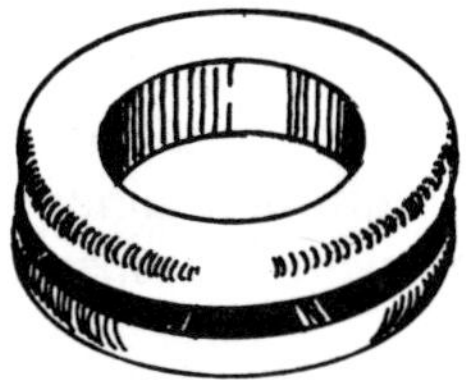

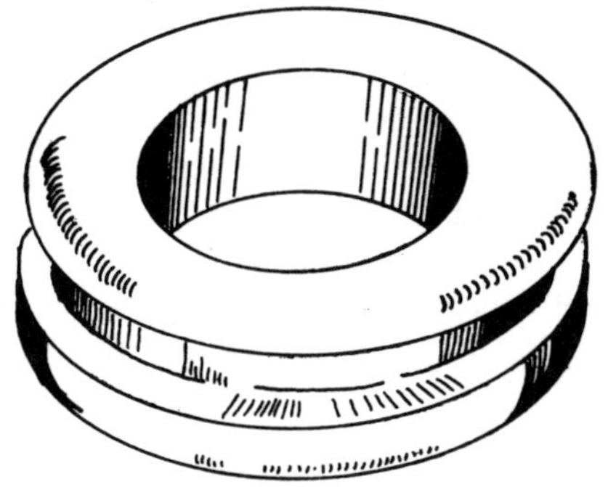

Fig. N.12. Grommets

UNIT 1. PROJECTS THAT UTILIZE WELDING FOR DECORATION

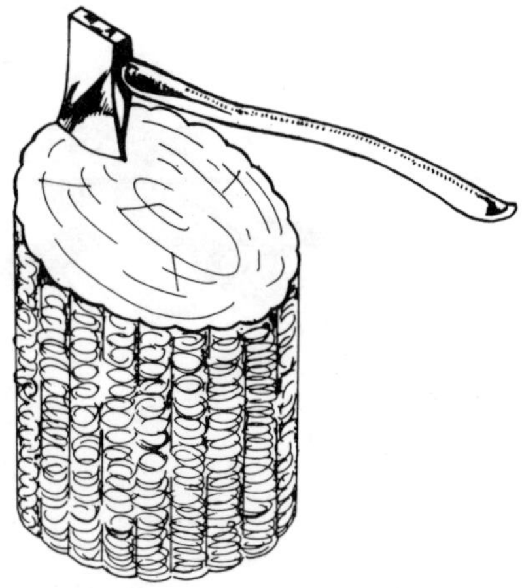

Fig. 1.1. Log paperweight

All of the projects in this unit utilize welding for decoration. The log projects use the weld bead to simulate the bark on a tree trunk. The desk pen holders use the weld simply for decoration.

LOG PAPERWEIGHT OR ORNAMENT

Materials required are one piece of 1½ in. black pipe 75 to 100 mm in length, one piece of 3/16 in. x ½ in. flat bar about 300 mm long, (Only about 20 mm of the flat bar is needed. However, it is easier to handle a longer section because it has to be hammered and ground.), and one piece of ⅛ in. oxyacetylene welding rod about 100 mm long.

This project can be made from scrap materials.

CHOICE OF WELDING PROCESSES

1. M.I.G. welding. 2. Electric arc welding (E7014 ⅛ in. electrode suggested). 3. Oxyacetylene welding (fusion). 4. Braze-welding to give the log a rich gold look. 5. Silver-brazing or braze-welding to join the axe head to the handle.

PROCEDURE

Run overlapping beads down the length of the pipe. (The welds do not have to be perfect. Take a look at the trunk of a large tree.) After welding cut one end of the pipe at 90° and the other at about 30° leaving a piece of pipe 65 to 75 mm long. Cutting both ends will get rid of the starts and finishes of the welding. At this point you will have to decide whether to fill the angled end of the pipe with metal or wood. If you decide to use wood it will have to be a piece of tree branch that has been allowed to dry out thoroughly. Using green wood or a plug of wood from a board will not work. Green wood will dry out and fall from the pipe. The plug of wood will not look realistic enough. It is suggested that you make your first attempt with a steel insert.

Cut a disc of steel 1³⁄₁₆ in. or 1¼ in. thick to fit the inside diameter of the pipe. Weld in flush and then grind to a smooth finish using a disc grinder. At this point the log is finished except for the saw cut to accommodate the axe head.

AXE

The axe head is shaped from a piece of ³⁄₁₆ in. x ½ in. flat bar or similar material. Flare out the end of the bar by hammering it on an anvil. Grind the sides to make the head even and balanced in appearance. Cut off about 20 mm of the bar. Shape a piece of ⅛ in. welding rod as shown and silver-braze or braze-weld to the axe head.

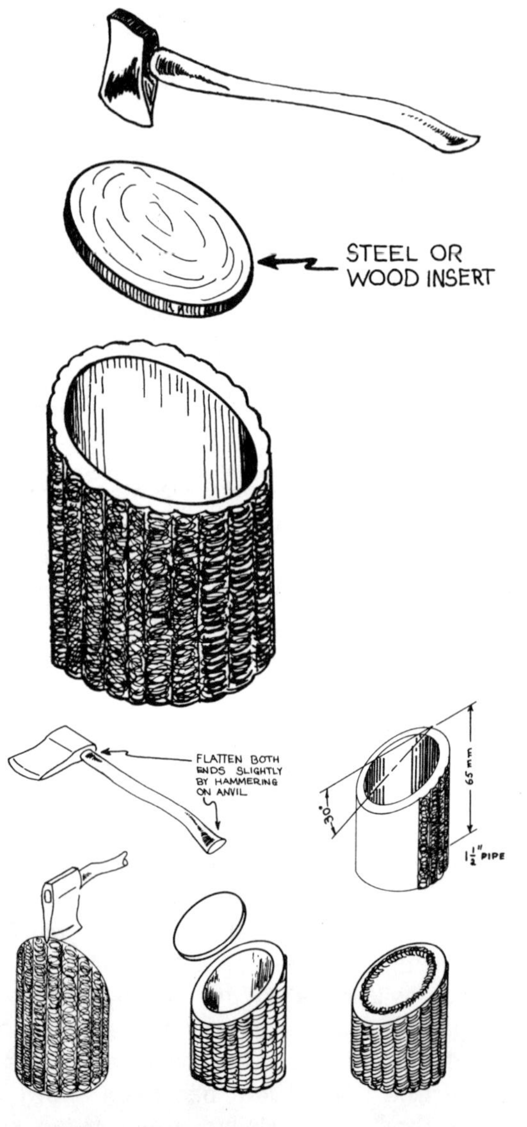

Fig. 1.2. Log paperweight

FINISHING

At this point you should polish the log with a wire wheel or wire brush. The power wire wheel will give the best finish. The axe should be rubbed with emery cloth or filed depending upon the type of finish you want. The axe could be left with a rough look similar to a real axe. To join the axe to the log, insert the axe head into a slot cut with a hacksaw in the log and put a *small* tack weld on the side of the log.

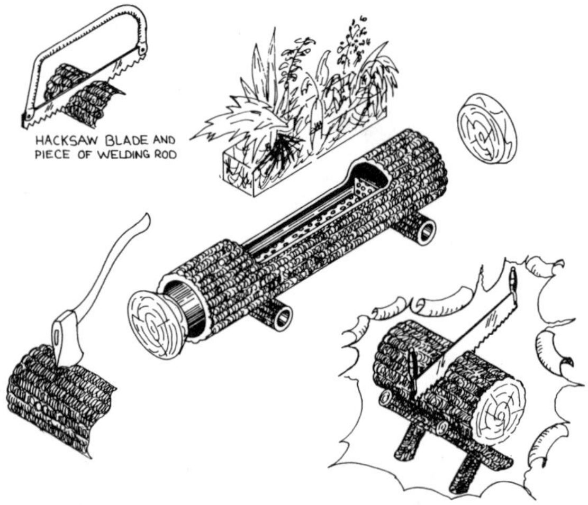

Fig. 1.3. Log planter

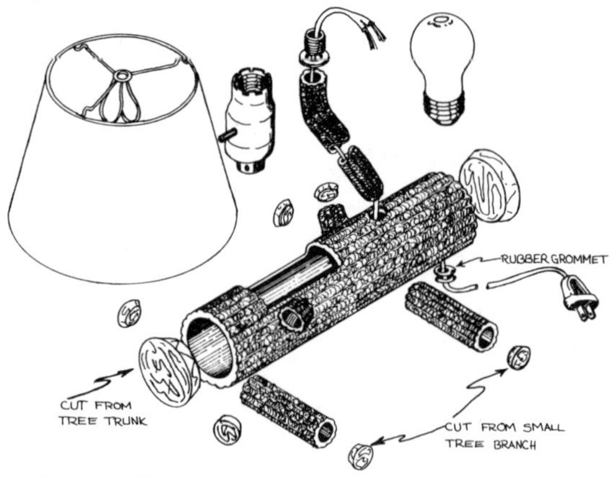

Fig. 1.4. Log lamp

LOG PLANTER

The log planter is best made from a piece of 3 in. pipe about 200 mm long. The feet are pieces of ½ in. pipe. The opening for the plants should be about 130 mm long and 40 to 50 mm wide.

The type of plant or decoration you decide to use will govern whether you need a piece of perforated metal inside. A piece of perforated metal will allow you to stick artificial flowers in the holes and support them. If you wish to use a live plant that will require watering consider making a small container from a piece of galvanized or tin plate.

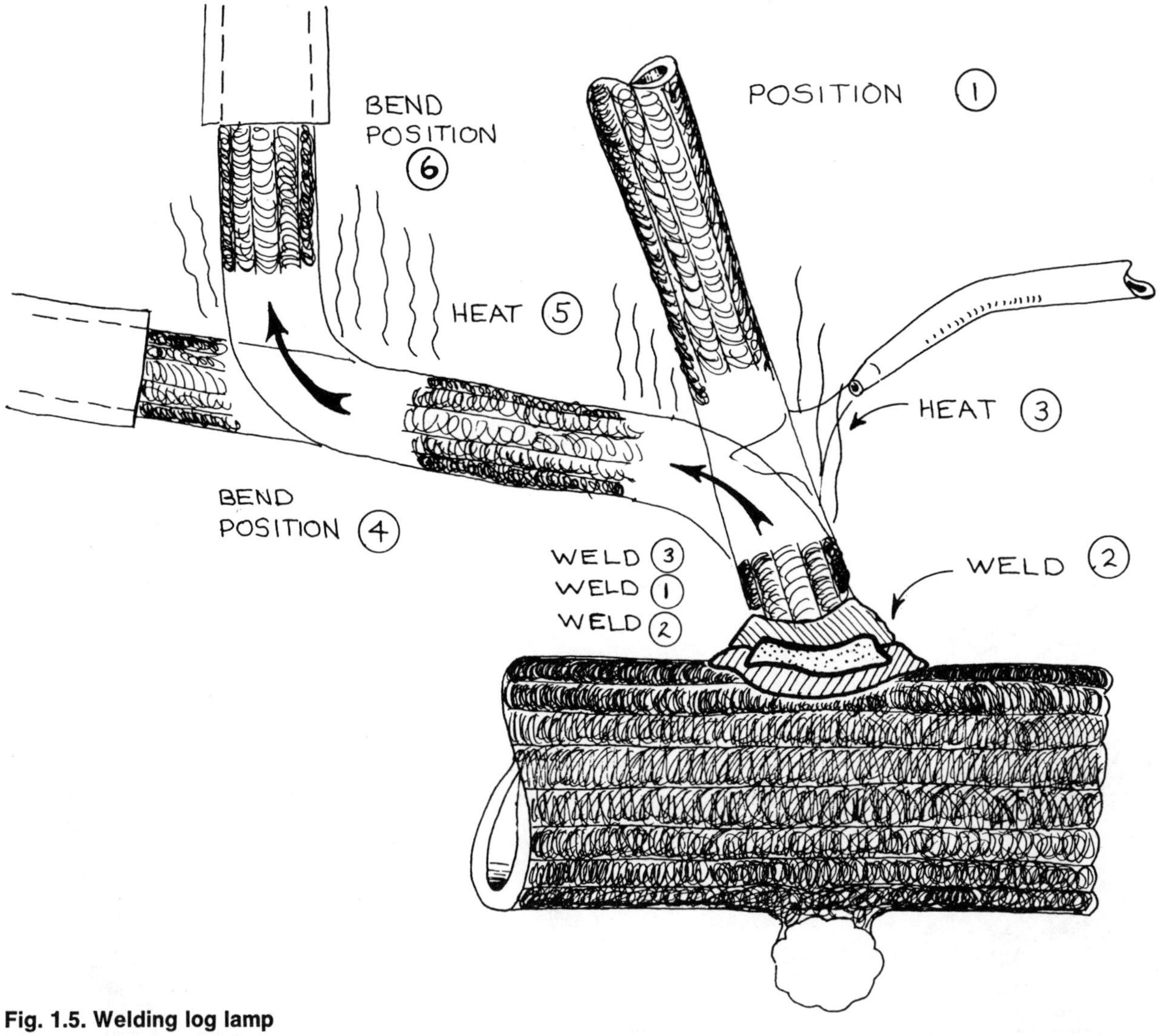

Fig. 1.5. Welding log lamp

LOG LAMP

Use 2 in. pipe for the main body and ½ in. pipe for the feet in making the log lamp. A piece of ½ in. pipe about 450 mm long with weld beads added will provide sufficient material for the branches. Place the main branch over the top hole at about a 70° angle to the body of the lamp. Weld as shown in Fig. I.5. using three welds or more to make the branch look as if it were growing from the main trunk. Realism is the key to this project. *Do not try to bend the long branch before it is welded to the trunk.* After welding the branch to the trunk heat the branch at the base and bend as shown. Do not make the bends too sharp; try to make them flow. The idea here is to end up with the end of the branch over the centre of the lamp base so that it is balanced in appearance. After bending the branch cut it 150-200 mm from the main body and add a short piece of ⅛ in. threaded pipe and a ⅜ in. washer if required. When welding the nipple and washer make sure that you do not plug the opening.

FINISHING

The log projects shown can be wire brushed then sealed with clear lacquer. They can also be made realistic in appearance by applying many coats of orange shellac to the brushed surface.

For variations consider welding the projects in brass, stainless steel, or aluminum if you have access to the various welding processes. *Design your own log projects such as log gun racks as shown.*

DESK PEN HOLDER

This project can be made with little difficulty. It is best made with steel at least 12 mm thick. The round base is welded from the outside in toward the centre. The vertical edge is welded last. The rectangular or square base is welded from one side to the other. After welding drill a hole for the machine screw that holds the pen base. Countersink the hole on the bottom to provide a recess for the head of the screw.

FINISHING

Wire brush the base before adding the pen, and seal with spray lacquer. Glue a piece of felt to the bottom of the base to prevent scratching. *Design your own penholders.* Consider shapes such as a triangle, star, oval, and so forth.

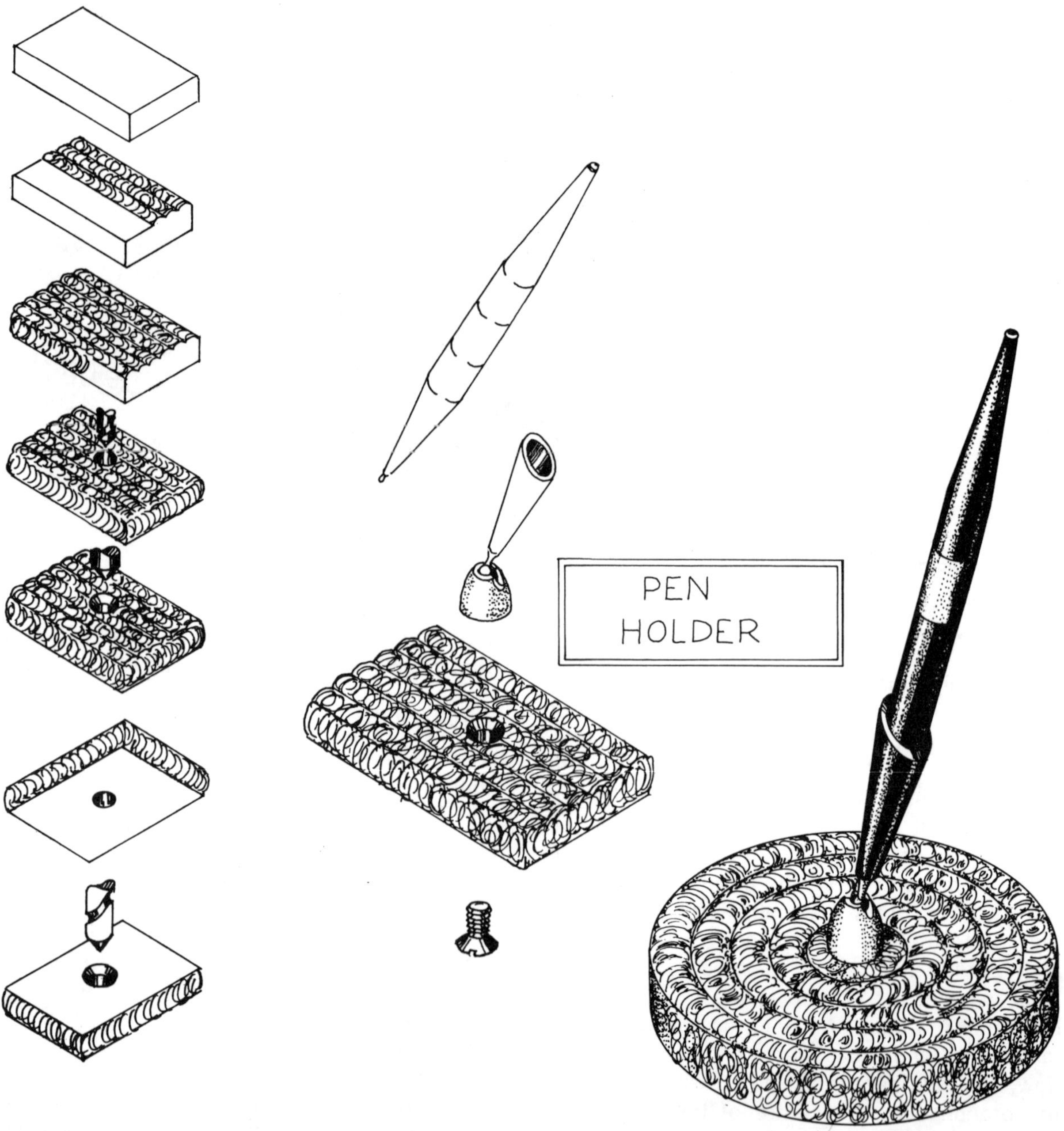

Fig. 1.6.

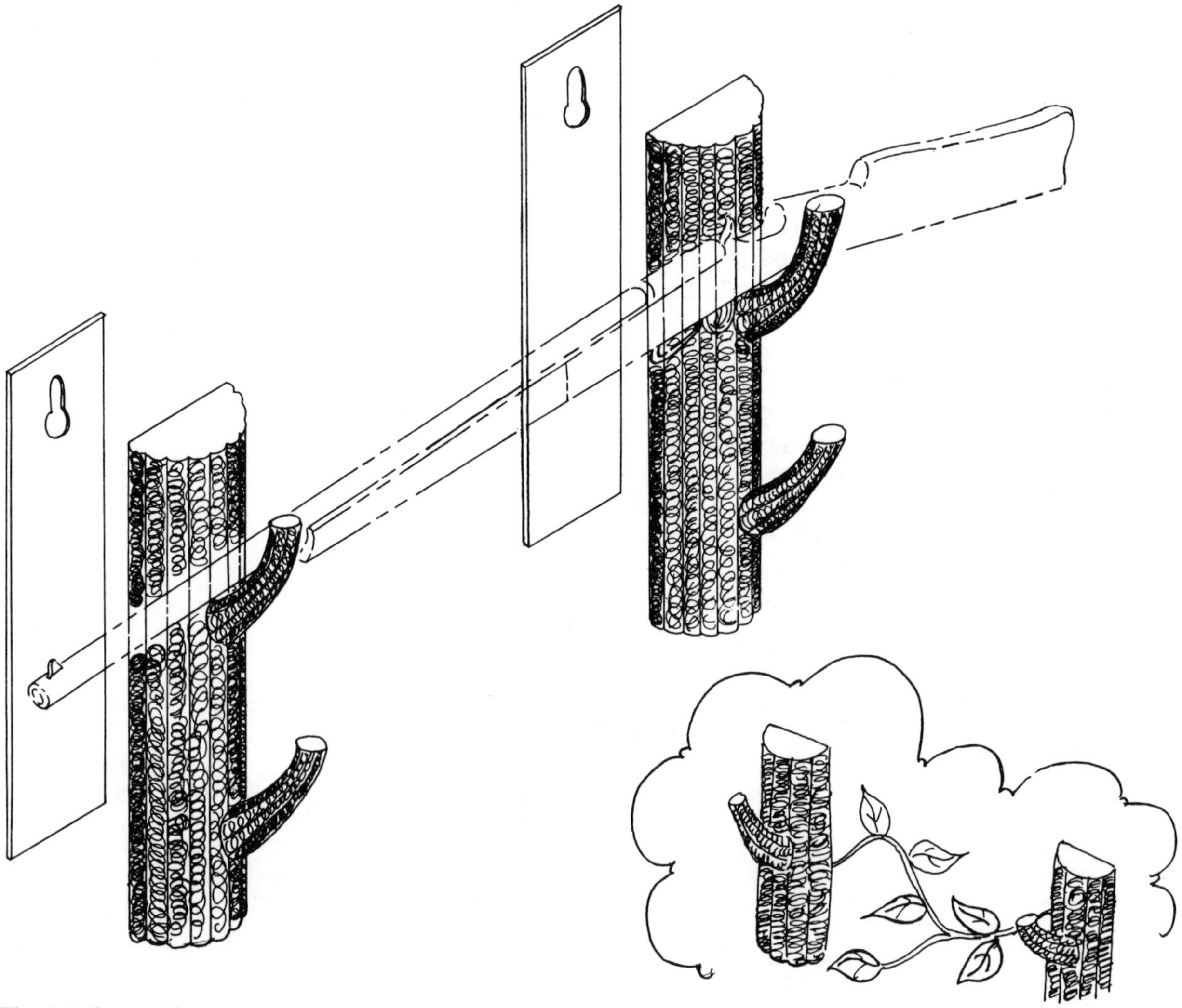

Fig. 1.7. Gun rack

UNIT 2. CUBES

The basic cube is a six-sided box that involves layout, cutting, and welding to complete. Drilling holes in the cube will provide the maker with a die.

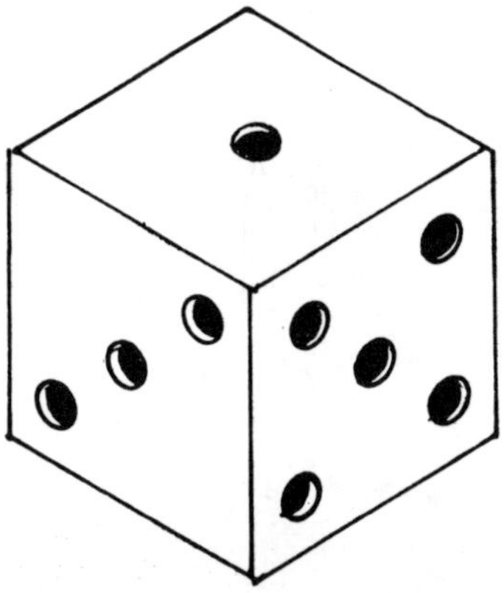

Fig. 2.1. Die

BASIC CUBE

Material required is one piece of 16 or 14 ga. sheet steel 50 x 300 mm (this will make a 50 mm cube). **Choice of welding processes** is oxy-acetylene welding (fusion).

PROCEDURE

Lay out six 50 mm squares as shown in Fig. 2.2. Drill all holes before cutting the squares apart. Tack weld all sides together before welding. Be sure that the sides opposite each other have holes totalling seven. For example, six should be opposite one. Before welding the joints of the cube decide how you wish the finished cube to look. Should the corners be sharp and clean? Should they be slightly rounded? Should the welds show?

If you wish to have sharp corners add filler rod as you weld, then grind smooth and sharp. For rounded corners do not use filler rod and either file later or leave as is.

USES FOR THE CUBE

The cube may be used as a pencil holder. Adding a desk pen will provide an attractive gift. Bookends may be made as shown in Fig. 2.4. The use of this project is limited only by your imagination. See Fig. 2.2. also.

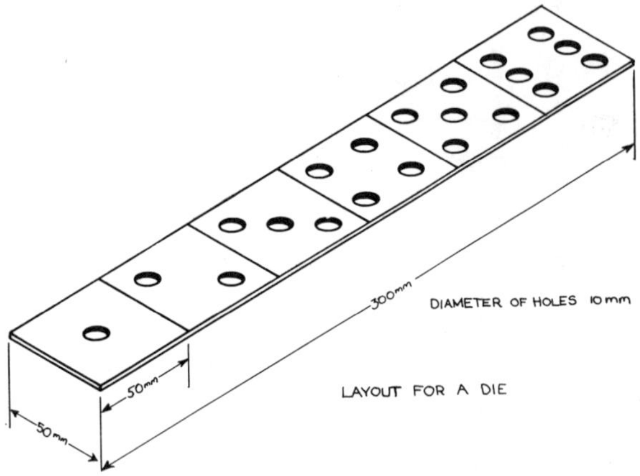

Fig. 2.2.

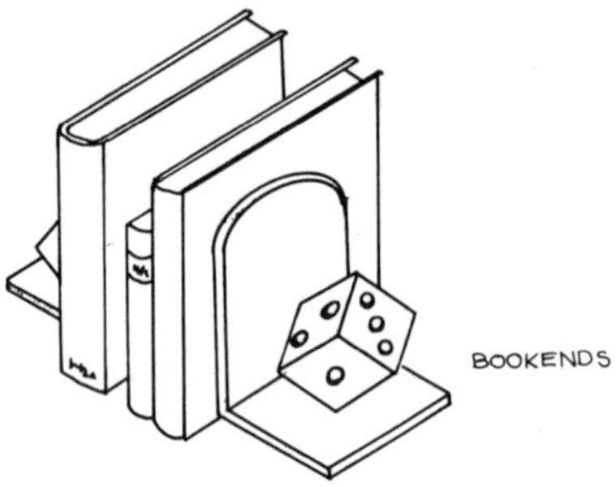

Fig. 2.4.

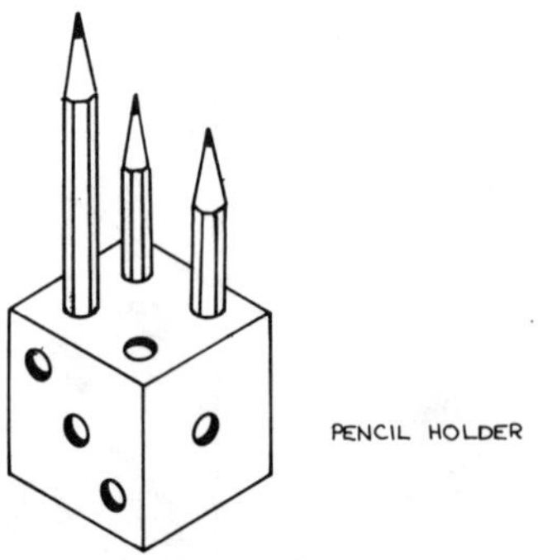

Fig. 2.5.

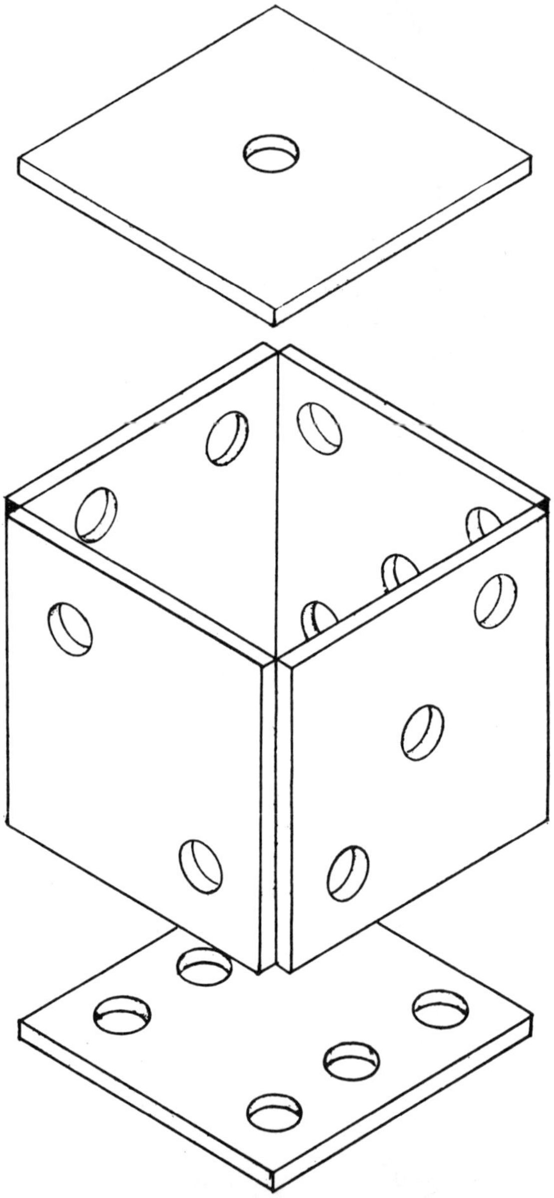

Fig. 2.3. Exploded view of die

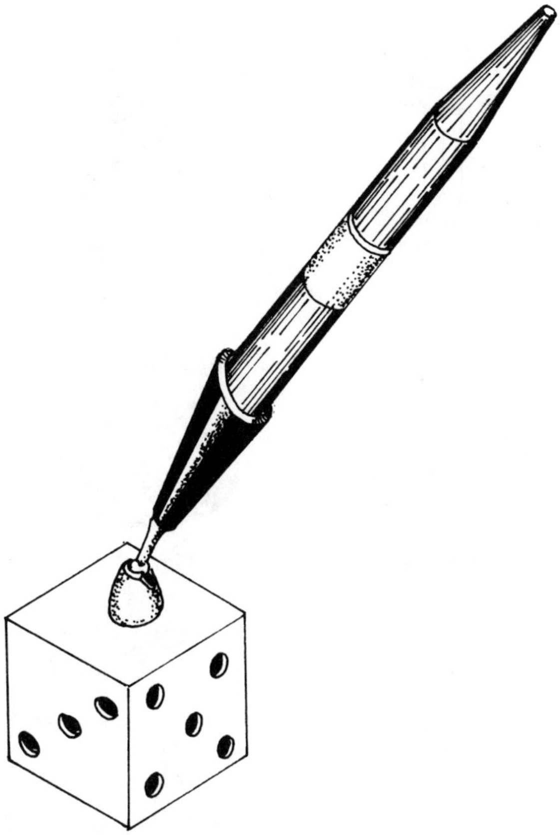

Fig. 2.6. Pen holder

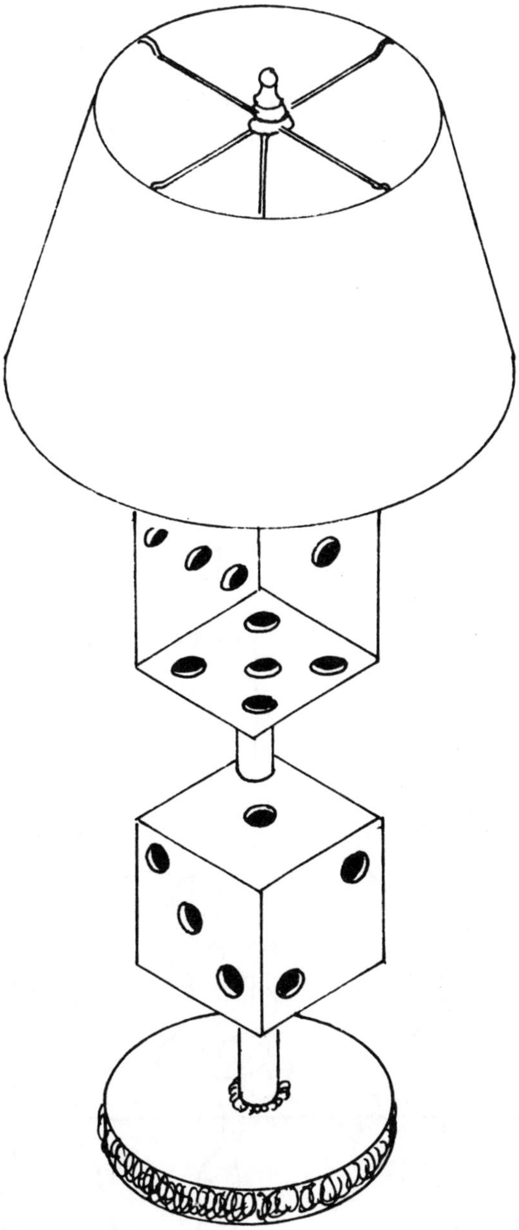

Fig. 2.7. Lamp

UNIT 3. CANDLES

The use of candles for decoration or supplementary lighting is very popular now. Candles can be obtained in all shapes and sizes, therefore, the candle holders shown in this unit range from the simple single candle holder for table or wall to the complex six or eight candle candelabra that would look good hanging over a dining room table or in a large stairwell. *Most candle holders may be adapted to accept wiring and electrical fixtures.* See the unit on lamps.

The basic candle holder consists of the cup

for holding the candle, the drip tray, and the body or base of the holder. Fig. 3.1. illustrates a simple holder. Before starting any of the candle holders shown study the material on the making of the various parts in the "how to do" section.

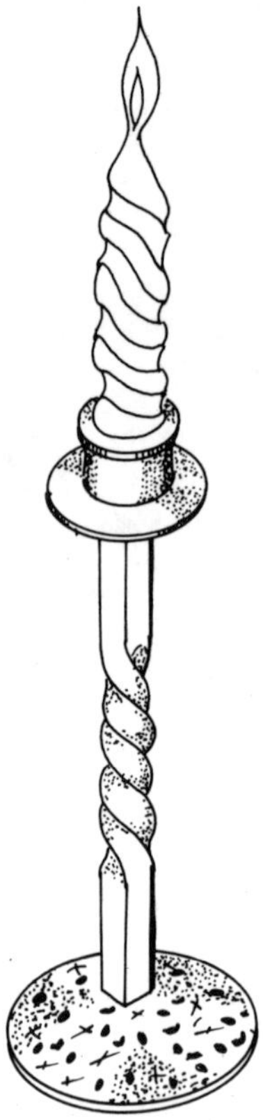

Fig. 3.1.

Fig. 3.2.

Fig. 3.3.

Fig. 3.4.

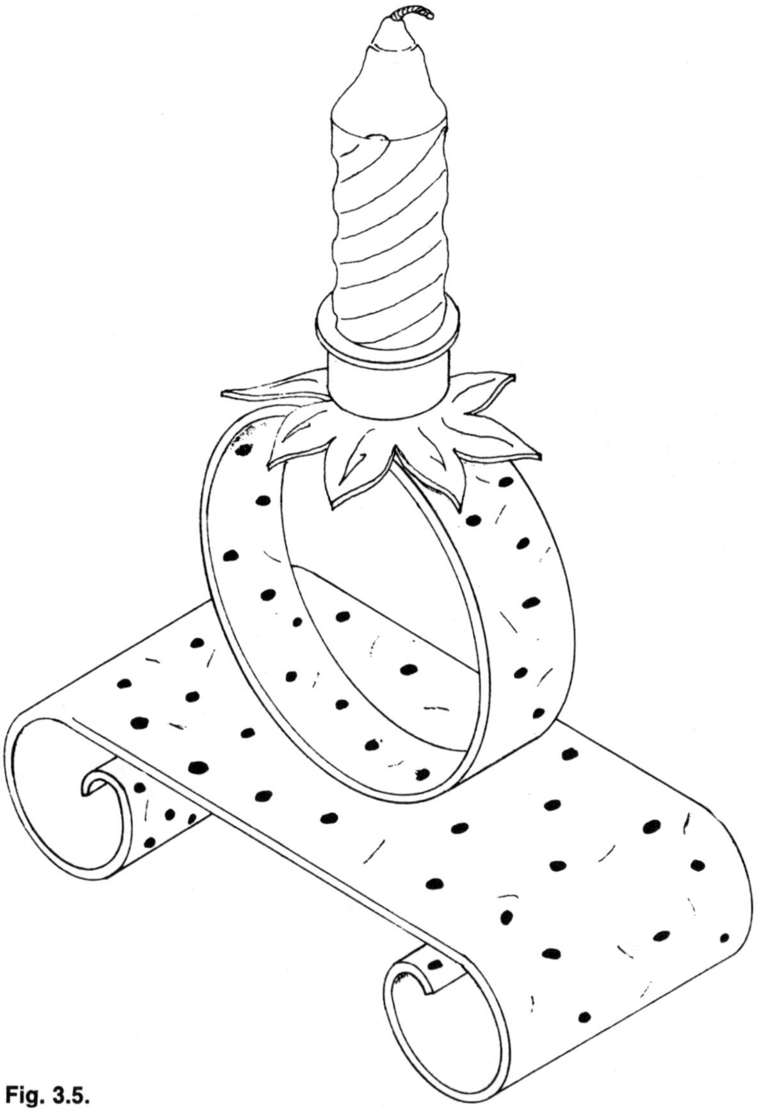

Fig. 3.5.

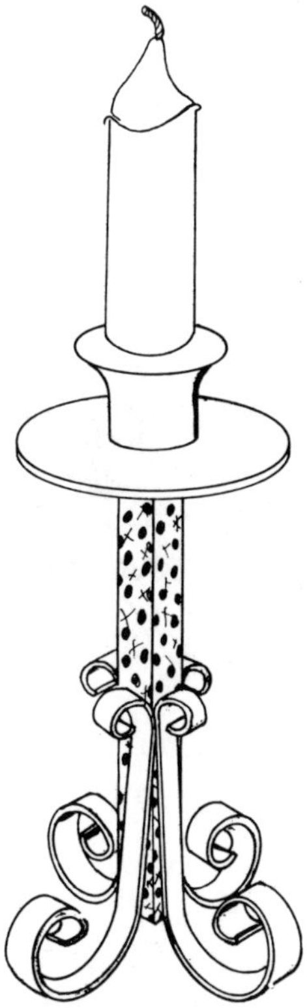

Fig. 3.6.

The many designs of candle holders shown require little instruction in fabricating other than to know the various dimensions and types of material used. Joining the candle holders may be accomplished by riveting, braze-welding, oxyacetylene fusion welding, silver-brazing, electric arc welding, and M.I.G. welding.

Fig. 3.1. (p. 100.) is 250 mm high. The stem is ⅜ in. square bar (twisted), the base, 100 mm diameter cup made from 16 or 14 ga. sheet metal. The candle holder is ¾ in. or 1 in. in diameter with 50 mm drip tray (dished).

Fig. 3.2. (p. 100.) is 250 mm high. The stem is made from a ¾ in. x ³⁄₁₆ in. flat bar hammered and formed into an "S" scroll. The top of the "S" is 65 mm; the bottom is 100 mm. The base of the figure is a 130 mm diameter disc (dished). The candle holder is ¾ in. or 1 in. in diameter with a 50 mm drip tray.

Fig. 3.3. (p. 100.) has the same scroll as Fig. 3.2. Vertical bars are ⅜ in. square bar twisted. The bars are approximately 100 mm long. Use ¾ in. or 1 in. diameter cups for the candles with 50 mm drip trays.

In Fig. 3.4. (p. 100.) the base is 1 in. x ³⁄₁₆ in. flat bar hammered. Each side is 150 mm long. The vertical bars are ⅜ in. square twisted. The longest is 200 mm, the next is 165 mm, and the shortest is 115 mm in length. Scrolls can be made with ³⁄₁₆ in. round bar (oxyacetylene welding rod) with a 65 mm scroll at one end and a partial scroll at the other.

In Fig. 3.5., the ring is 150 mm in diameter and may be made from 1 in. x ⅛ in. flat bar. The base is 150 mm long and 75 mm wide. Use ¾ in. or 1 in. diameter cup with 50 mm drip tray.

In Fig. 3.6. the overall height is 230 mm. The centre post is ½ in. x ½ in. square bar. The scrolls are made from ½ in. x ⅛ in. flat bar with 65 mm scrolls at one end and a partial scroll at the other. The candle holder is 1½ in. in diameter with a 75 mm drip tray (dished).

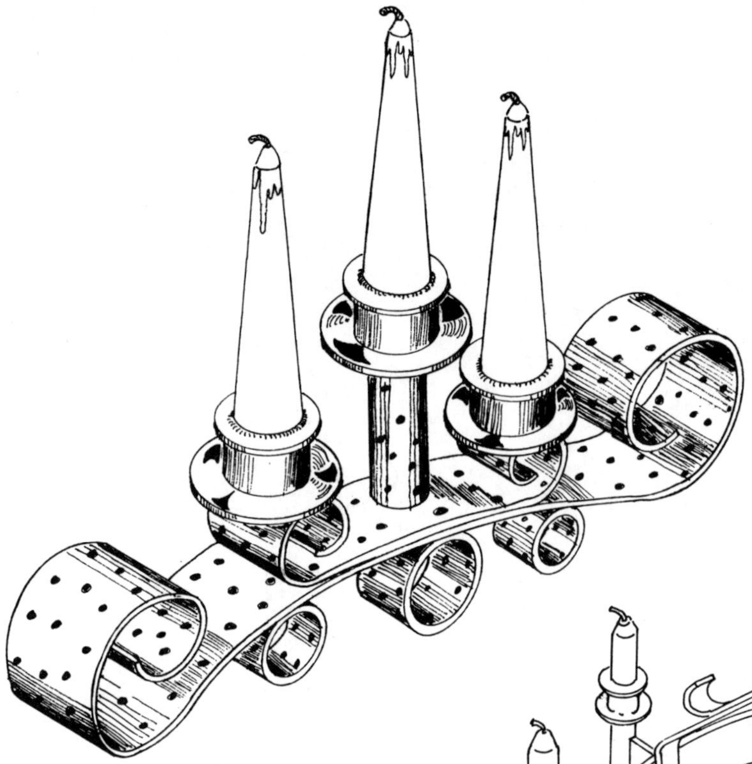

Fig. 3.7.

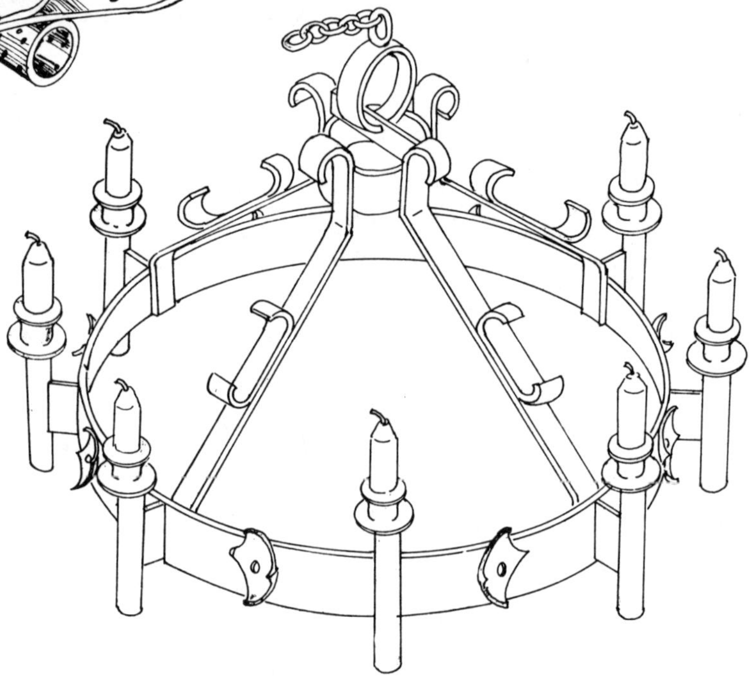

Fig. 3.8.

The overall width of the candle holder in Fig. 3.7. is 355 mm. The material for the base is ⅛ in. x 1½ in. flat bar hammered and scrolled. The centre post is a piece of ⅝ in. tube 50 mm long. The second part of the base is made from ⅛ in. x 1 in. flat bar hammered and scrolled. The candle holders are ¾ in. or 1 in. diameter cups with 50 mm drip trays.

DESIGN YOUR OWN CANDLE HOLDERS

The basic designs shown above may be altered by changing the type and shape of materials used. The dimensions may also be changed. Whatever you change in your designs, try to retain balance and flow.

The candelabra shown in Fig. 3.8. can be made from relatively inexpensive materials such as flat bar and tubing. The largest item in this project is labour in terms of hours required to hammer the surfaces of the rings and to hand rub the surfaces after assembly. There is no short cut method of obtaining the finished appearance. Any use of grinders or buffers does not leave an authentic finish.

Materials required are numbered in Fig. 3.9. (p. 103.) 1. Scrap piece of chain. The length will depend upon location. 2. 3⁄16 in. x ¾ in. x 165 mm flat bar. A piece of 2 in. pipe may be substituted. 3. 3⁄16 in. x ¾ in. flat bar cut to fit ring 2. 4. ⅛ in. x 2 in. flat bar. This is the same material that is used for the main ring 7. This ring can vary from 75 mm to 150 mm in diameter. 5. ⅛ in. x ½ in. flat bar 180 mm long. 4 required. 6. ⅛ in. x ½ in. flat bar 500 mm long. 4 required. 7. ⅛ in. x 2 in. flat bar. 70 in. should be sufficient. We stripped a piece of ⅛ in. sheet steel 8 in. wide. This gave us enough material for 4 units. *At all* *times think of substitute materials that may be used.* 8. ¼ in. x 1 in. flat bar. Eight pieces 40 mm long. 9. A scrap piece of 16 or 14 ga. sheet steel sufficient to make 8 diamond shaped pieces 40 x 65 mm. 10. ¾ in. or 1 in. tubing. Eight pieces 120 mm long. 11. Candle holders. See the section on candle holders. 12. Drip trays. Before you decide to make items 11 and 12 you should consider the type and size of candles you intend to use. You may also have to change the diameter of item 10 if you decide to use large diameter candles. Consult the section on candle holders.

CHOICE OF WELDING PROCESSES

1. M.I.G. welding. 2. Electric arc welding. 3. Oxyacetylene welding (fusion). 4. Oxyacetylene welding (braze-welding).

PROCEDURE

Hammer both sides of the ⅛ in. x 2 in. flat strip to a "Swedish Iron" appearance. Roll the strip on a set of sheet metal rolls to obtain a ring about 550 mm in diameter. Cut off the excess material and retain for making the smaller ring. Weld the joint in the ring. Do not grind the weld unless it is too high. Placing one of the candle holders over the weld will hide the joint.

Roll the smaller ring and repeat the above operations. Form a partial scroll on item 6 (⅛ in. x ½ in. x 20 in.). Do not make the other bends at this time. Mark off ring 4 and tack weld all 4 flat bars to the ring. Make sure that the flat bars are even and square to the ring. At the opposite ends of the bars place a chalk mark about 45 mm in from the ends. Use a square to locate the centre of the ring. Heat the bar about 25 mm above the ring and bend outward until the measurement from the centre of the ring to the mark on the bar measures 275 mm. Repeat the operation for the other bars. Finally heat and bend all 4 bars at the chalk marks making sure that the bent sections are parallel with the ring. Lay this assembly aside for now.

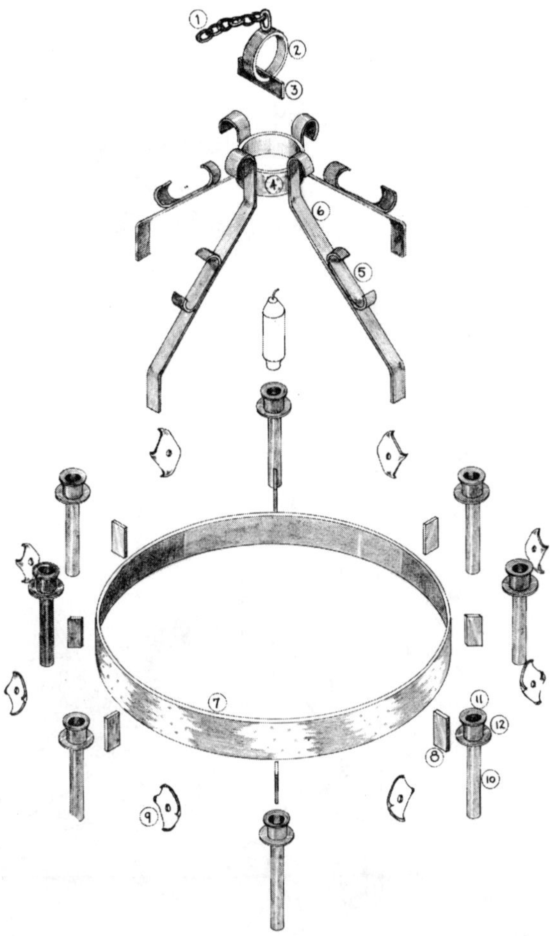

Fig. 3.9. Exploded view of candelabra

Fig. 3.10.

Fig. 3.11.

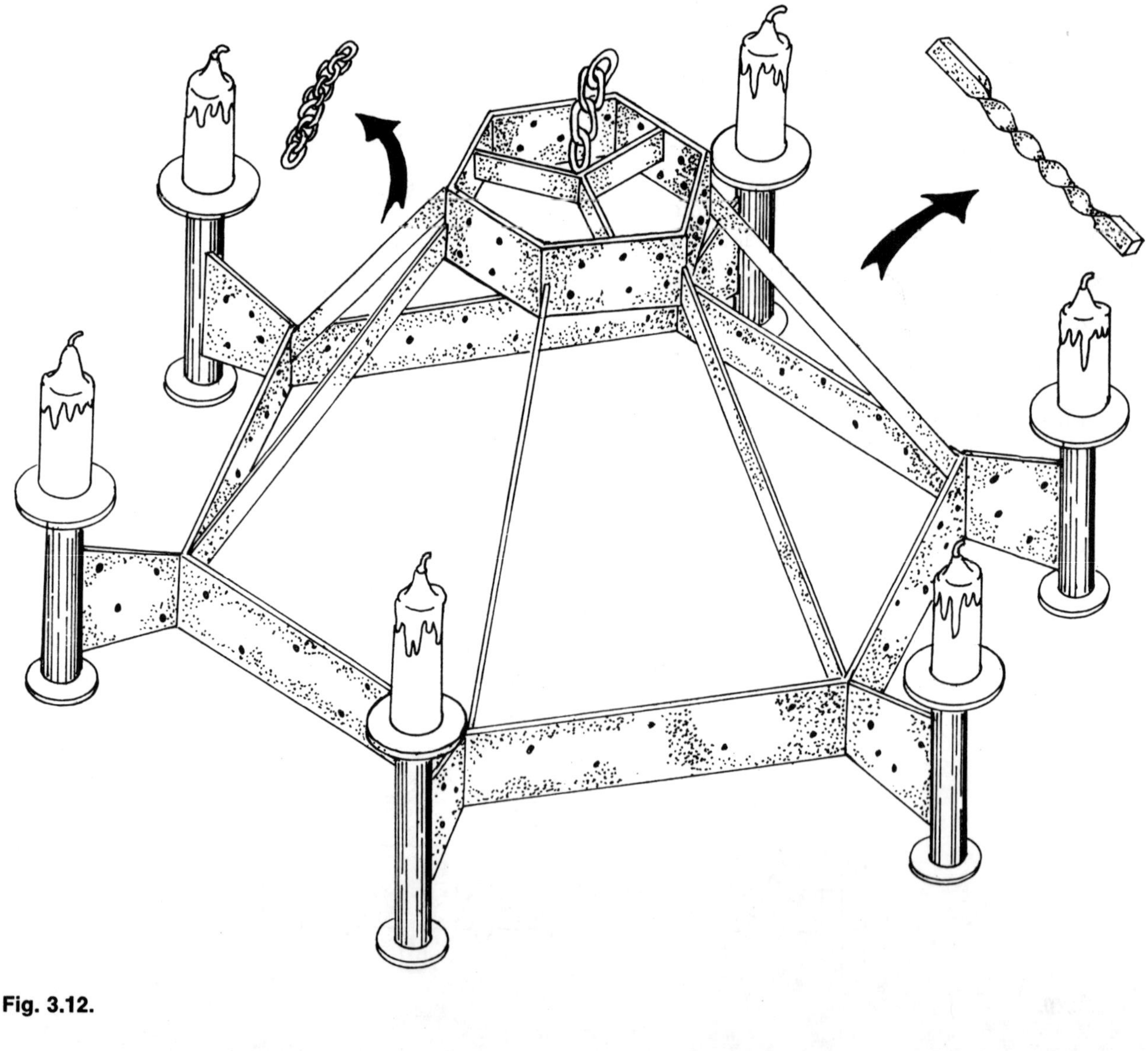

Fig. 3.12.

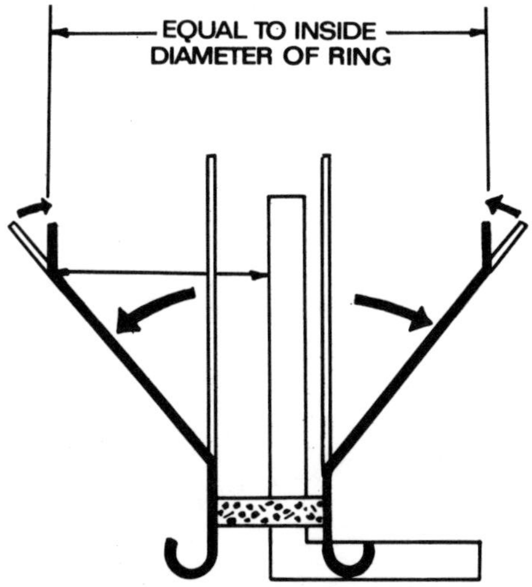

Fig. 3.13. Assembling candelabra

CANDLE HOLDERS

The candle holders used on this particular project were made with 1 in. diameter holders, 50 mm drip trays, and ¾ in. tubing. Cut 8 pieces of ¾ in. tubing 125 mm long. Seal one end of each by welding a small disc of metal over the end. Join the drip trays to the tubing, then the candle holders to the trays. Use whatever welding process you have available.

ASSEMBLY

Mark off the main ring into eight equal parts and weld one of item 8 at each intersection. Weld one candle holder assembly to each piece of flat bar. Make sure that all assemblies are vertical and square to the main ring. Place the assembly on a flat surface and check that all candle holder tubes are touching the surface. Measure the space between the bottom of the ring and the surface of the table at various points.

At this point you can join the two subassemblies together. Clamp the four bars (item 6) to the main ring. See Fig. 3.8. for positioning. At this point you will probably have to juggle and adjust the sub-assembly to make it sit parallel with the main ring as in Fig. 3.10. Be very particular at this point otherwise the project will not hang properly. When you have adjusted the assembly to your satisfaction weld the flat bar (item 6) to the main ring.

ORNAMENTATION

Items 5 and 9 may now be added to the project. Before scrolling items 5 drill a ¼ in. hole in the centre of each. After scrolling plug weld items 5 to 6. Consider adding 4 more scrolls on the inside of each 6.

Lay out eight diamond shaped pieces 40 mm x 65 mm on a piece of 16 or 14 ga. sheet steel. Drill a ¼ in. hole in the centre of each before cutting apart. Hammer each diamond shape then plug weld it to the main ring. A raised weld here looks much like a rivet. To shape the diamonds heat them with a torch until red then bend the vertical points with pliers. The horizontal points are bent by forcing a screwdriver between the main ring and the point and prying out while the material is hot. At this point it is only necessary to add items 1, 2, and 3 to complete the project. Upon completion of the assembly check that it is square and hangs level.

FINISHING

Paint the entire project with flat black paint. Allow the paint to dry then rub all hammered surfaces with sand paper. After sanding the surfaces to your choice of appearance spray the entire project with clear lacquer to seal the surfaces.

This is a complex project and if you are satisfied with the end result consider making Fig. 3.12. using the same dimensions or, better still, design your own. For ideas visit your local stores that sell lighting fixtures. Most fixtures can be adapted for candles.

The variety of projects in this section range from the very simple to some complex types. By changing types of material and various dimensions it is possible to have an endless selection of candle holders. When deciding on dimensions for any of the projects shown always consider width and height for balance and a pleasing appearance.

SUGGESTIONS

The remaining projects in this unit have been added to stimulate your interest. We have taken the liberty of adding our suggestions for dimension and types of material. Remember that they are only suggestions.

The candle holders in Fig. 3.14. (p. 106.) may be made from a 75 mm disc of 22 ga. sheet steel cut with aircraft snips. The body could be ¼ in. square bar twisted, the base a 100 mm diameter disc dished. Overall height could be 250 mm. The bottom arm could be 410 mm overall, the centre arm 230 mm, and the top arm 100 mm. All candle holders are sitting on 25 mm arms. Consider adding drip cups or trays.

The back plate for the project shown in Fig. 3.15. (p. 106.) could be oval or circular and could be made from 1/8 in. sheet steel hammered. Consider a 150 mm circle or an oval 200 mm high and 125 mm wide. The arms could be made from 3/8 in. square bar also hammered. The candle cups are 3/4 in. in diameter using 50 mm drip trays.

The project in Fig. 3.16. could be either a table centrepiece or hang from a ceiling. Consider a diameter of 410 mm with a height of 250 mm to the four centre candle holders. The scrolls could be made with 1/4 in. brass welding rod.

Consider a diameter of 355 mm and a height of 355 mm for Fig. 3.17. (p. 107.) Scrolls could be 100 mm and 65 mm. Material could be 1/4 in. round or 3/16 in. x 1/2 in. flat bar. The centre post could be 5/8 in. tubing.

For Fig. 3.18 (p. 107.), consider all suggestions as in Fig. 3.17., but increase the diameter to 410 mm.

The project in Fig. 3.19. (p. 107.) could be made using all dimensions from the project in Fig. 3.9.

As was suggested at the beginning of this unit candle holders come in all shapes and sizes. One could take three small "C" scrolls and make possibly a dozen different candle holders. The possibilities are unlimited. Graph paper with candle cups in position should start you off on the designing of your own exclusive wall sconces, candelabras, or simple candle holders for use as decoration on tables, fireplace mantles, and so forth.

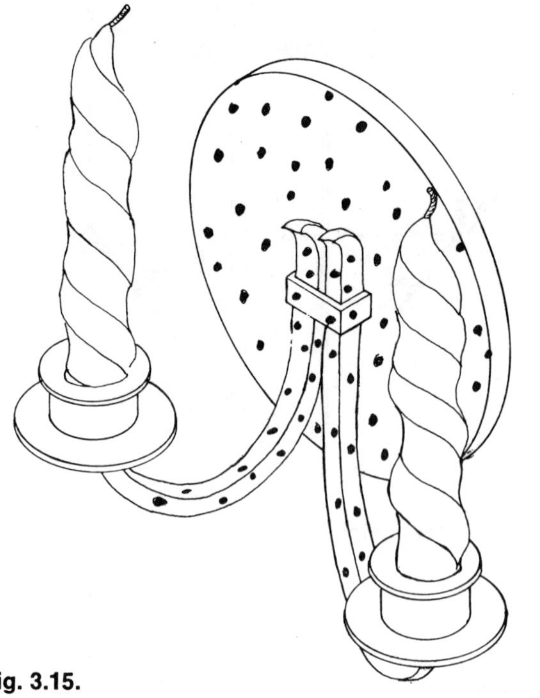

Fig. 3.15.

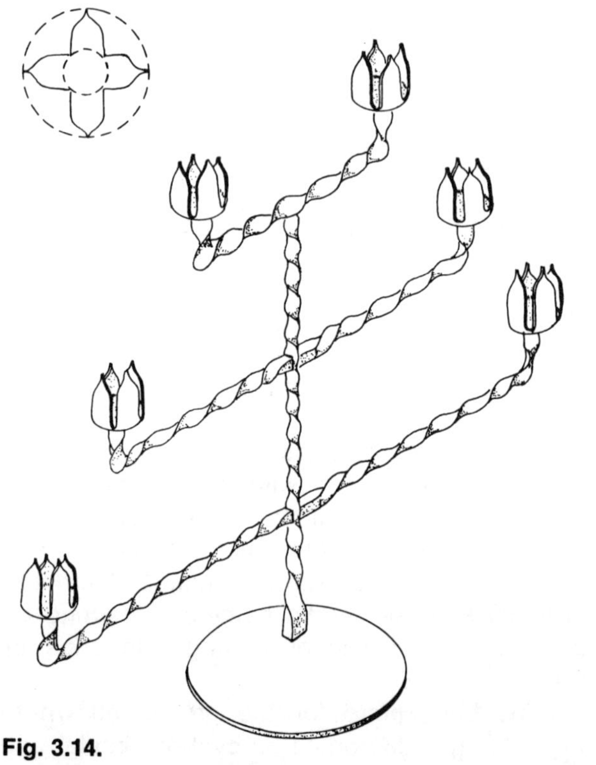

Fig. 3.14.

Fig. 3.16.

Fig. 3.17.

Fig. 3.18.

Fig. 3.19.

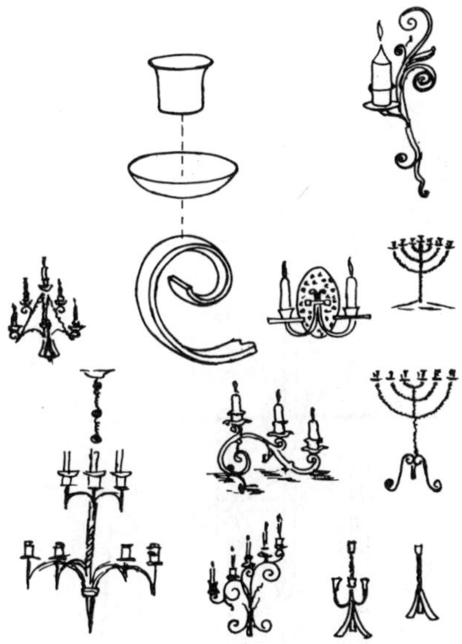

Fig. 3.20.

Fig. 3.21.

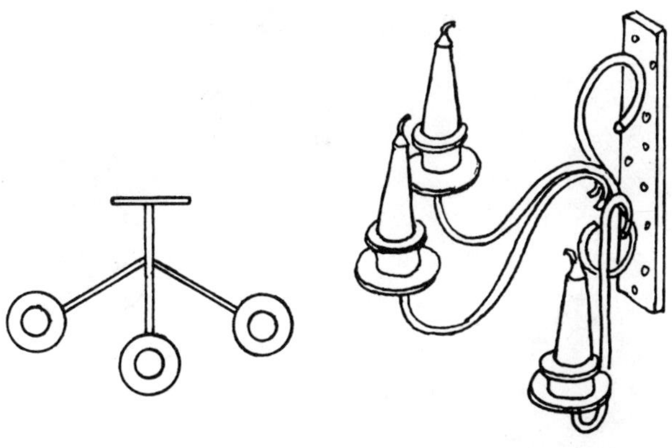

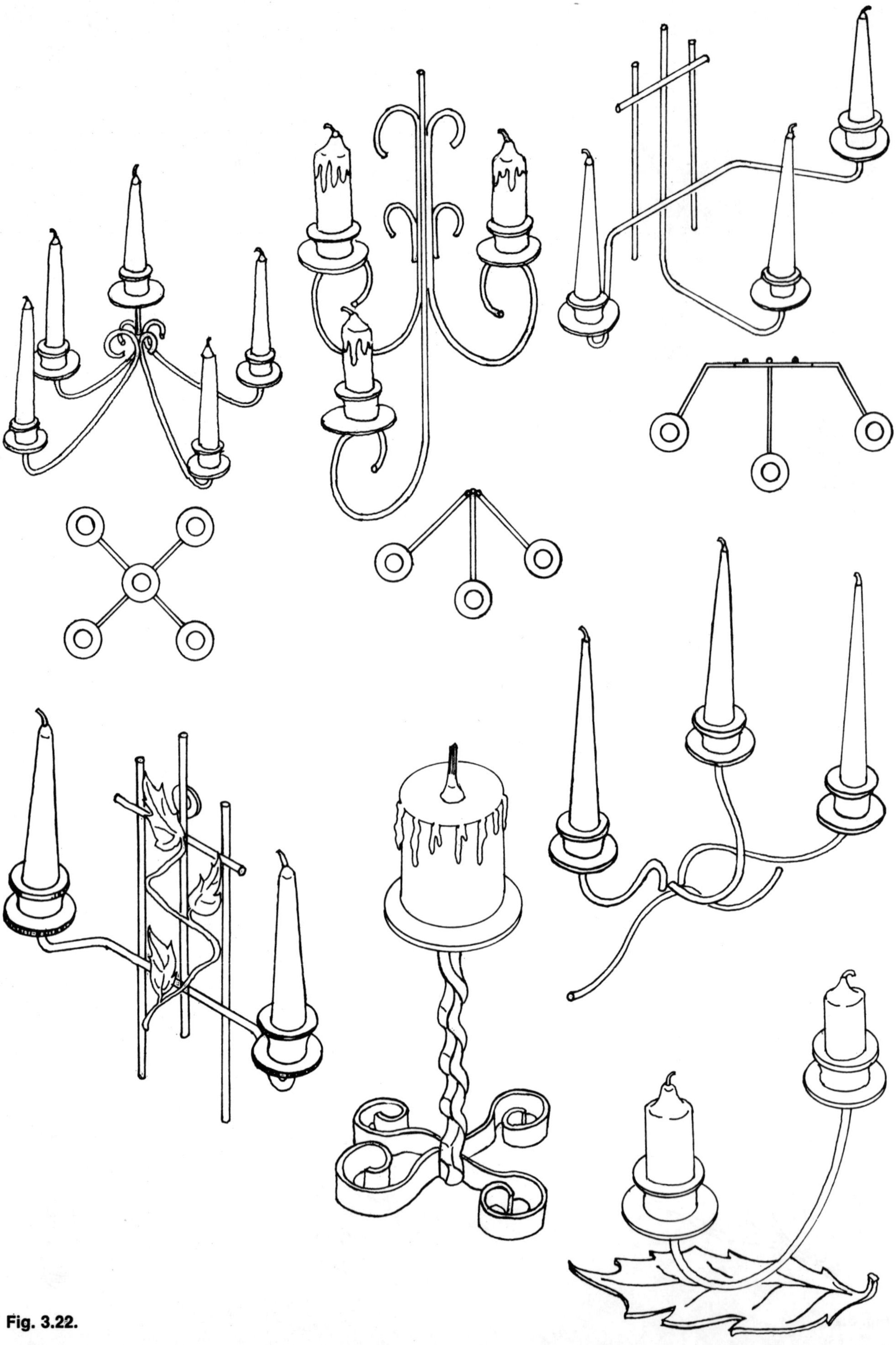

Fig. 3.22.

UNIT 4. LAMPS

Since the projects in this unit all require wiring and use electricity, the following suggestions apply to all projects. 1. Holes drilled to allow passage of wiring should be filed smooth on both sides. 2. Where possible rubber grommets (3/8 in. diameter) should be fitted in all holes carrying wires. 3. On larger projects such as the chandelier, wiring should be supported at intervals by means of clips. 4. The Underwriters' knot should be used where wiring may be subjected to strain. See Fig. 4.1.a. 5. If you are not sure about doing your own wiring ask for assistance from a qualified electrician. 6. Do not take short cuts with this type of project. Follow the electrical code in effect for your area of residence.

Lamps come in various types and shapes. They all, however, contain the basic materials such as a bulb or bulbs, a cord or wiring, and a connection to obtain power such as a plug. See Fig. 4.1.b.

The design of lighting fixtures today is unlimited. Materials such as plastic, glass, and cloth are all used. All of the projects in this unit are made with metal. None have any parts that cannot be made with basic tools found in most school shops. Some parts can be fabricated more easily using sheet metal equipment. However, with a little thought and ingenuity it is possible to change the design and substitute shapes and materials that you can form with available tools.

TABLE LAMPS

Because the table lamps shown are simple to construct only suggestions as to materials and

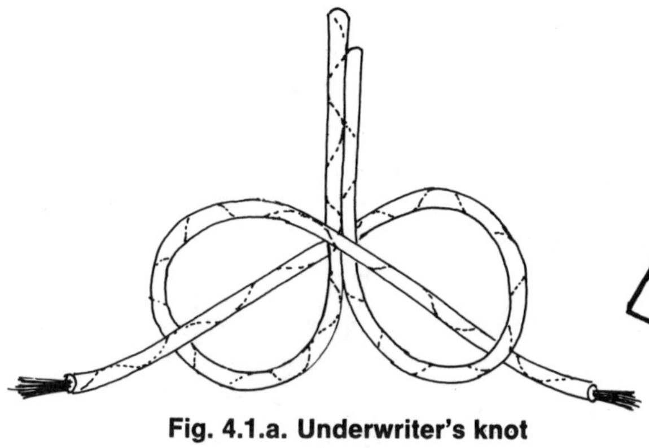

Fig. 4.1.a. Underwriter's knot

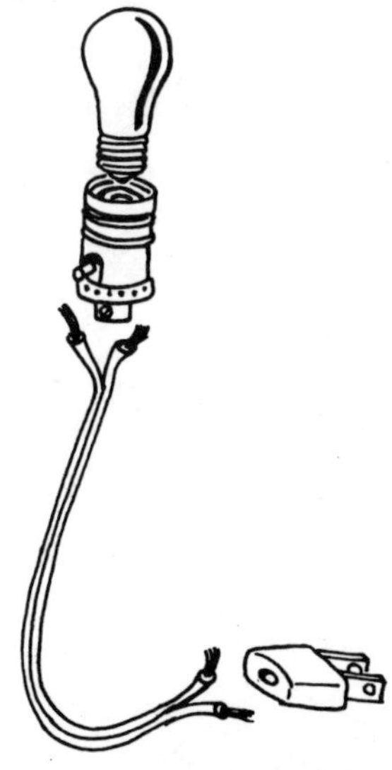

Fig. 4.1.b. Basics for a lamp

Fig. 4.2. Table lamp

dimensions are given. Personal taste will determine the type and shape of lamp shade that you will select for the various lamps shown. Give some thought to the shape, size, type of material, and the colour of the shade you intend to use. A lamp is only as attractive as its overall appearance.

The lamp in Fig. 4.2. can be made from steel or copper tubing. The curve at the base could be 150 mm in diameter and the height to the socket 200 to 250 mm. Most sockets fit $\frac{1}{8}$ in. threaded pipe. The leaves can be cut from sheet metal and silver-brazed or braze-welded to the stem.

The outer frame of the lamp in Fig. 4.3. can be $\frac{3}{8}$ in. square bar with the scrolls made from $\frac{3}{8}$ in. x $\frac{1}{8}$ in. flat bar or $\frac{3}{16}$ in. round such as welding rod. The lamp could be a square with 125 mm sides and a height of 300 mm to the end of the tubing. Consider twisting the square bar.

In Fig. 4.4., the lamp can be made using a 125 mm to 150 mm disc (dished) for a base. Overall height to the socket could be 355 mm.

Fig. 4.3.

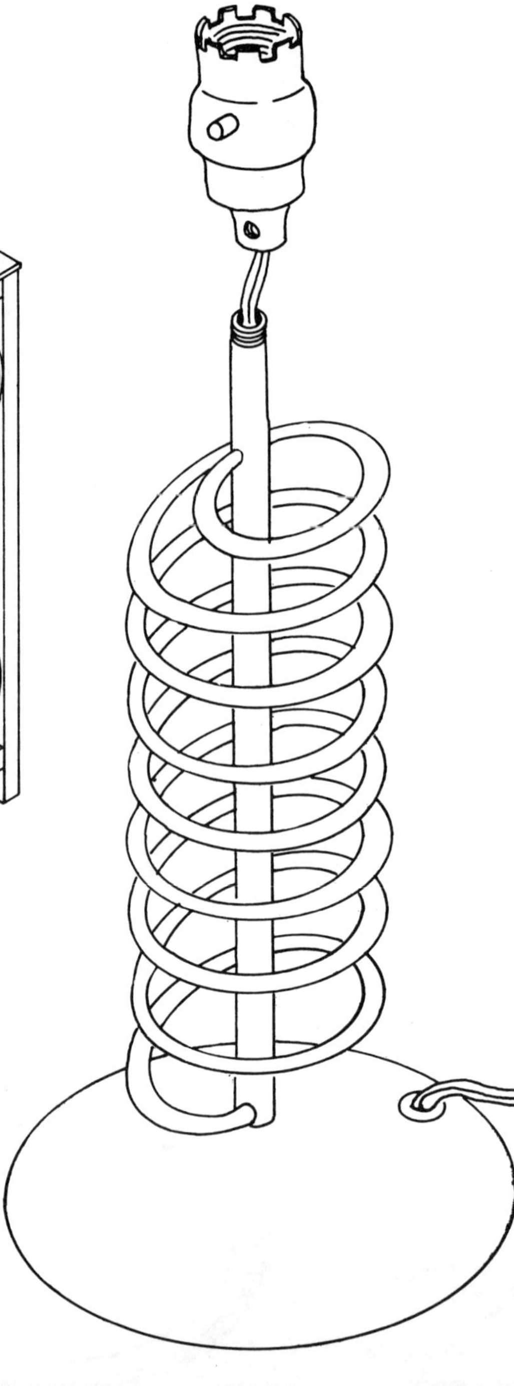

Fig. 4.4.

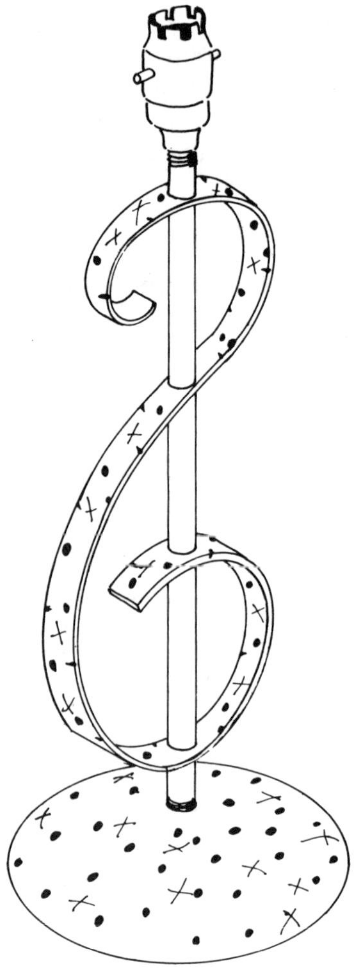

Fig. 4.5.

The coil should be 100 mm in diameter. Consider using brass rod for the coil and painting the ⅛ in. pipe flat black.

The design in Fig. 4.5. (p. 110.) presents a challenge. It looks simple, but skill is needed to drill all the holes in line to permit the ⅛ in. pipe to pass through. Consider ¼ in. x 1 in. flat bar for the scroll. The base could be 150 mm in diameter.

In Fig. 4.6., use ⅛ in. IPS pipe for the centre piece and paint flat black. Make four scrolls with a 65 mm scroll on one end and a 100 mm on the other using 3⁄16 in. brass welding rod. Height to the top of the pipe could be 300 mm.

Use the same dimensions as for Fig. 4.3. in Fig. 4.7. Choice of materials could be from any of the projects mentioned.

SWAG LAMP

This project is designed to use a swag lamp kit containing 4.6 metres of electrical cord with plug attached, 3.7 metres of decorative chain, lamp socket, ceiling hooks, and all hardware. These kits can be purchased in antique gold with black chain and fittings and tri-light sockets, or whatever styles seem appropriate.

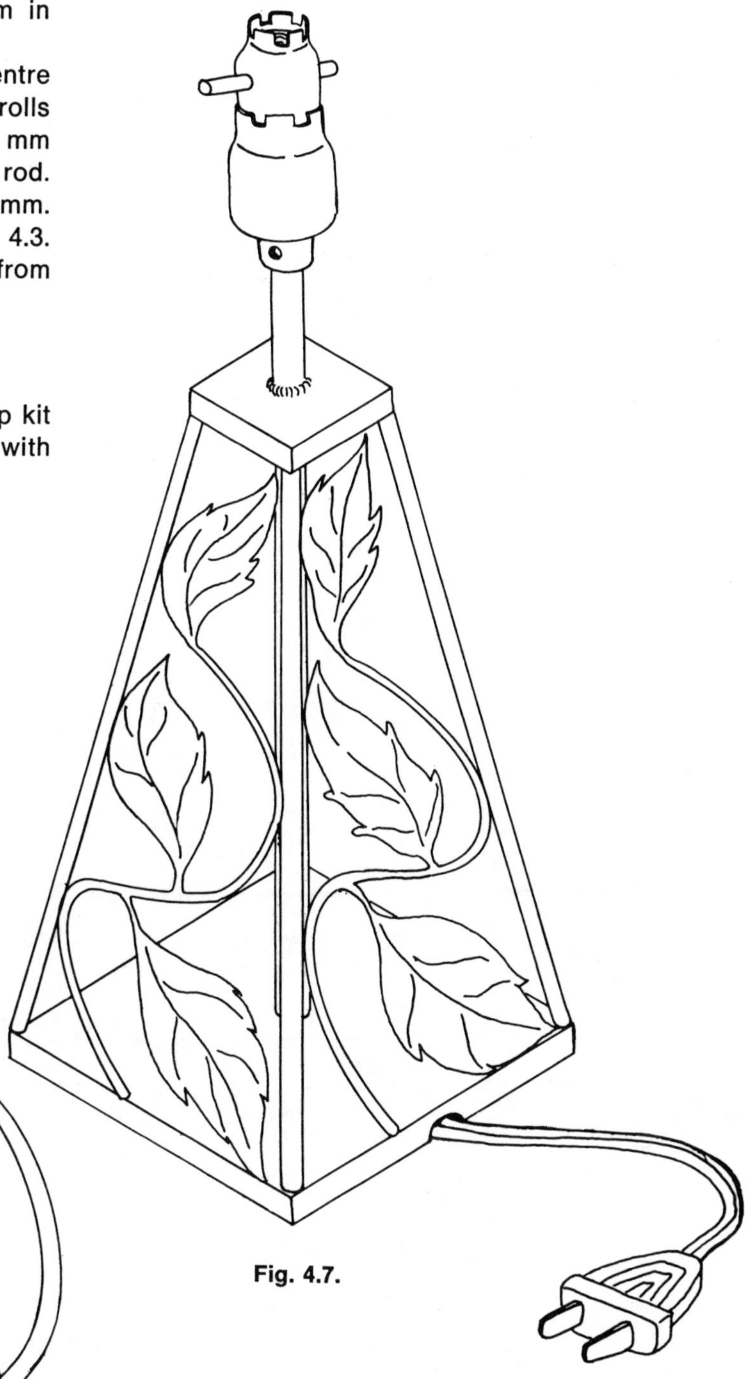

Fig. 4.7.

Fig. 4.6.

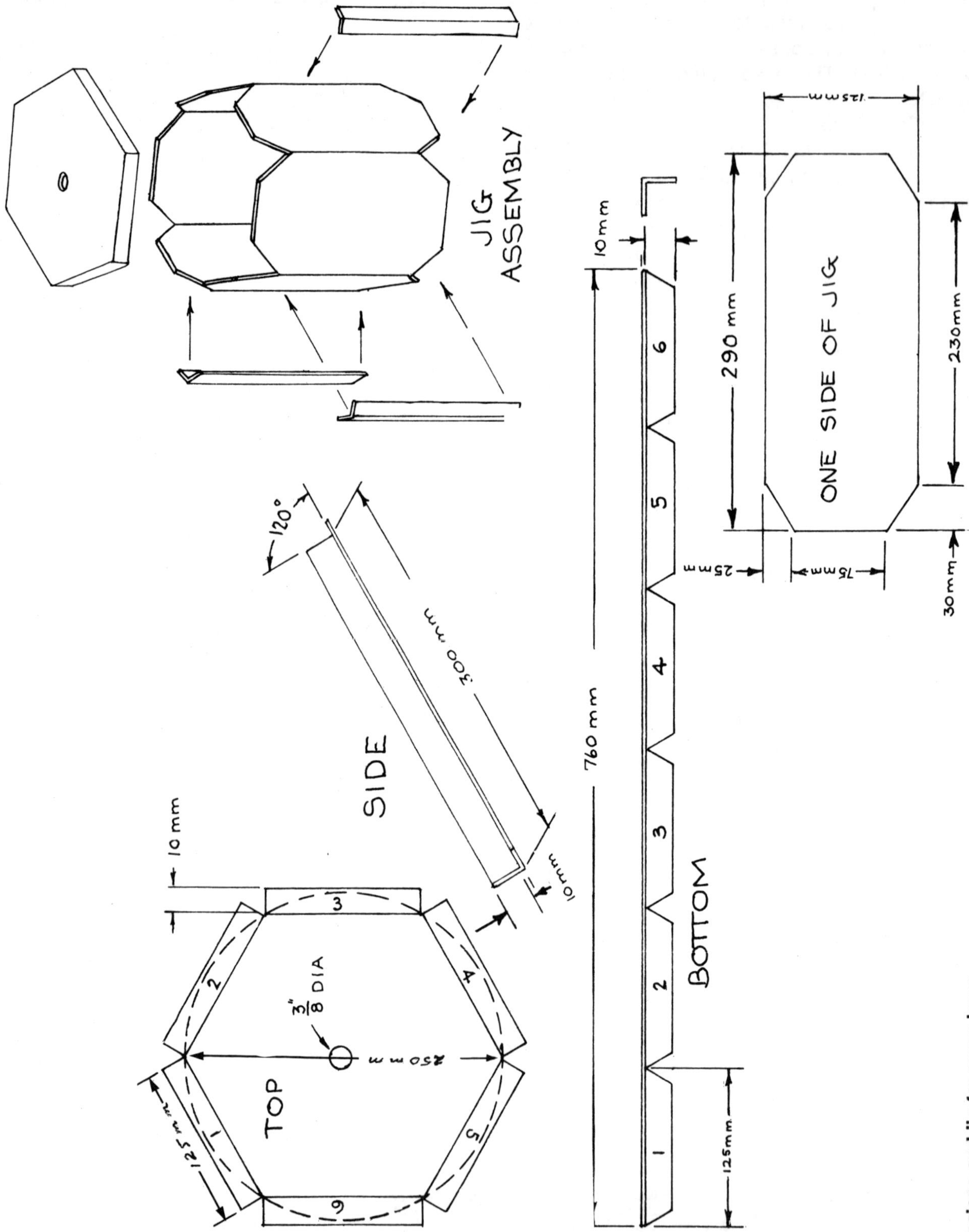

Fig. 4.8. Dimensions and jig for swag lamp

Fig. 4.8. (p. 112.) shows all the dimensions and also a jig that may be fabricated from 14 ga. sheet steel if you intend to make more than one lamp. Use 22 ga. sheet steel for making the lamp body. Spot welding is the best method of joining. If you do not have access to a spot welder consider gluing the joints.

LAYOUT

Top: lay out the top as shown and fold the edges on a pan brake. If a pan brake is not available use a hardwood block and a hammer to form the sides. Drill or punch a ⅜ in. hole in the centre. See Fig. 4.9. Sides: shear 6 strips of 22 ga. sheet steel 25 mm wide by 300 mm long. Bend each to 120°. Bottom: cut a strip of 22 ga. sheet steel 25 mm wide by 750 mm long. Mark off in 125 mm lengths. Cut notches with snips then form to the same shape as the top. It is not necessary to spot weld the joints. When the sides are spot welded to the bottom it is held in shape.

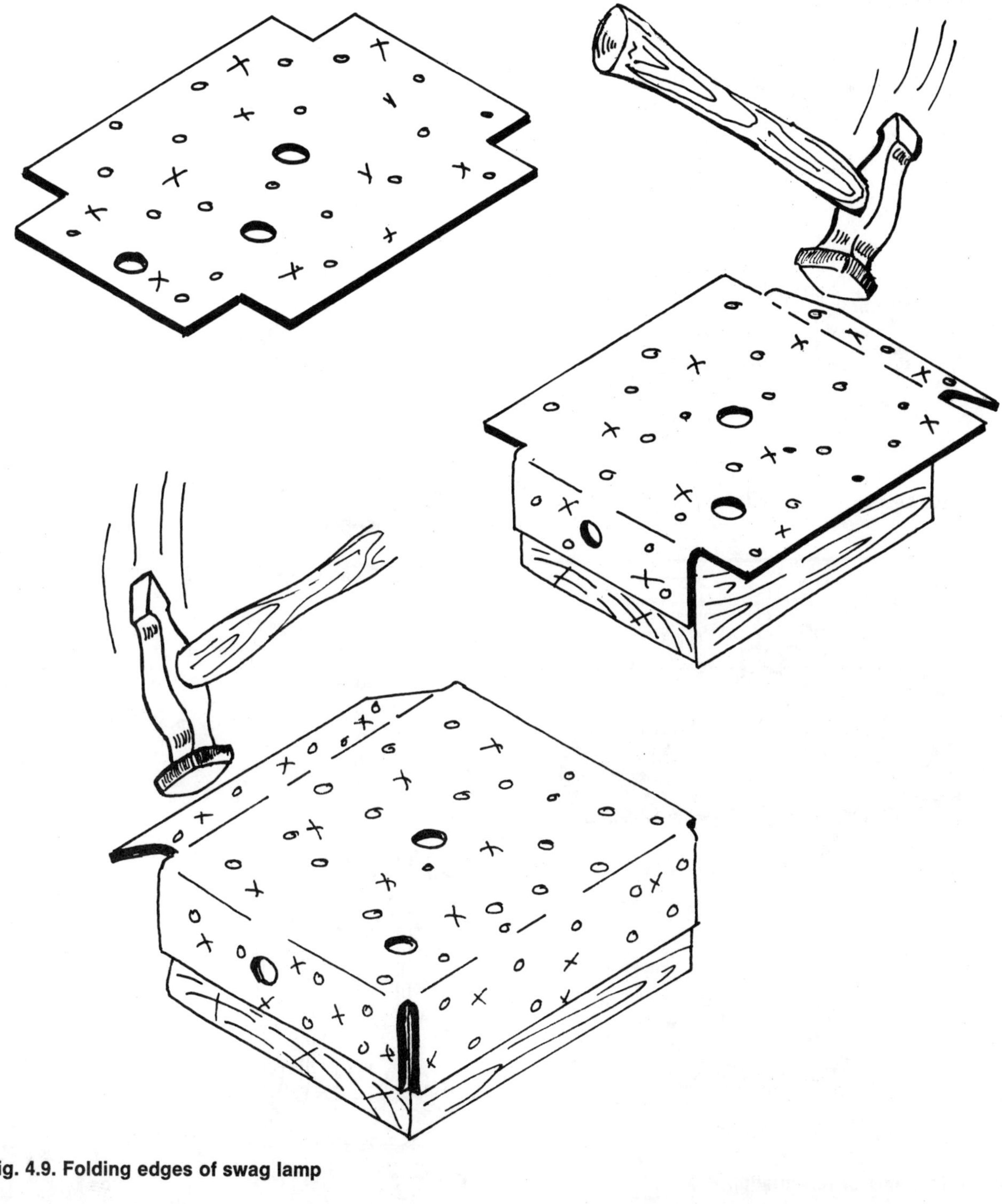

Fig. 4.9. Folding edges of swag lamp

ASSEMBLY

Spot weld each upright to the top. Make sure that each is square before joining. When all six uprights have been added join the bottom to complete the frame.

At this point it is necessary to cut and add the clips to hold the plastic or glass in place. Each side of the lamp requires a clip at the top and bottom. Cut 12 clips 6 mm wide by 20 mm long from 22 ga. sheet steel. Bend each at right angles and spot weld to the frame. See Fig. 4.10. Use a scrap piece of plastic as a spacer when adding the clips.

ORNAMENTATION

As you can see in the illustrations the ornamentation may be varied. The scrolls on top and bottom of the lamps were made with 3/16 in. welding rod. The decoration on the sides may be on the panels of acrylic or plastic or may be made from 1/8 in. welding rod. See Fig. 4.11. for suggestions as to sizes for ornamentation. All ornamentation was silver-brazed to the lamp body. Soft solder may also be used. Ornamentation may be used only on the top, or top and bottom according to preference and taste.

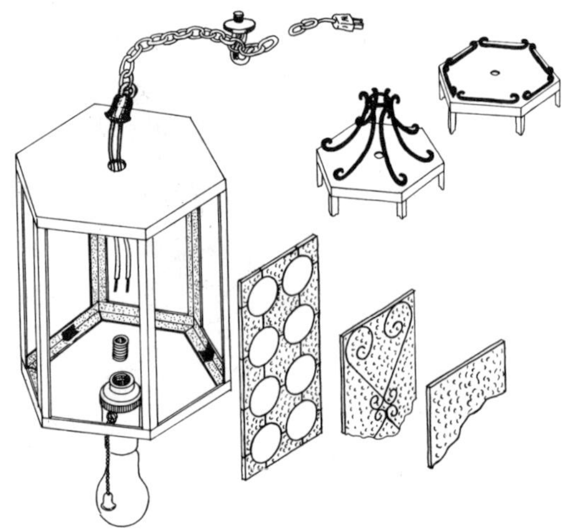

Fig. 4.10. Assembling swag lamp

Fig. 4.12. Swag lamp

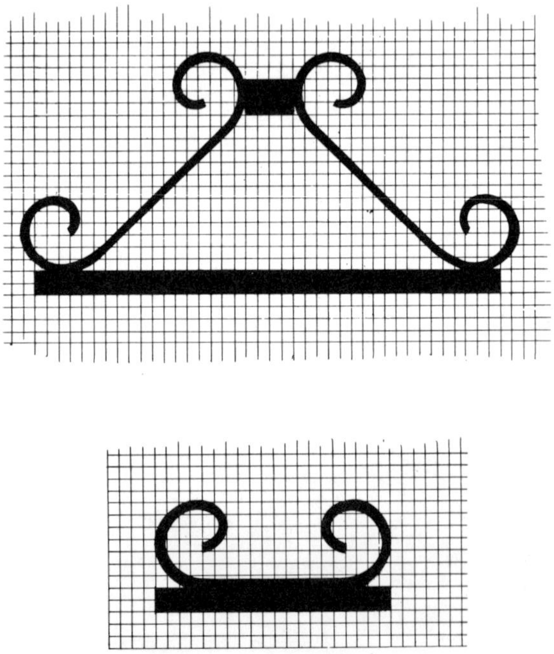

Fig. 4.11. Lamp ornamentation

Fig. 4.13.

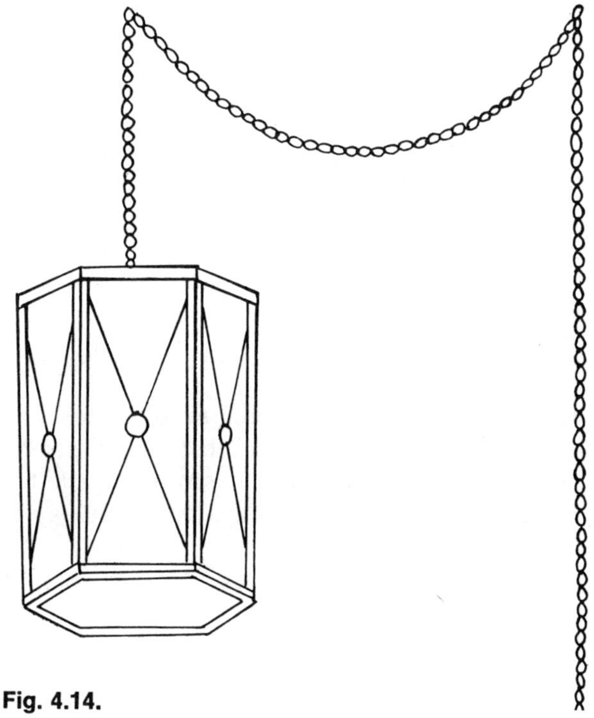

Fig. 4.14.

Fig. 4.15.

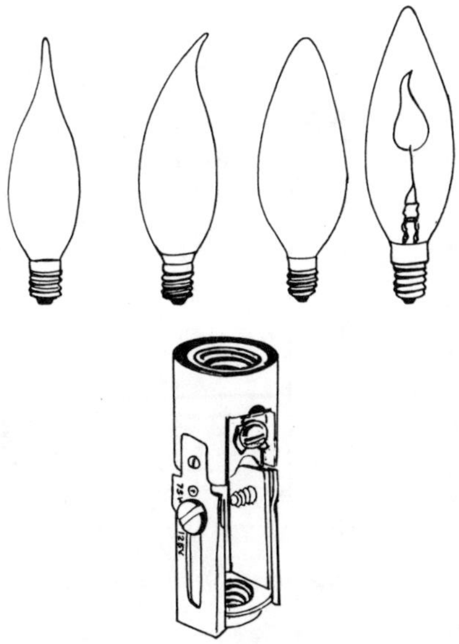

Fig. 4.16. Socket and ornamental bulbs

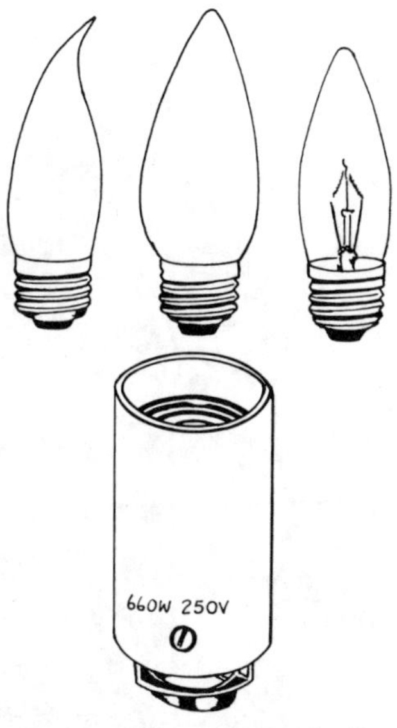

Fig. 4.17. Socket and ornamental bulbs

CHALLENGE

If you have access to a T.I.G. welder consider making your lamp body with 24 oz. sheet weight copper and the ornamentation with 3⁄16 in. brass welding rod silver-brazed to the body. Polished and lacquered this lamp looks superb.

The remainder of the projects in this unit use the decorative type of socket, plastic cover, and ornamental bulbs as shown in Fig. 4.16. and Fig. 4.17. The socket in Fig. 4.16. is a candelabra socket. It is available in a variety of heights; adjustable sockets are also available. The bulbs used in this type of socket must be the candelabra or small base type. They are available in an almost unlimited selection of shapes and colours. The more decorative the bulb the greater the cost. They are available also in various wattages from the 3 watt flicker flame on the right to the pinch shape with up to 60 watts. They come in clear, frosted, and white finish.

The socket shown in Fig. 4.17. is the standard or regular base type. The ordinary type of household bulb will fit this socket, but the bulbs shown look much better with a plastic cover over the socket. The plastic covers have to be purchased separately and come in white, white with either plain or gold candle drip appearance, or black with gold drip. The choice between the two sockets shown is based upon which size blends in with the design of the lamp you intend to make.

WALL LAMP

These lamps provide attractive supplementary lighting and are usually used in pairs. The one shown in Fig. 4.18. has a hammered "Swedish Iron" finish. It is also attractive with a plain finish painted either flat black, gold, or copper. There are only three major parts to this lamp excluding the electrical parts, the box or back, the scroll, and the tray or cup.

BOX

Lay out the pattern shown in Fig. 4.19. on 16 ga. or 14 ga. sheet steel. Notch or cut out the corners. Before bending the sides hammer the surface to the desired appearance. See the section on finishing metal. When folding the sides remember to fold away from the hammered side. See Fig. 4.9.

SCROLL

Hammer both sides of 460 mm length of $\frac{3}{16}$ in. x 1 in. flat bar. Form a 100 mm scroll on one end and a 65 mm scroll on the other end. Bend the rest of the scroll over an anvil horn to form a progressive curve.

TRAY

The tray used was 75 mm in diameter and dished. You may wish to substitute a candle drip tray purchased from a store. Check the section on holders and trays for candles for information. A $\frac{5}{16}$ in. hole is needed in the centre of the tray. A $\frac{3}{8}$ in. long $\frac{1}{8}$ in. IPS nipple is then centred over the hole and silver-brazed into place.

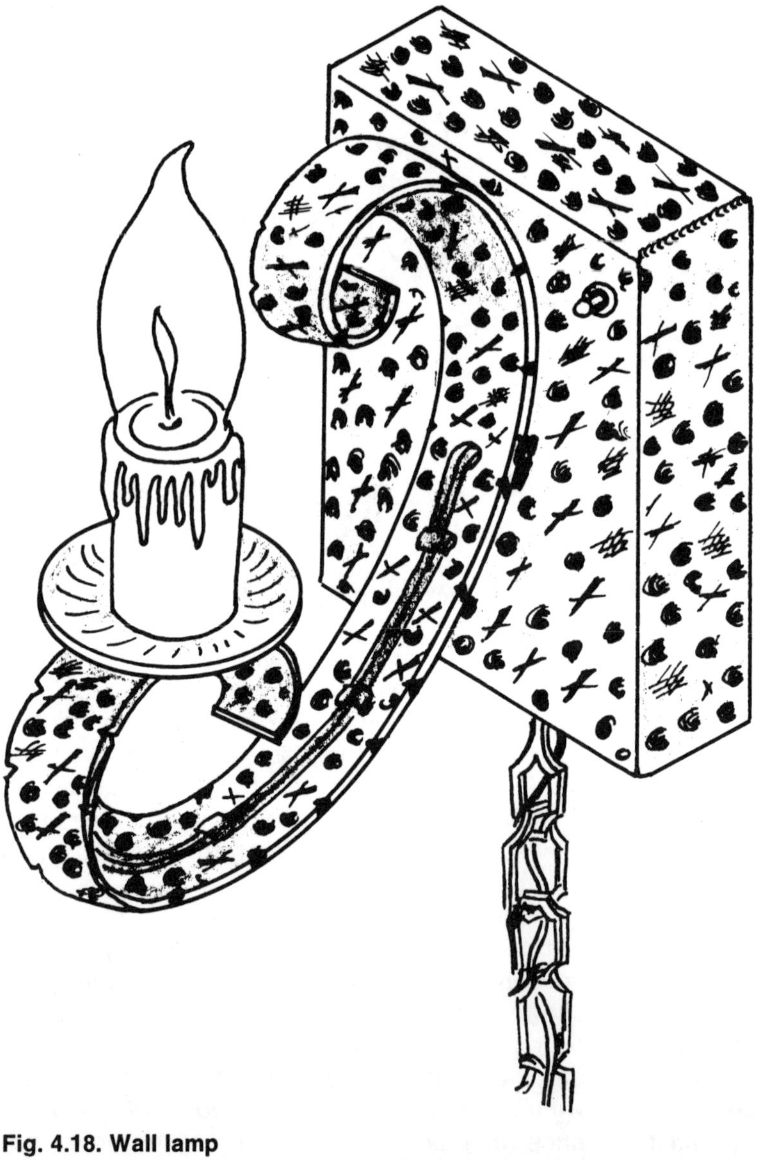

Fig. 4.18. Wall lamp

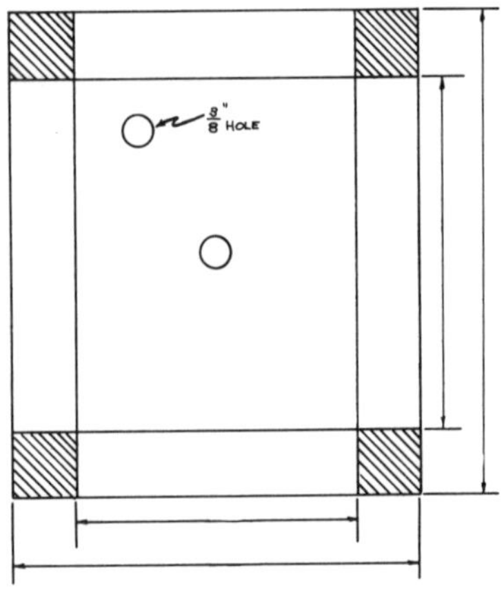

Fig. 4.19. Wall lamp box layout

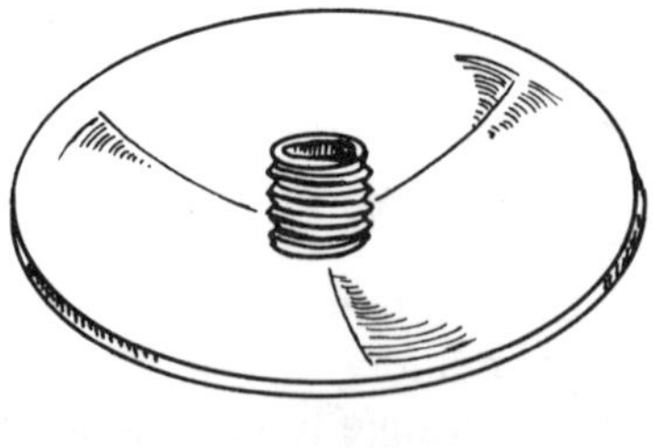

Fig. 4.20. Wall lamp tray

ASSEMBLY

Centre the scroll on the box and join with a small weld on either side of the scroll. With the box placed against a wall place a mark on the top of the bottom scroll. Drill a 5/16 in. hole at the mark. Place the tray on the scroll making sure that the holes are in line. Weld the tray to the scroll making sure that the tray is at 90° to the box. Drill a 5/16 in. hole through the centre of the scroll and the box. Cut a strip of 22 ga. sheet steel 6 mm wide and 70 mm long. This will provide 4 clips for holding the electrical wire to the curve of the scroll. Spot weld or tack weld the clips to the scroll. Do not bend the clips at this point. Drill a 5/16 in. hole on the bottom side of the box and insert a rubber grommet.

The chain shown is optional. A small half circle of 1/8 in. welding rod was welded to the bottom of the box and a length of decorative chain added. This helps to cover the wire that runs down to the electrical outlet. Also optional but nice to include is a single pole rotary type switch. If you wish to add the switch, drill a 3/8 in. hole 3/4 in. in from one of the corners to accommodate the switch. See Fig. 4.21. for wiring details. All that remains now is to add the small plate for hanging the lamp as in Fig. 4.21.

FINISHING

Paint the complete lamp then rub off the paint to leave a silvery surface with black indentations. Add the socket (standard base type), wire the lamp, and add the plastic cover and bulb of your choice. Consider the alternate designs shown. By substituting various parts from each you can fashion your own design.

CHANDELIER

If you have already constructed the large candelabra shown in Unit 3 you may wish to construct a similar project that utilizes electric light rather than candles. Fig. 4.24. (p. 118.) shows what this project looks like when fitted with electrical sockets and bulbs. The construction is similar except for the details shown in Fig. 4.25. (p. 118.) and Fig. 4.26. (p. 118.)

Drill 5/16 in. holes in the main ring where all holders attach to the main ring. Use 1/2 in. tubing in place of the pieces of flat bar. Be sure to drill a hole in each upright tube. Silver-braze a 1/8 in. IPS nipple 3/8 in. long to each drip tray. Add clips to support the wiring on the inside of the main ring. See Fig. 4.24.

It is suggested that you use the adjustable candelabra socket with 100 mm plastic covers. When construction has been completed wire all sockets but one as in Fig. 4.25. The remaining socket is wired with sufficient wire to go around the main ring, up one support arm, through the chain, and to the ceiling fixture as in Fig. 4.24. The wires from the other sockets can now be joined to the main wires. *Remember, make all electrical connections according to local code.*

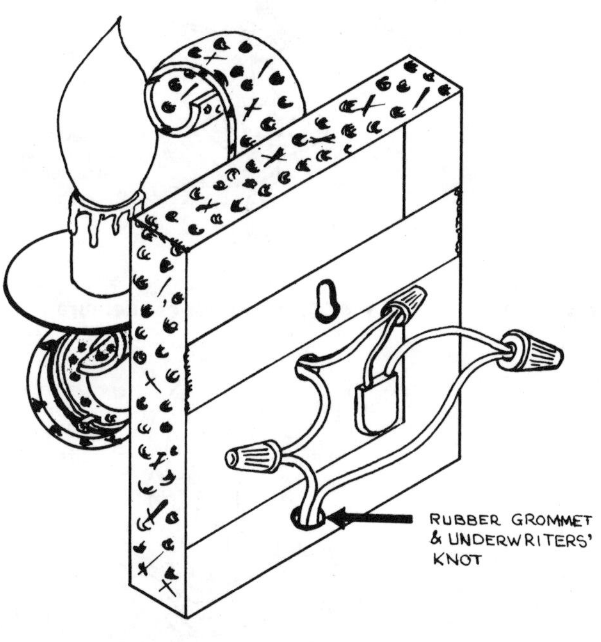

Fig. 4.21. Wiring details

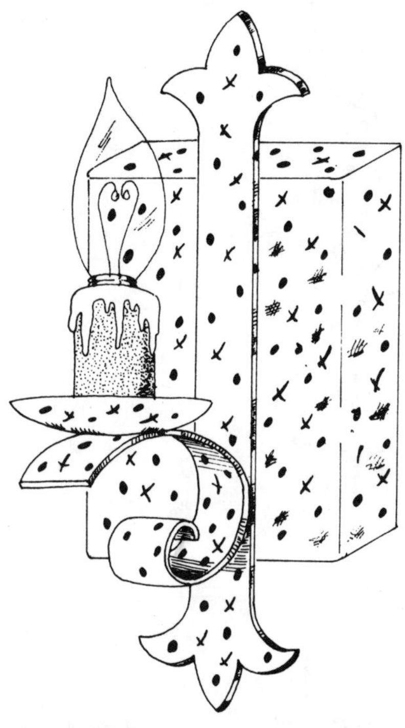

Fig. 4.22. Wall lamp

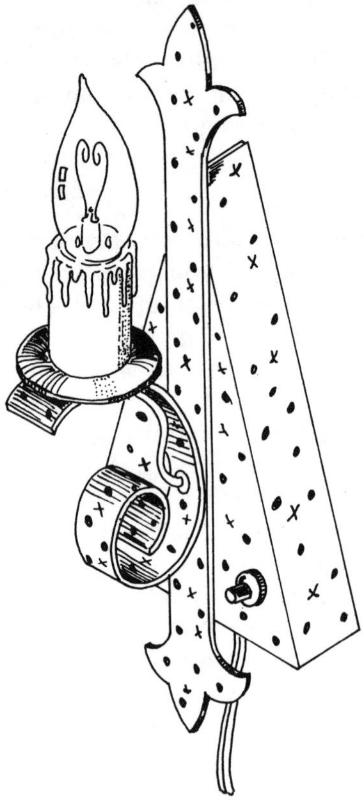

Fig. 4.23. Wall lamp

Fig. 4.24. Chandelier

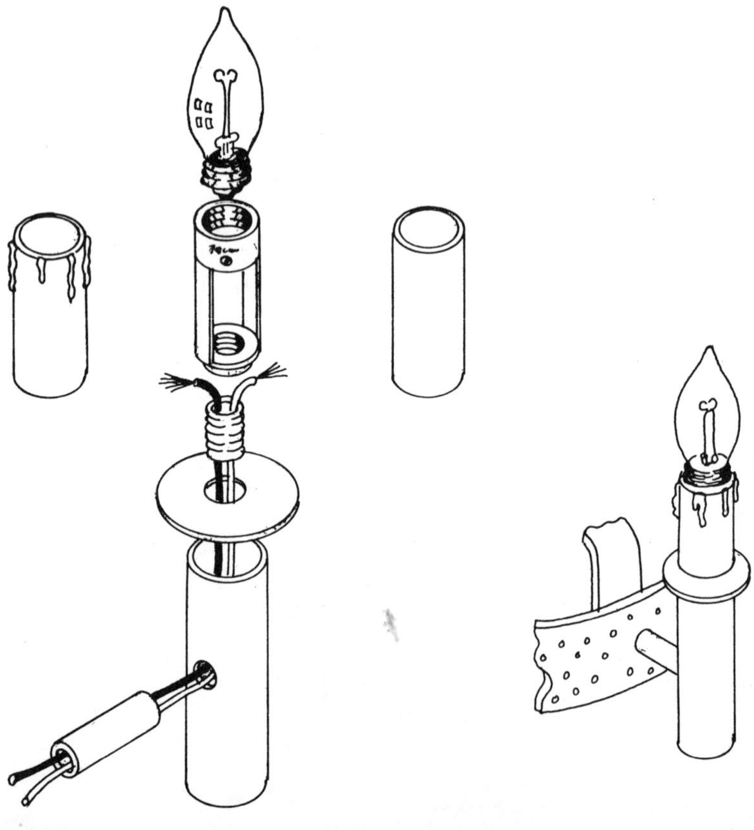

Fig. 4.25. Electric fixture for candelabra

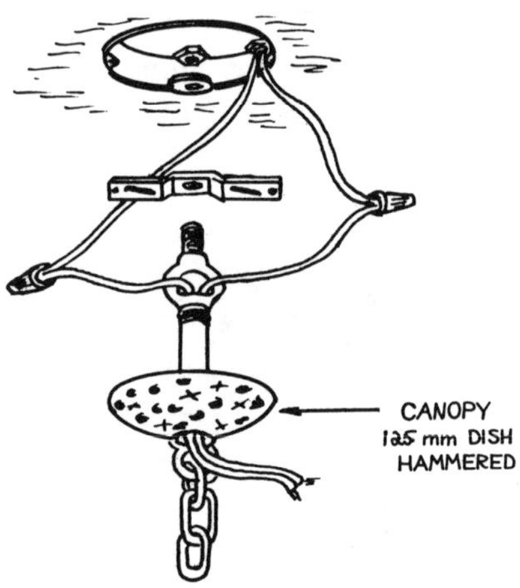

Fig. 4.26. Wiring for candelabra

Fig. 4.27. Alternate design

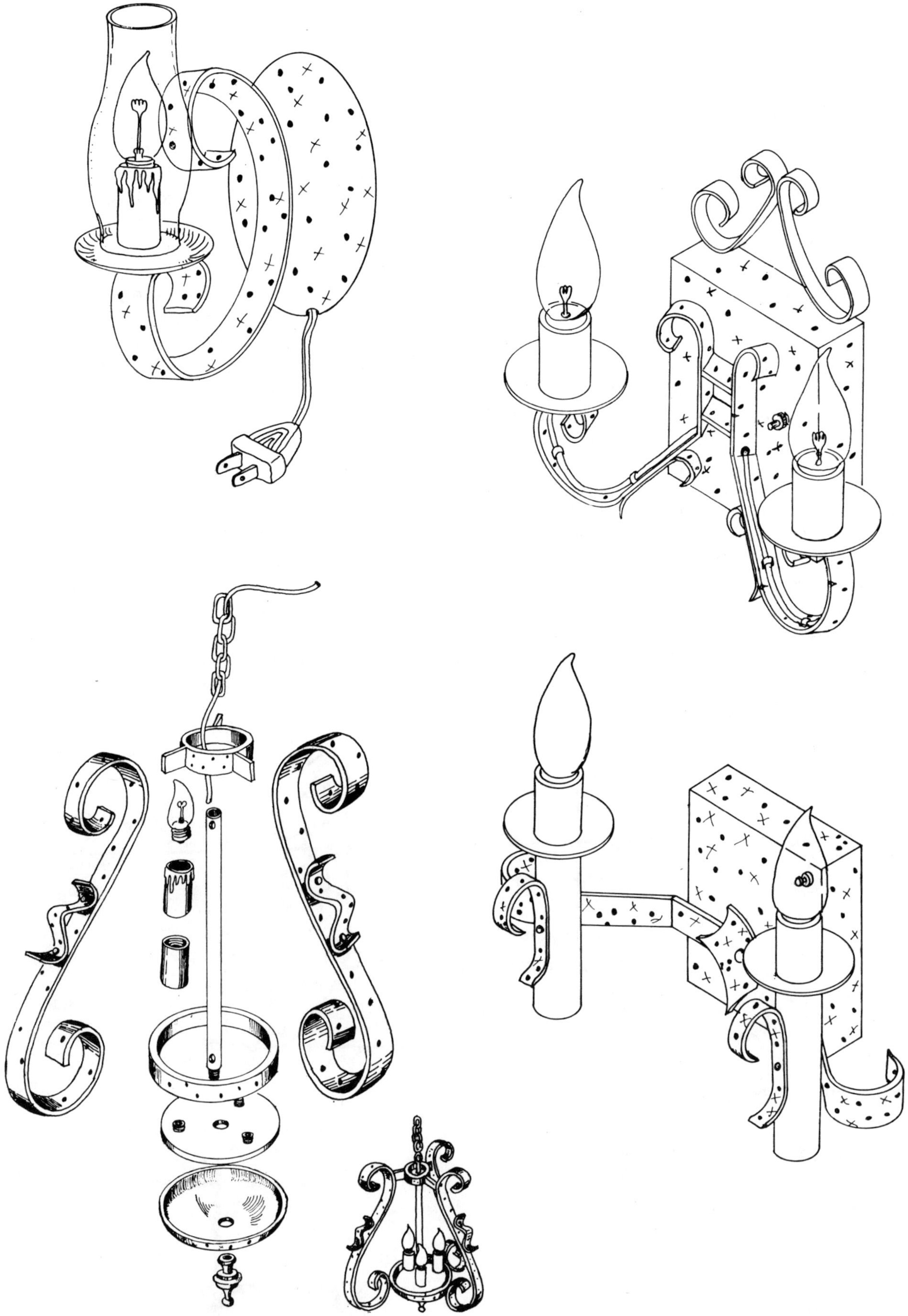

Fig. 4.28. Wall lamps and exploded view of ceiling light

UNIT 5. FIREPLACE ACCESSORIES

A fireplace is simply a hole in the wall until equipped with the accessories with which to maintain a fire safely. The list shown below includes all items for holding the burning logs, poking the fire, moving the ashes, brushing the hearth, picking up burning logs, containing sparks, and so forth.

The firebasket holds the burning wood off of the hearth and lets the draft keep the fire burning. Logs used to be placed across the back bars of the andirons.

The firescreen can be made in a variety of shapes. The screen can have metal curtains that can be drawn open or closed similar to window drapes. Fig. 5.8., Fig. 5.9., Fig. 5.10., and Fig. 5.12. show a variety of screens.

Andirons serve mostly to decorate the otherwise bare fireplace opening. They may be placed either in front of or behind the firescreen. See Fig. 5.17., Fig. 5.18., and Fig. 5.19. for various designs.

Log or wood holders are convenient for carrying wood into the house and can be decorative enough to set them beside the hearth.

The poker is of course necessary for moving the fire around to keep it burning properly. It is a functional item but it can be made attractive by twisting the shaft, hammering the surface, and making a decorative handle.

The brush is self-explanatory. It should match the poker in appearance.

Tongs can be made in a variety of designs. Their purpose is to grip the log so that you can control and place it where you wish. The poker, brush, and tongs usually are a matched set and hang either on a stand or on brackets fixed to the wall.

FIREBASKET

The firebasket has first to be functional. If it is made from stock that is too wide the heat from the fire will quickly burn and oxidize the metal. Round bar or square bar ½ in. thick is usually used for construction.

Fireplace openings are not all made to standard dimensions. The opening may be as large as 1.8 m wide and 1.8 m high. The im-

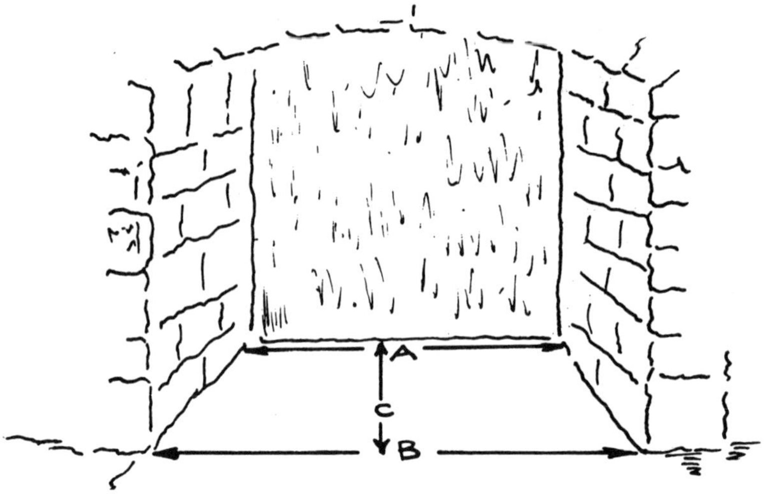

Fig. 5.1. Fireplace opening

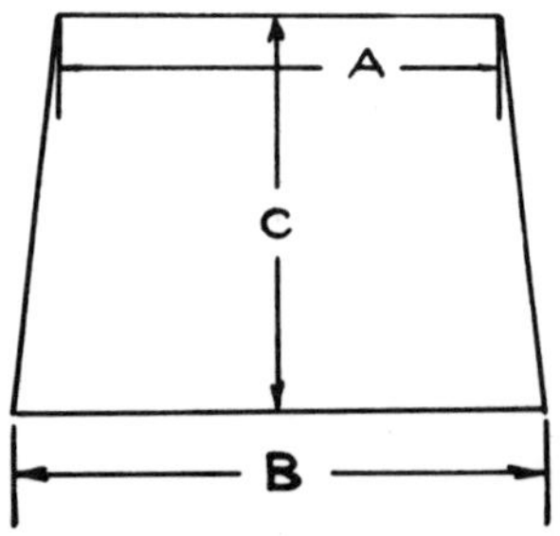

Fig. 5.2. Plan view of fireplace opening

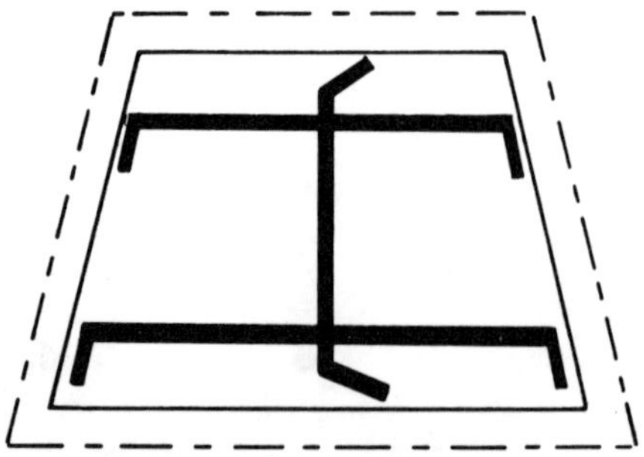
Fig. 5.3. Firebasket is smaller than opening

portant thing is to make the basket fit the opening. Obtain the measurements shown in Fig. 5.1. (p. 120.) The width across the front of the opening, the depth from front to back, and, if the opening is tapered, the width across the back. Fig. 5.2. (p. 120.) shows a plan view of such an opening.

If you make the firebasket the full size of this opening the log may roll to the front of the basket and the smoke may come out into the room instead of going up the chimney. It is suggested that you make the basket slightly smaller than the opening. Fig. 5.3. (p. 120.) shows by means of a dotted line the opening shown in Fig. 5.2. The solid line is 50 mm in from each side making it 100 mm shorter in both depth and width. If the fireplace opening is quite small you may have to make the basket the size of the opening in width but do make it slightly smaller in depth. On the centre line of the same drawing we have laid out one bar of the basket and also the two bars for the legs. When you figure out how many bars are required for your basket allow from 40 mm to 50 mm between each bar.

Notice that Fig. 5.3. shows a tapered opening. Most fireplace openings are this shape. This does not mean that you have to make the basket tapered. The tapered basket simply utilizes the entire opening. In fact, it is easier to make the basket with the front and back dimensions the same.

The basket can be made with a bar all around the top as in Fig. 5.6. The basket shown in Fig. 5.5. can also be made with side guards. The amount of effort that one puts into making a firebasket will depend upon whether one wishes the basket to be functional or merely decorative.

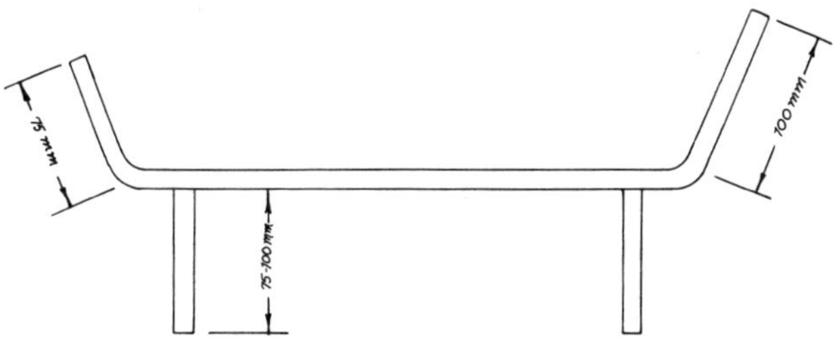

Fig. 5.4. Side view of firebasket

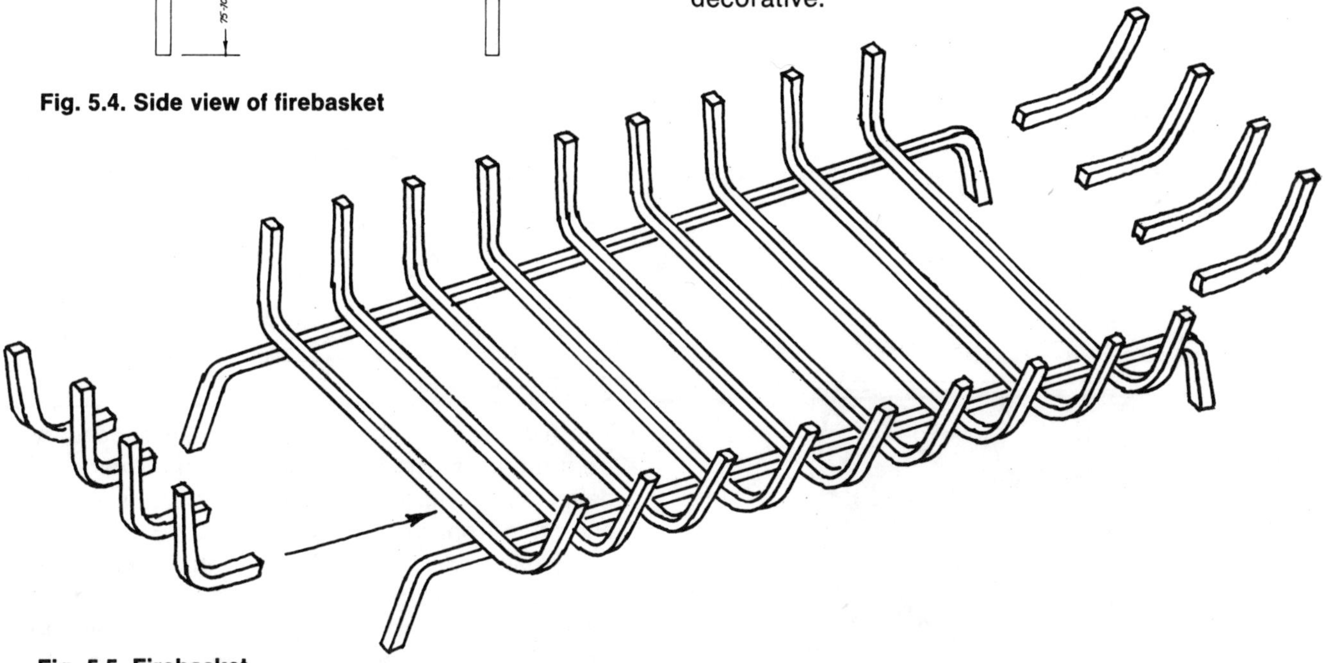

Fig. 5.5. Firebasket

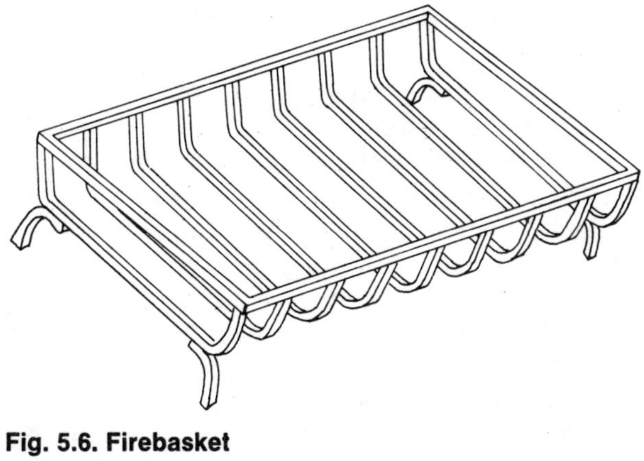

Fig. 5.6. Firebasket

Fig. 5.7. Firebasket

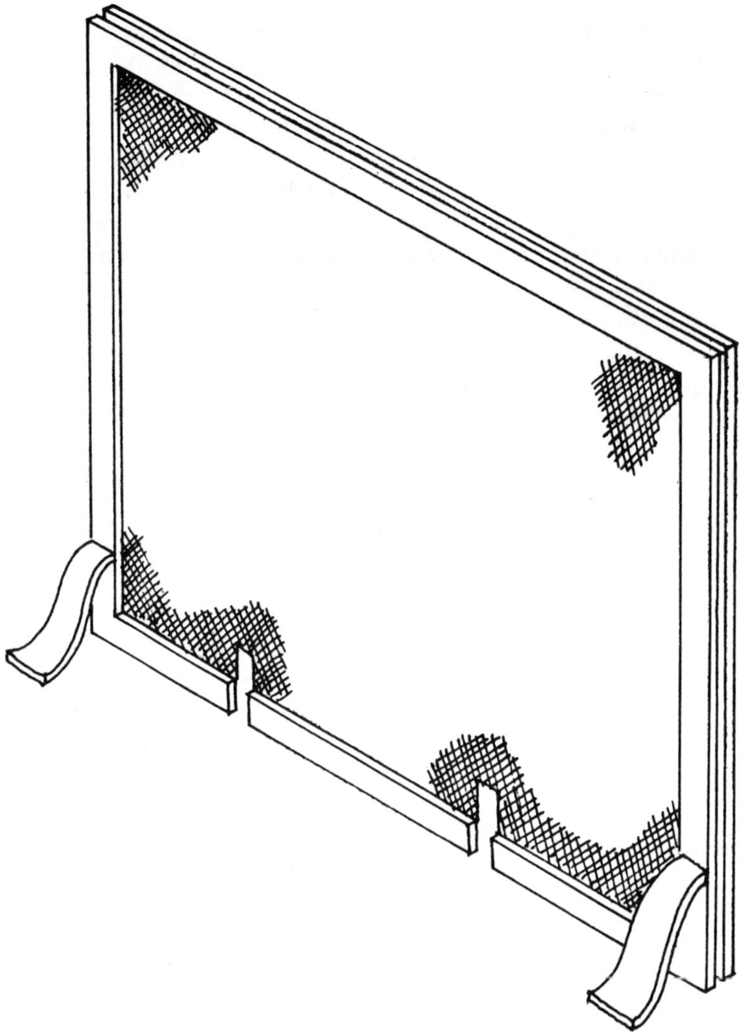

Fig. 5.8. Firescreen

Fig. 5.9.

Fig. 5.10.

FIRESCREEN

The firescreen is the most important accessory from the point of view of safety. The fire should never be left unattended unless there is a fire-screen in front of it.

Firescreens can be divided into two groups. One is the type of screen that must be moved away from the opening in order to get at the fire. The other more convenient type is the screen that can be opened in the same manner as one pulls open window drapes or curtains. The fixed screen, however, can be more decorative. See Fig. 5.9. and Fig. 5.10. for examples.

The fixed screen can be made using ⅛ in. x 1 in. flat bar welded at the corners to form a frame. With two frames the screen material can be riveted between the frames. Riveting is recommended over welding as the screen has to be stretched as the frame is put together. The decorative designs shown in Fig. 5.9. and Fig. 5.10. can be cut from 22 ga. sheet steel and riveted to the screen.

The frame of the draw type of screen can

be formed from 16 ga. sheet steel. If thicker steel such as 14 ga. is available it should be used. Use strips 75 mm wide and the length needed to go around the perimeter of the fireplace. Bend the material on a sheet metal brake to a 90° angle. Cut to length to form a top, a base, and two sides as in Fig. 5.12. Cut the corners at 45°, then weld together. If 16 ga. steel is used it will be necessary to stiffen the frame. Make a frame of ¼ in. x 1 in. flat bar as in Fig. 5.13. (p. 124.) and add it to the main frame using a few small tack welds.

Make two scrolls from 3⁄16 in. x ¾ in. flat bar and add them to the bottom of the frame to aid the frame in standing. Two small scrolls made from ⅛ in. x ½ in. flat bar and placed over the top joints will hide the welds.

It is now necessary to make the rods that will support the screening. The screens should be able to overlap each other at the middle of the frame to make sure that all sparks are contained. Two rods made from ¼ in. round bar and shaped as in Fig. 5.14. (p. 124.) should do the job. To complete the firescreen add a small decoration to each piece of screen that can be used as a handle to draw the screens. See Fig. 5.15. (p. 124.)

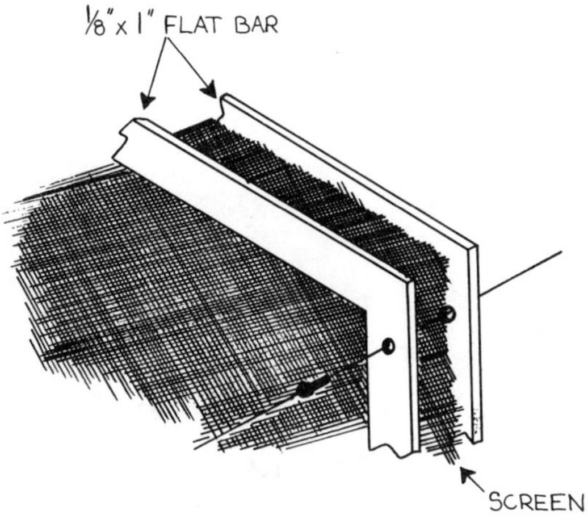

Fig. 5.12. Firescreen

Fig. 5.11. Assembling frame

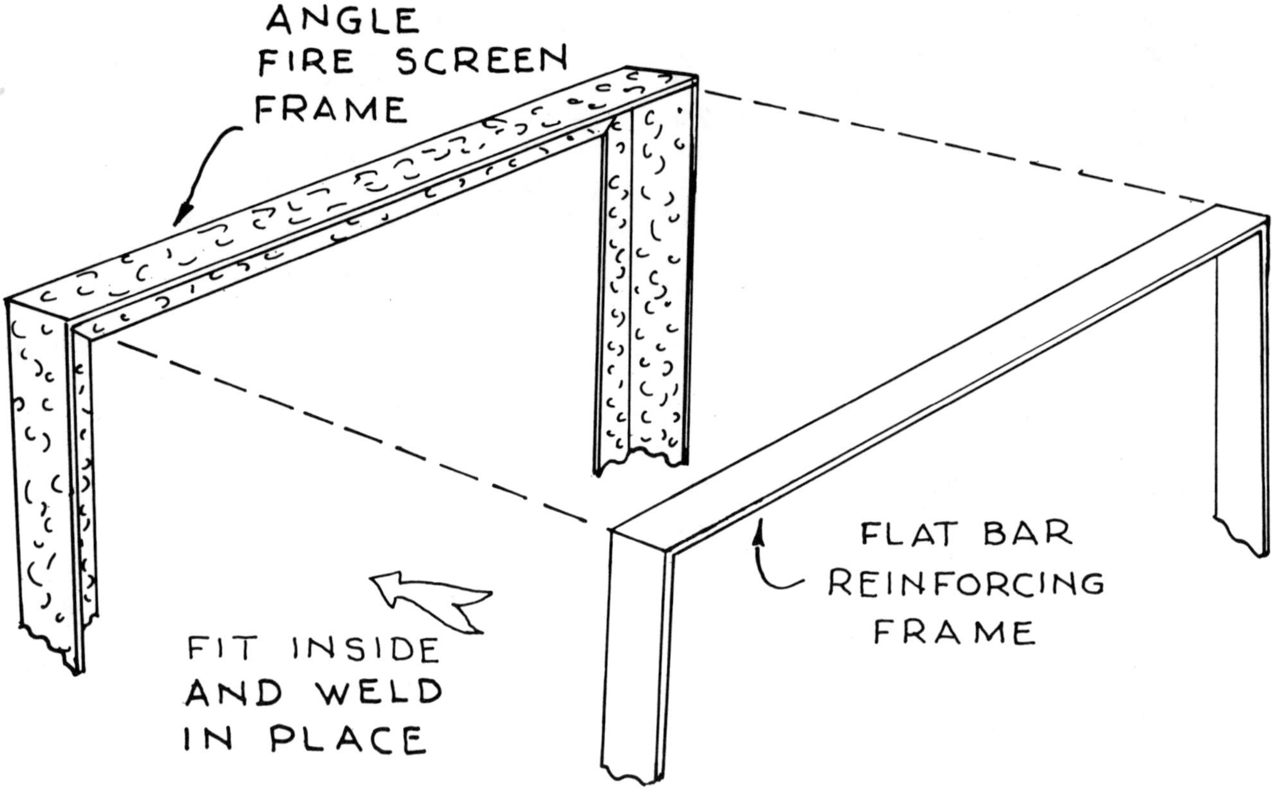

Fig. 5.13. Reinforcing firescreen frame

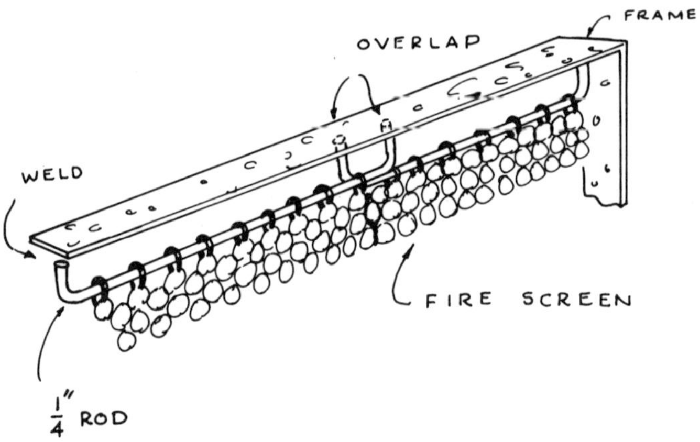

Fig. 5.14. Rods to support screen

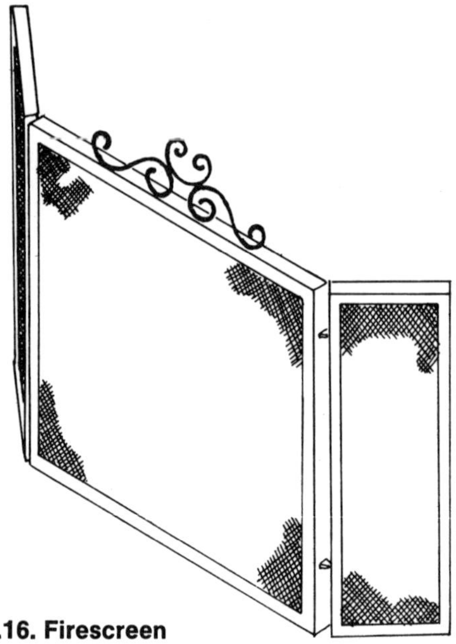

Fig. 5.16. Firescreen

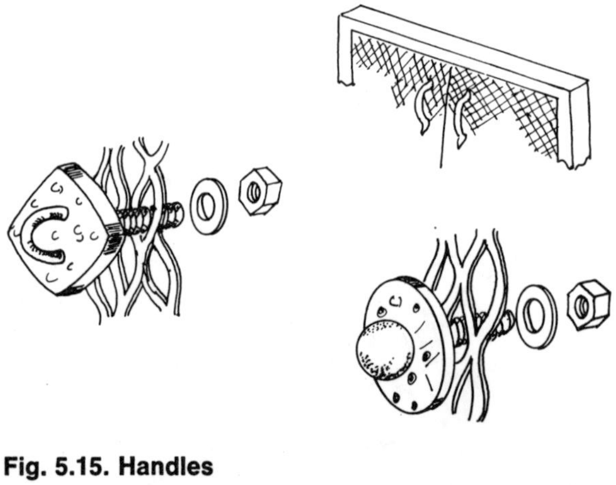

Fig. 5.15. Handles

ANDIRONS

Today andirons are used mainly for decoration. A firebasket is used in their place to support the logs. As can be seen in the illustrations, andirons can be plain and simple or very ornate. The back leg of the andiron can be made using ½ in. or ⅝ in. square bar. The main leg or front of the andiron was originally forged or shaped by hand. Today it is difficult to find someone who has these skills.

An easy way to make andirons that appear similar to those made many years ago is to use the small castings which decorate ornamental railings and columns. Fig. 5.20. shows a few of the more common shapes available. It is suggested that you try to design your own andirons or attempt to interpret the designs shown by substituting the ornaments shown. Check the section on accessories for further information on finials and pickets.

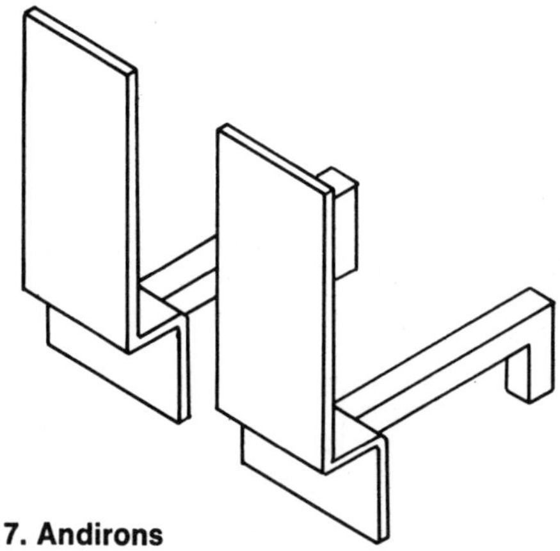

Fig. 5.17. Andirons

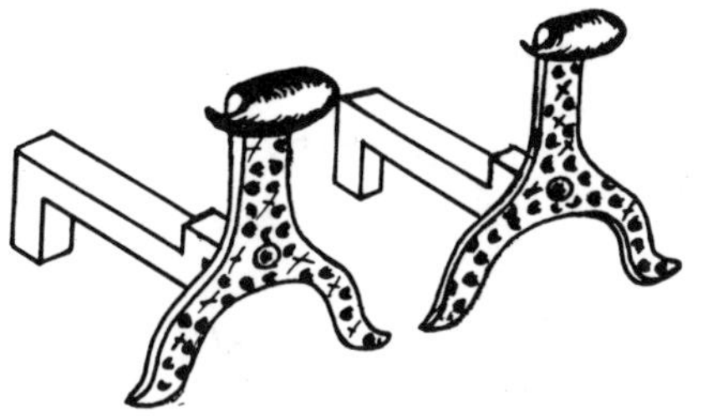

Fig. 5.18.

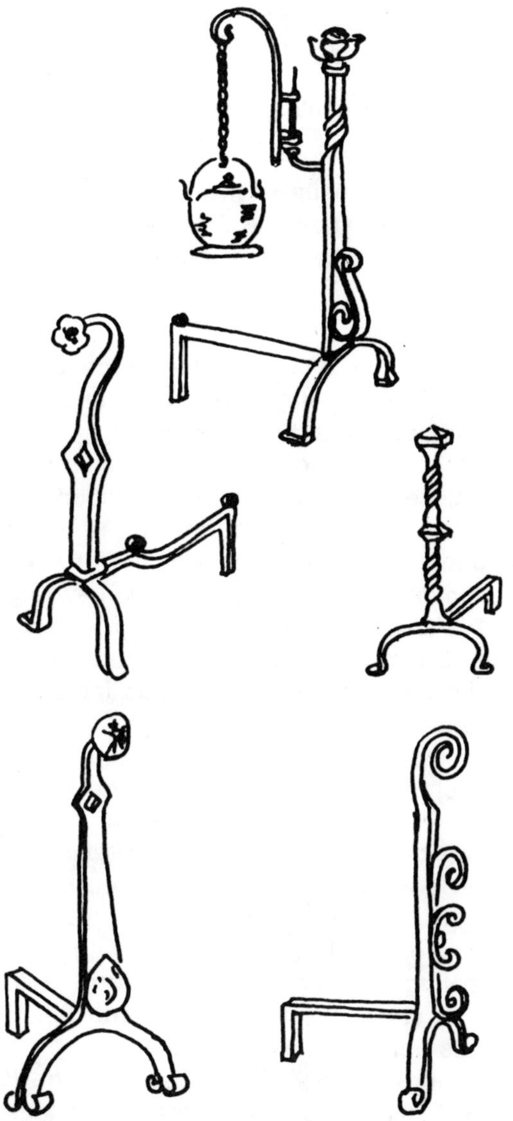

Fig. 5.19.

Fig. 5.20. Ornamental castings

Fig. 5.21. Log holder

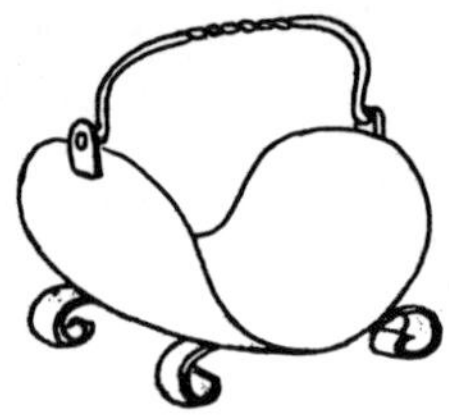

Fig. 5.22. Log holder

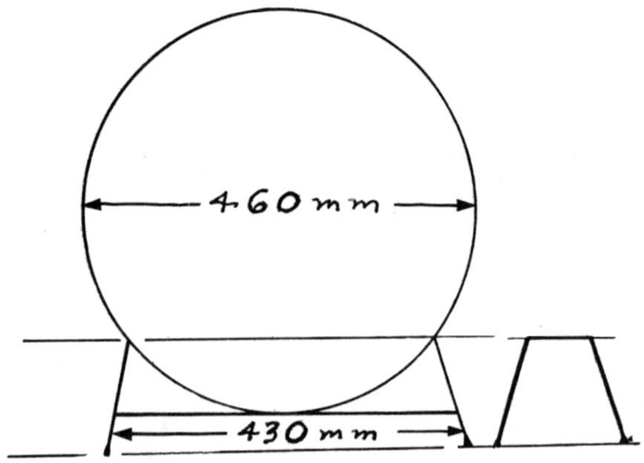

Fig. 5.23. Details of Fig. 5.21.

LOG HOLDERS

A log or wood holder should match the rest of the fireplace set. If you decide that the firescreen is to be made with brass then all accessories should be made using brass. If a modern design is chosen then all accessories should have a modern look.

Fig. 5.21. shows a simple modern log holder. Compare it with the holder shown in Fig. 5.22. Each will enhance a fireplace providing that it matches the overall appearance.

MODERN LOG HOLDER

This log holder is very simple in appearance and is relatively easy to make. It is made from one size and type of material and there are only three different shapes to make. Materials required are 2 pieces of 3/8 in. round bar 1500 mm long, 2 pieces of 3/8 in. round bar 430 mm long, and 2 pieces of 3/8 in. round bar 560 mm long. Roll the 2 pieces 1500 mm long into circles 460 mm in diameter. Weld the joint in each ring. Bend the 560 mm pieces as shown in Fig. 5.23. Lay the two rings together and tack weld at one spot. Pull the rings apart about 150 mm at the side opposite to the weld. Add the legs and braces as shown in Fig. 5.23. Finish welding, then paint the entire log holder flat black.

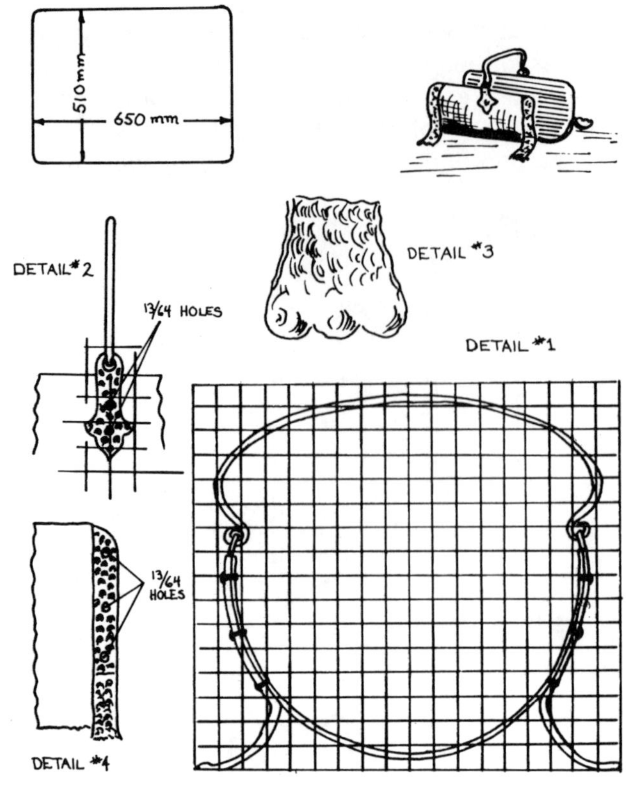

Fig. 5.24. Details of old log holder

OLD TYPE LOG HOLDER

This log holder has the appearance of those used in colonial days. The legs and handle brackets look like hinges made by the local blacksmith. This project requires some forging. Material required are 16 ga. sheet steel 650 mm x 510 mm, about 610 mm of 3/8 in. round bar, about 1.3 m of 1/4 in. x 1 in. flat bar, and 205 mm of 1/8 in. x 1 in. flat bar. Lay out detail 1 shown in Fig. 5.24., using squares of about 25 mm. This will be close to actual size. Roll the steel sheet to the radius indicated by the squares.

Hammer each end of the 1/8 in. x 1 in. flat bar to obtain the shape shown in detail 2. Drill two 1/2 in. holes then cut the flat bar apart. The two pieces should be symmetrical. Round each end where the hole is and you now have the hinge parts of the handle.

Draw out both ends of the 3/8 in. round bar by heating and hammering to form a taper. Bend as shown in detail 1 and add the hinge sections before closing the ends.

The legs are cut and shaped from the 1/4 in. x 1 in. flat bar. Be sure to make two legs left hand and two right hand. The legs should be hammered and shaped as in details 3 and 4. Try hammering the flat bar over a block of lead.

Drill 13/64 in. holes to attach both the legs and the handle hinges. Assemble the project using rivets or plug weld through the holes. Try for authenticity with this project.

The log basket shown in Fig. 5.22. can be made from a 460 mm circle of 16 ga. sheet steel (hammered), 3/8 in. square bar for the handle and 3/16 in. x 1 in. flat bar for the legs.

POKER, BRUSH, TONGS, AND SO FORTH

All of these items are usually made as a set; keep the length, type of material, and handle design the same. All three items can stand in a corner of the hearth, but will look much neater hanging on a stand or on wall brackets. The length of all of the items will again depend upon the size of the particular fireplace. They may be as short as 380 mm or as long as 1200 mm. Consider 3/8 in. square bar for the smaller ones and 1/2 in. square bar for the longer items.

The handles are usually the most decorative part of the set and often the top of the stand has the same shape incorporated in its design. Fig. 5.25. shows a variety of shapes. The work end of the poker can also be decorative. It does not have to be just a point. See Fig. 5.25. for ideas.

The shovel pan is usually formed from 16 or 14 ga. sheet steel. A selection of shapes are shown in Fig. 5.26. (p. 128.) Tongs can vary greatly in design even though the function is the same. The handles at the top are either moved outward or inward in order to make the opposite ends grip the wood. Fig. 5.27. (p. 128.) shows a variety of designs and shapes.

The brush head should be obtained before you begin making your set. Consider using a painter's dust brush or even part of a floor broom. The brush head may be round or rectangular in shape. The head of the brush may be covered with metal such as 22 ga. sheet steel or may simply be painted black. You should notice that little specific information is given here, only suggestions and ideas. Put together a choice of handles, lengths, and designs to make your exclusive set of fireplace accessories.

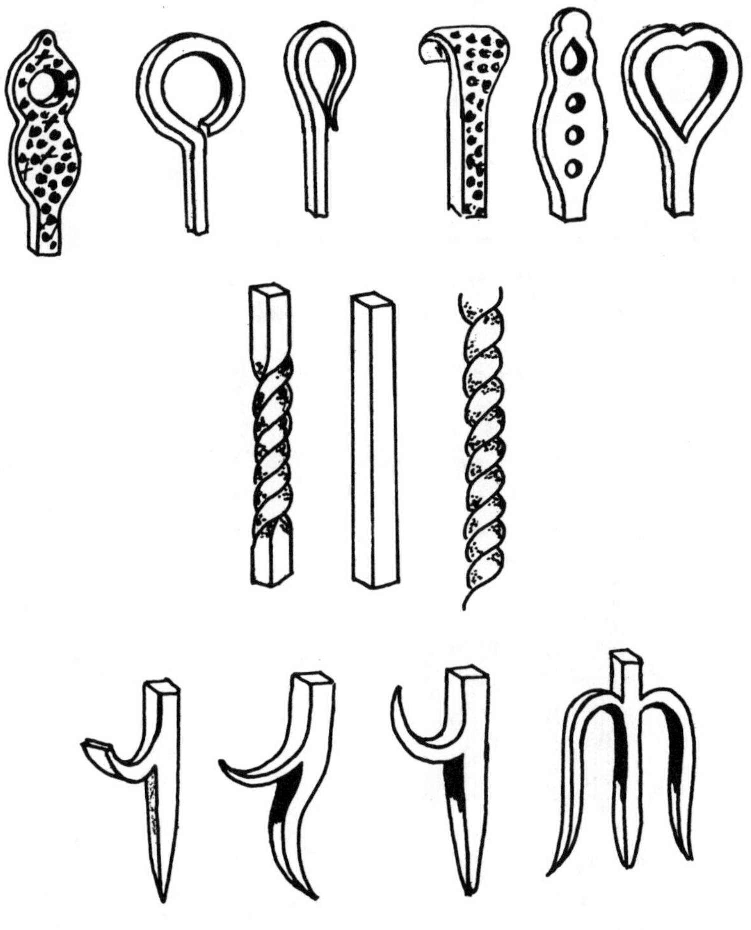

Fig. 5.25. Pokers

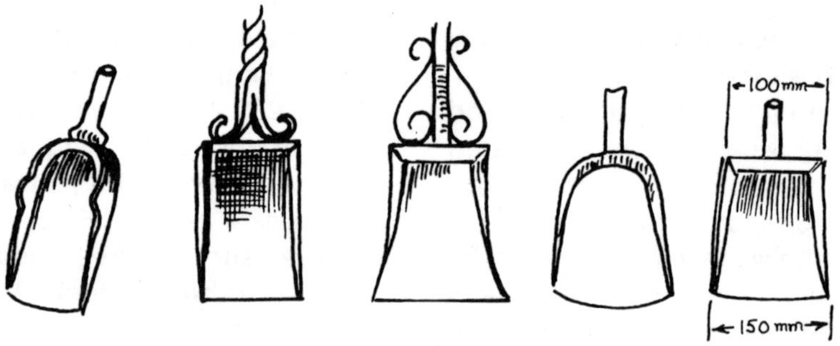

Fig. 5.26. Shovel pans

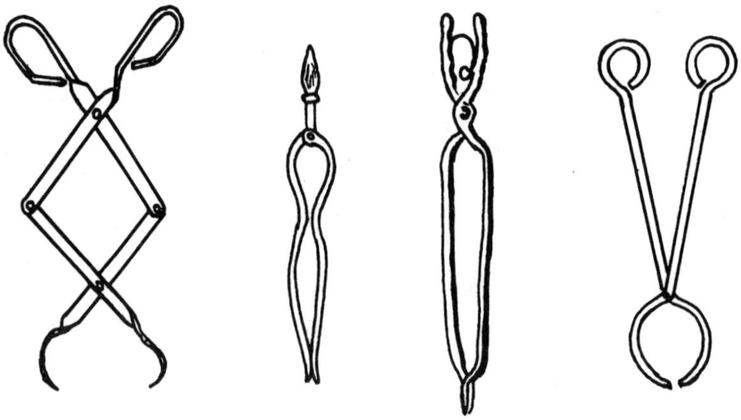

Fig. 5.27. Tongs

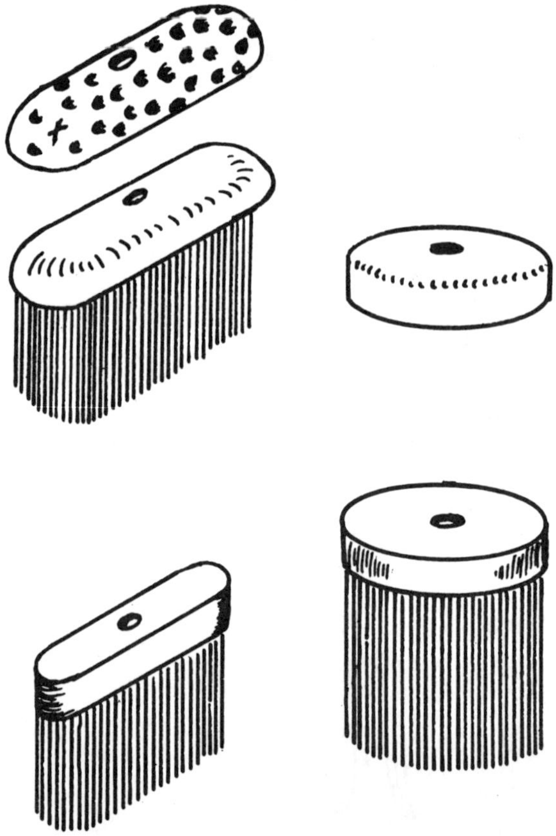

Fig. 5.28. Brush heads

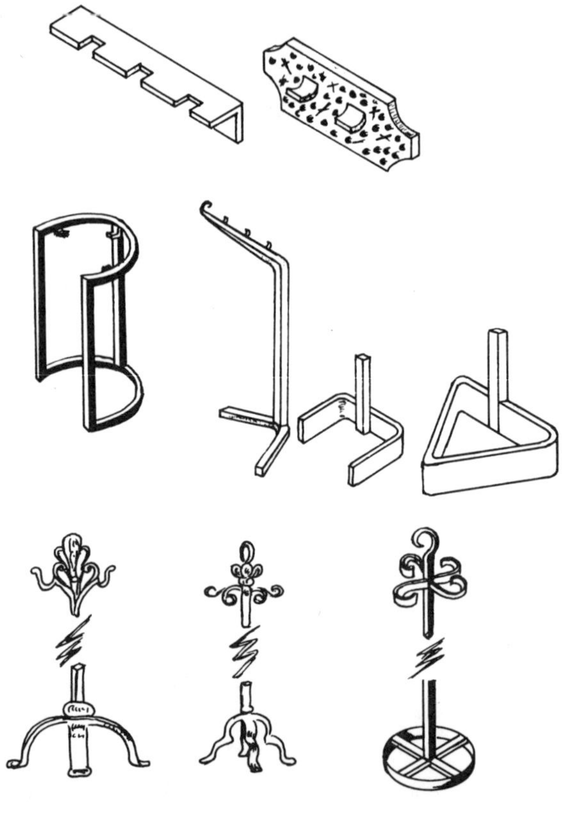

Fig. 5.29. Stand and wall brackets

UNIT 6. TABLES

Tables are excellent projects for students to make. They can be small and simple, such as the one in Fig. 6.1. or large and complicated, such as the one in Fig. 6.2. They can be square, round, hexagonal, or cube shaped. They can be made using flat bar, angle iron, square tubing, and so forth.

Fig. 6.1.

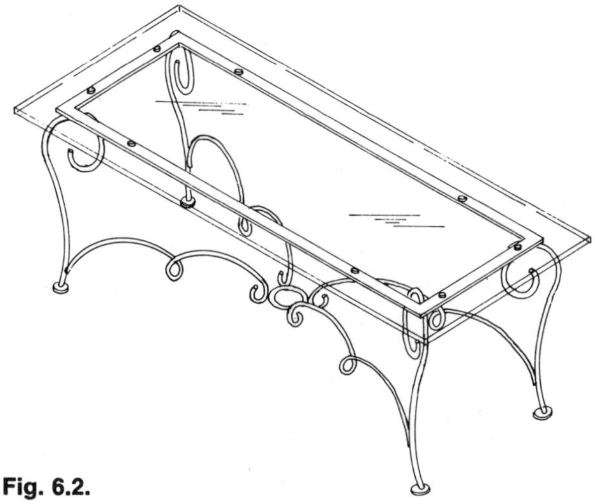

Fig. 6.2.

OCCASIONAL TABLE

Materials required are one piece of 3/4 in. x 3/4 in. angle iron 1.5 m long, one piece of 5/16 in. round bar 1.6 m long, and one piece of 1/4 in. round bar 740 mm long. Material lengths allow about 40 mm to 50 mm for cutting and waste. Choices of welding processes are 1. oxyacetylene welding (fusion), 2. electric arc welding, and 3. oxyacetylene welding (braze-welding).

PROCEDURE

For the top, make a frame with angle iron 355 mm square. It does not matter whether it is 355 mm square on the inside or outside measurement. Check the section on measurement and layout for information on angle iron corners. Be sure to weld the corners on the outside of the angle iron. Welding on the inside prevents the top from seating properly.

Cut four pieces of 5/16 in. round bar 380 mm long for the legs. All four pieces must be the same length. Mark each leg at 215 mm from one end. Heat with a welding torch and bend as in Fig. 6.3. When heating the legs, heat only a small section to obtain a neat angle. Place all four legs on top of each other and check that they are identical in shape.

At this point the legs may be tack welded to the frame. *Do not weld the legs to the frame at this point.* Each leg should be placed on the corner of the angle iron with the bend pointing toward the opposite corner of the angle iron frame. The leg should be held so that the end of the leg away from the frame is directly over the corner of the frame as in Fig. 6.4. You will need help to do either the tack welding or to hold the legs. After tack welding all four legs to the frame it is necessary to square the legs so that the table will be level.

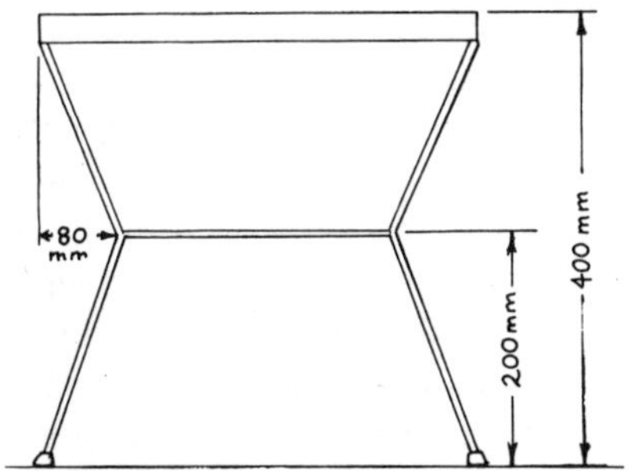

Fig. 6.3. Occasional table

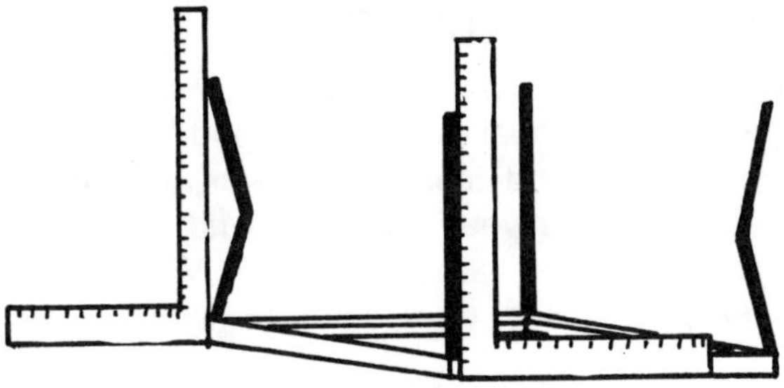

Fig. 6.4. Square legs both ways

ORNAMENTATION

Using the 1/4 in. bar make a ring about 75 mm O.D. See the section on making scrolls and circles for information. The ring does not have to be exactly 75 mm. Use the remainder of the bar to make four cross members as shown in Fig. 6.5. (p. 130.) Lay the ring and cross members on a flat surface. Tack weld together, check for alignment, then weld. Tack weld the ornamentation to the legs then weld all joints after checking whether the entire piece is square and level.

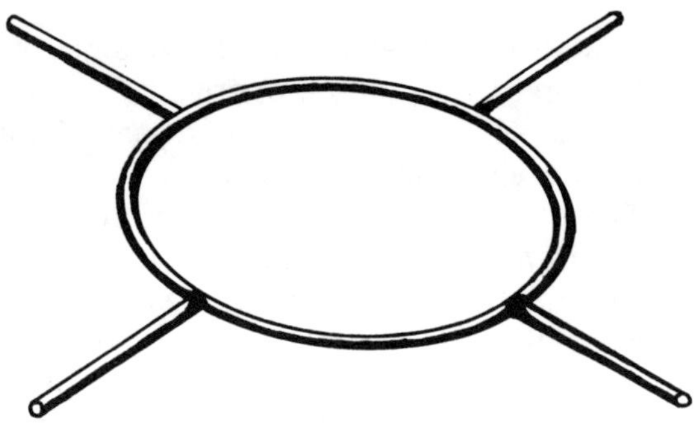

Fig. 6.5. Ornamentation for table

FINISHING

Rub all surfaces with a wire brush to remove scale and so forth, then prime and paint. The top fitted to this table is a scrap piece of arborite or formica-covered plywood ¾ in. thick. Arborite can often be obtained in the same way that scrap glass can be obtained. See the section on accessories. Arborite manufacturers often sell scraps for a token fee or give them away. To seat the top into the angle iron it is necessary to reduce the thickness of the top by running it over a table saw equipped with a dado head. Remove ⅛ in. of the thickness ¾ in. wide around the perimeter of the top and the top will seat flush with the top of the angle iron. Fit the top as tightly as possible for a better looking finish. Add four rubber or plastic tips to the legs to complete the project. Design your own ornamentation for this table. Other ideas are shown in Fig. 6.6.

SMALL COFFEE TABLE

This project can be used as a small coffee table, an occasional table, or an end table. If it is used as an end table, the height should be increased to between 460 mm and 585 mm. Materials required are 5.4 m of ¾ in. square tubing, and 12.8 m of ⅛ in. x ⅝ in. or ¾ in. flat bar. The tubing provides the complete frame of the table and the flat bar is used for all of the ornamentation. Choices of welding processes are 1. M.I.G. welding and 2. oxyacetylene welding (fusion).

PROCEDURE

To make the frame, cut four pieces of tubing 545 mm long and four pieces 420 mm long all with 45° corners. Be as accurate as possible when making the cuts. Using two pieces 545 mm and two pieces 420 mm, assemble a frame and tack weld together. *Do not weld any joint until the entire project is assembled.*

If necessary file the joints to obtain the best fit possible. Poorly fitted joints will require extra weld material which will certainly result in distortion of the frame. When one frame is assembled build the second frame on top of the first using it as a template.

Cut four pieces of ¾ in. tubing 370 mm long with square ends. Again make sure that all are the same length. If necessary file the ends square. Assemble the two rectangles with the 370 mm pieces between to form a cube. Tack weld and square. At this point you can weld the joints and then disc grind them smooth.

ORNAMENTATION

Cut 12 pieces of flat bar 370 mm long. These pieces will serve as frames for the scrolls. Cut 24 pieces of flat bar 345 mm long for the scrolls. Make a 65 mm scroll at one end of each of the

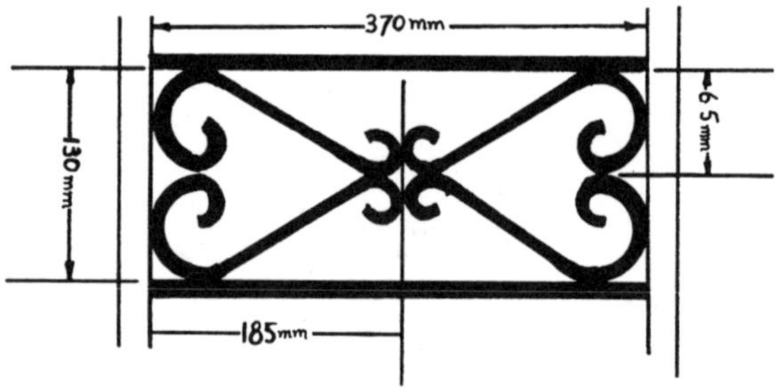

Fig. 6.7. Ornamentation for coffee table

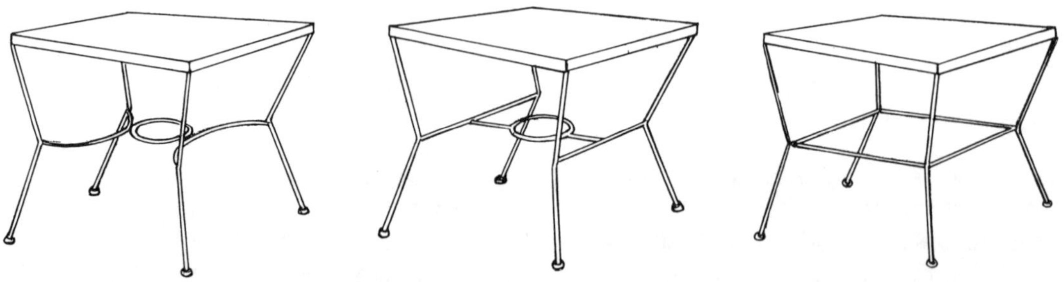

Fig. 6.6. Alternate designs, occasional table

twenty-four 345 mm long pieces. Make a partial 65 mm scroll on the opposite ends so that the overall length of each scroll measures 185 mm. The best way to do this is to work out a trial and error process while making the first scroll then place a mark on the scroll die holder where you finished up. Pull all other scrolls round to this mark.

After all the scrolls have been made and checked against each other you can begin to assemble the scroll sections. A simple jig will help here in holding the sections for tack welding. See Fig. 6.8. Assemble four scrolls and two frame pieces as in the illustration. When joining the scrolls make all of the tack welds on one side only. It is not necessary to weld these sections. The tacks will be sufficient to hold them.

When all six sections have been assembled they can be installed in the cube section as in Fig. 6.9. Lay the cube on a flat surface and place one scroll section in the end so that it is centred. Tack weld all four spots where the uprights meet the square tubing. Once again the tack welds are all that is needed to hold the pieces together. Turn the cube end for end and repeat. Repeat for the sides placing the scroll sections 75 mm in from each end as in Fig. 6.9.

FINISHING

Rub all scroll section with emery cloth before painting. The top used for this table was a section of glass from an old television set. The table dimensions were selected to match the dimensions of the glass, which is hardened and cannot be cut.

Design your own table. The construction of the cube table lends itself to many uses.

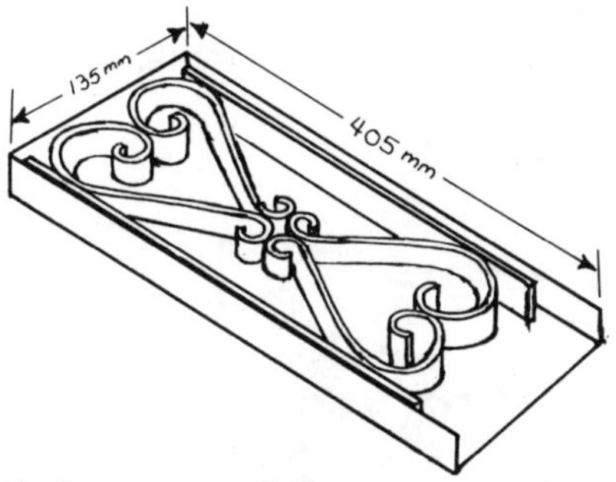

Fig. 6.8. Jig for ornamentation

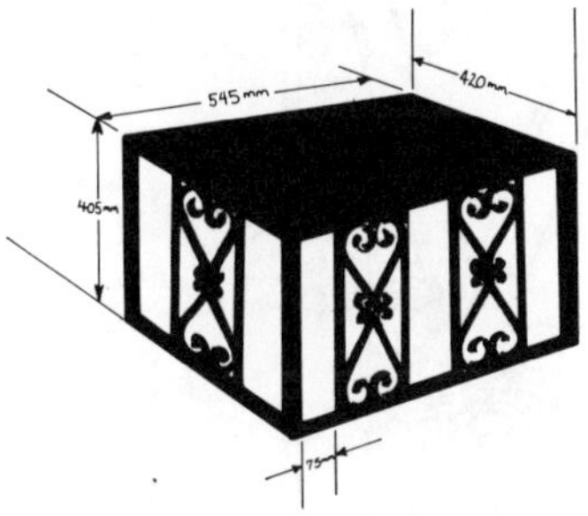

Fig. 6.9. Installing ornaments in cube

Fig. 6.10. Alternate ornamentation

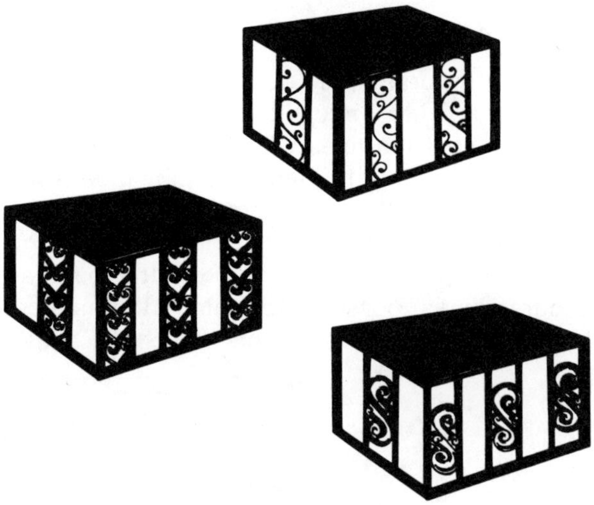

Fig. 6.11. Alternate ornamentation

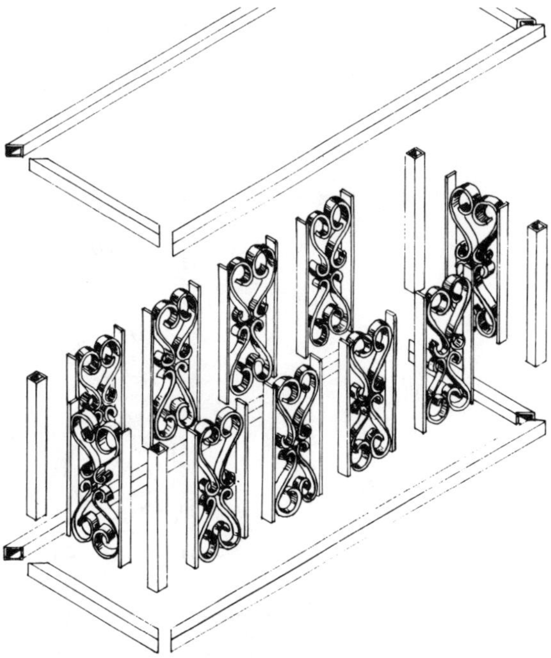

Fig. 6.12. Alternate design, cube table

Lengthening the cube to 1.2 m or even 1.5 m will turn it into a coffee table. Alternate ornamentation can be seen in Fig. 6.10. and Fig. 6.11. (p. 131.)

HEXAGONAL OR DRUM-SHAPED TABLE

This table is made of square tubing and flat bar. Materials required are 2.6 m of ¾ in. square tubing and 12.7 m of ⅛ in. x ⅝ in. or ¾ in. flat bar. The tubing provides the top and the bottom of the table. The flat bar is used for the ornamentation which also is the main support for this table. Choices of welding processes are 1. M.I.G. welding and 2. oxyacetylene welding (fusion).

PROCEDURE

To make the frame, lay out a 430 mm diameter circle and divide it into six equal parts as in Fig. 6.14. Connect the lines with bars the thickness of the tubing. Cut 12 pieces of tubing with the ends matching the angles on the layout. This will provide the pieces of tubing for both top and bottom of the table. Try to cut the joints as accurately as possible to obtain the neatest joints you can make. Joints which fit neatly do not require welding on the inside angle of the joint. If the joints do not fit well, you will need to weld them to hide the gaps. Do not weld inside the joints of one of the rings if you intend to inset the top. After welding is finished, the top and bottom of the finished scroll assembly should be ground with a disc grinder so that the corners are sharp and smooth.

ORNAMENTATION

The ornamentation on this table is similar to the cube-shaped coffee table. See Fig. 6.7. and Fig. 6.9. Cut the flat bar to provide 12 pieces 370 mm in length and 24 pieces 345 mm in length. The 370 mm pieces are used for the uprights and the 345 mm pieces become the scrolls.

ASSEMBLY

Place the top and the bottom on a flat surface standing vertically. Lay one section of scroll work between them and tack weld together after making sure that everything is parallel and square. Turn the table over and add another section. Repeat until all six sections are in place. Finish welding the table. Do not overweld.

Fig. 6.13. Hexagonal or drum-shaped table

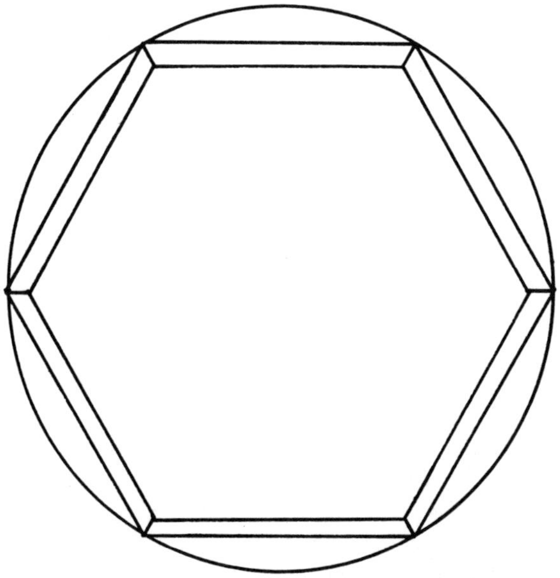

Fig. 6.14. Frame of hexagonal table

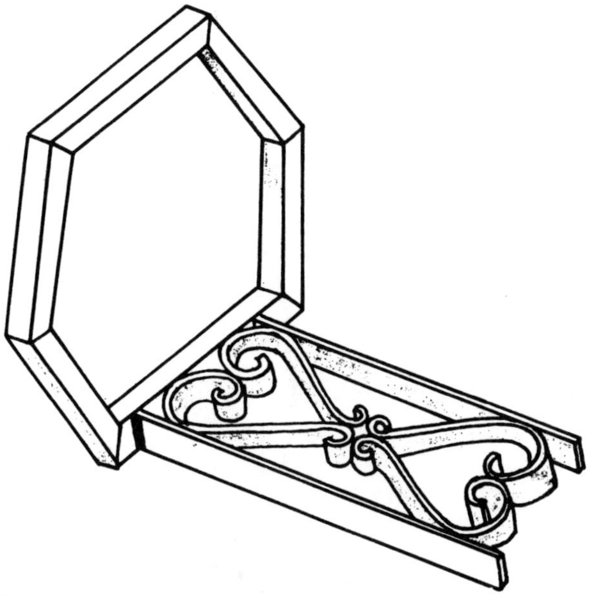

Fig. 6.15. Assembling hexagonal table

Fig. 6.16. Small chess table

The top for this table can be either glass or arborite-covered plywood. The glass should sit on top of the frame and should be cut to match the outer dimensions of the top. The arborite should be cut to fit tightly into the top so that it is level with the top of the tubing. We used a scrap piece from a counter top.

To finish the table, sand, prime, and paint in the colour of your choice. The shape of this table matches the shape of a swag lamp shown in the lamp section. Design your own table. How would this design look as a large coffee table 1500 mm in diameter? How about an eight-sided table 600 mm in diameter? Remem-

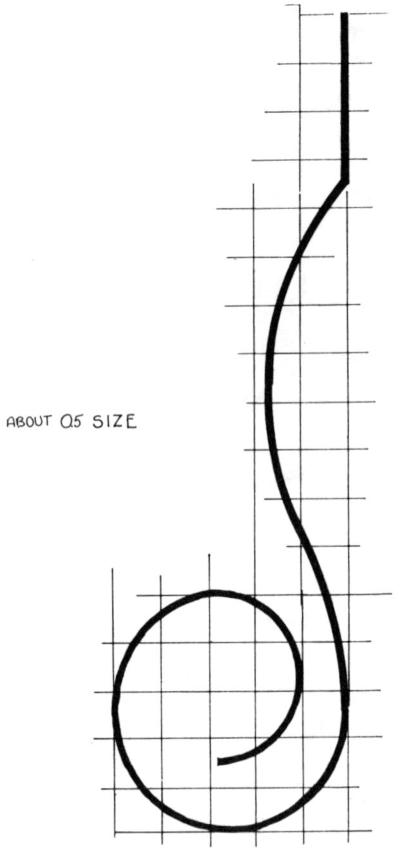

Fig. 6.17. Leg of chess table

ber that a coffee table is usually 300 mm to 460 mm high and end tables are 460 mm to 585 mm high.

SMALL CHESS TABLE

The frame of this table may be used with a variety of tops. The top shown in the illustration (p. 133.) was made as a chess board with no thought of its being used as a top for a table. The frame of the table was designed to match and blend with the wooden top. Materials required are, item 1, 4 pieces of 3/16 in. x 3/4 in. flat bar 330 mm long; item 2, 4 pieces of 1/8 in. x 5/8 in. flat bar 400 mm long; item 3, 4 pieces of 3/16 in. x 3/4 in. flat bar 700 mm long; and item 4, 1 piece of 1/8 in. x 5/8 in. flat bar 565 mm long. Choices of welding process are 1. M.I.G. welding, 2. oxyacetylene welding (fusion), 3. oxyacetylene welding (braze-welding), and 4. electric arc welding.

PROCEDURE

Make a flat frame 330 mm square using the material from item 1. You may have to shorten two of the pieces of flat bar depending upon how you put the corners together.

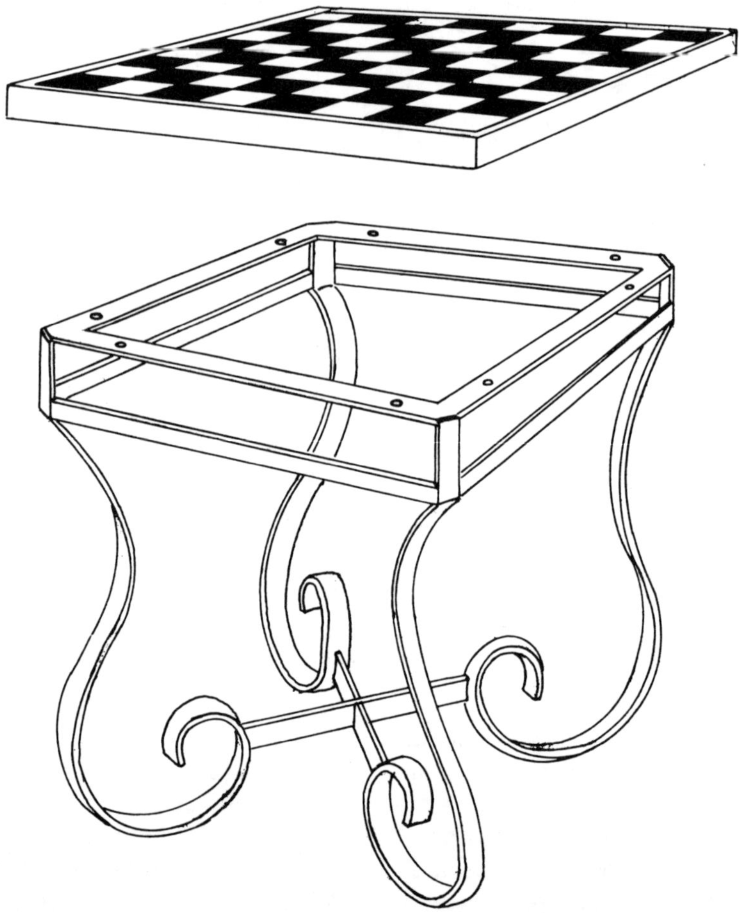

Fig. 6.18. Alternate design, chess table

Form a 130 mm scroll on one end of each of item 3. Bend the rest of the flat bar as in Fig. 6.17. (p. 134.) This will provide a flowing effect in the legs. If you wish to use an opaque projector to project Fig. 6.17., you will have a full-sized image of the legs if you create an image with 1 in. squares. Using item 4 form a ring 180 mm in diameter.

ASSEMBLY

Before assembling the table cut off the sharp points of the top frame. This will also provide a location point for each of the legs. Tack weld each leg to the frame making sure that each points toward the opposite corner.

Place the 180 mm ring between the scrolls and tack weld it to all the scrolls. Make sure that the joint on the ring is hidden by one of the scrolls. Double check that the ring is parallel with the top frame. Add item 2 to the legs as in Fig. 6.16. Weld the project then grind all welds showing before painting. *Try to place all welds so that they are hidden when the table is standing on the floor.* Alternate designs are shown in Fig. 6.18. (p. 134.) and Fig. 6.19. In Fig. 6.19. the legs are placed facing each other and the corners are not cut on the top frame. The ornamentation in both examples is made from 3⁄16 in. x 3⁄4 in. flat bar.

PATIO OR DINING ROOM TABLE

This project should be considered carefully before you decide to make it. It is an ambitious project and should not be attempted unless you are prepared to persevere and overcome the problems of fabrication and assembly that will certainly crop up. You should also be prepared to make at least four chairs (see the unit on chairs and seats) that are needed to make use of this project.

You should decide at this point what type of top you intend to use. See the section on accessories for information. We used a piece of 1⁄4 in. plate glass removed from a front window in a house. We also used 3⁄4 in. square tubing for supporting the glass top. You may substitute 3⁄4 in. x 3⁄4 in. x 1⁄8 in. angle iron. For the top use 4.75 m of 3⁄4 in. square tubing, 4.75 m of 1 in. x 1⁄2 in. x 1⁄8 in. channel, and 0.54 m of 3⁄4 in. square tubing. For the legs use 8 pieces of 1⁄2 in. round bar 760 mm long and 4 1 1⁄2 in. diameter washers.

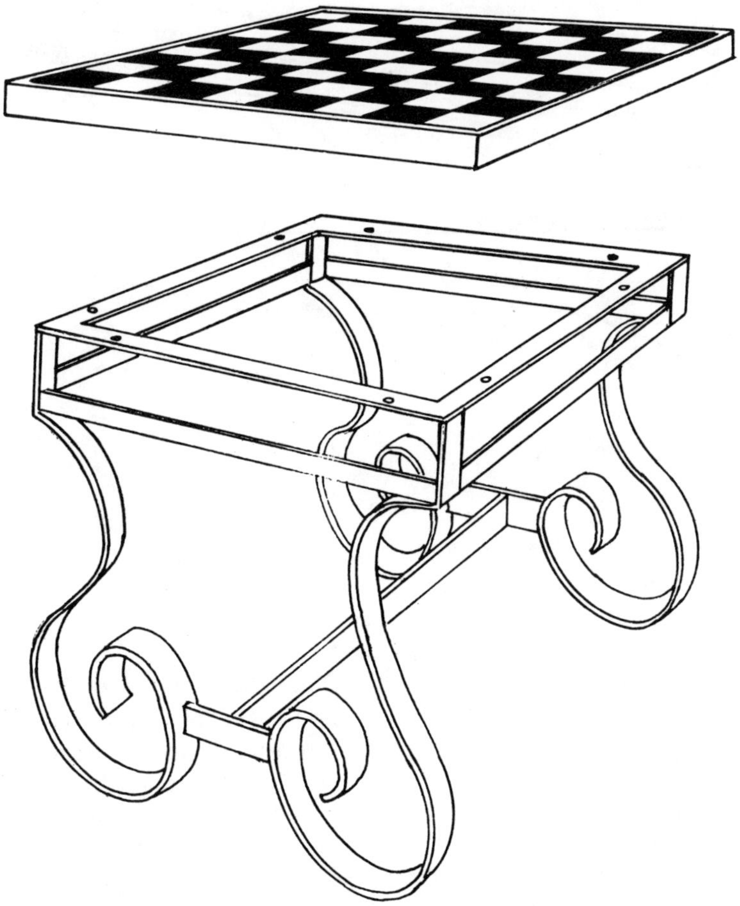

Fig. 6.19. Alternate design, chess table

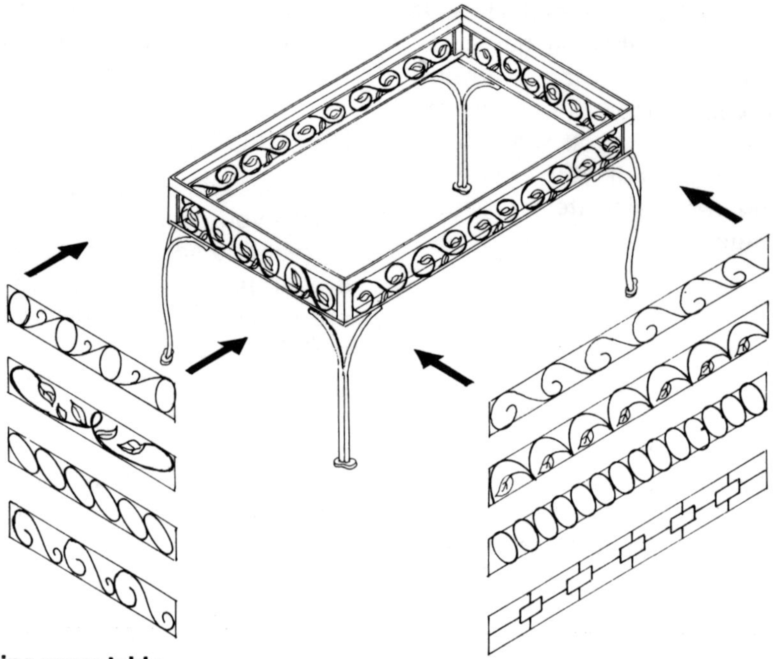

Fig. 6.20. Patio or dining room table

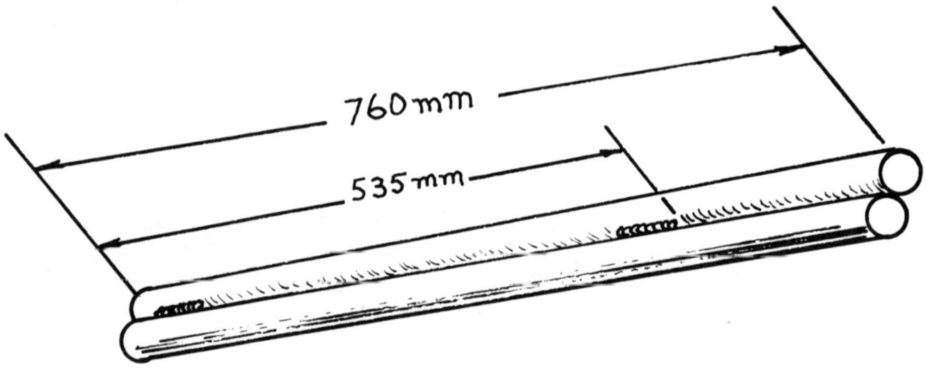

Fig. 6.21. Leg of patio or dining room table

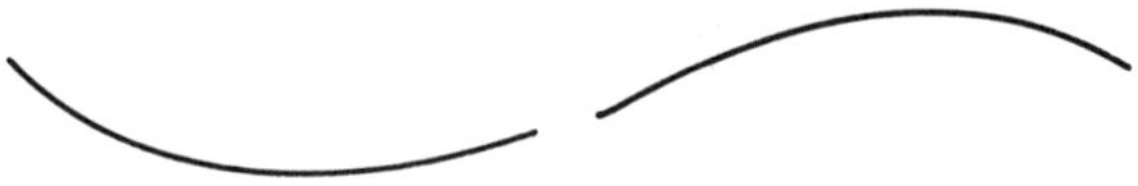

Fig. 6.22. Cyma curve

PROCEDURE

Cut the square tubing using 45° corners to form a frame 1500 mm x 865 mm. Form an identical frame using the channel. The corners of the channel must also be cut at 45°. Cut 4 pieces of ¾ in. tubing 125 mm long from the 0.54 m length. Make sure that they are identical in length and have square ends. File the ends if necessary. Assemble the top from the two frames with the 125 mm tubing pieces as spacers in the corners. The square tubing will be above the channel in the finished table. After checking for accuracy and squareness weld the assembly.

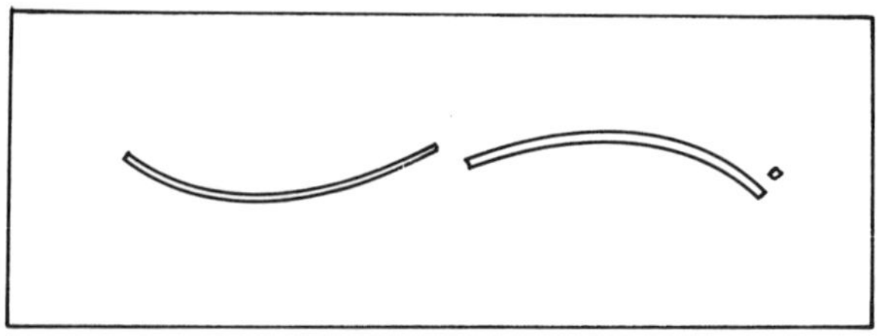

Fig. 6.23. Forming jig

LEGS

Place two pieces of ½ in. round bar together and weld as in Fig. 6.21. Repeat for the other three legs. Do not make long welds; 25 mm of weld in each place will be sufficient.

At this point it is necessary to build a forming jig to aid in shaping the legs to form a cyma curve. Fig. 6.22. shows a cyma curve consisting of a convex and a concave curve. Both curves have the same radius. To make our jig we used

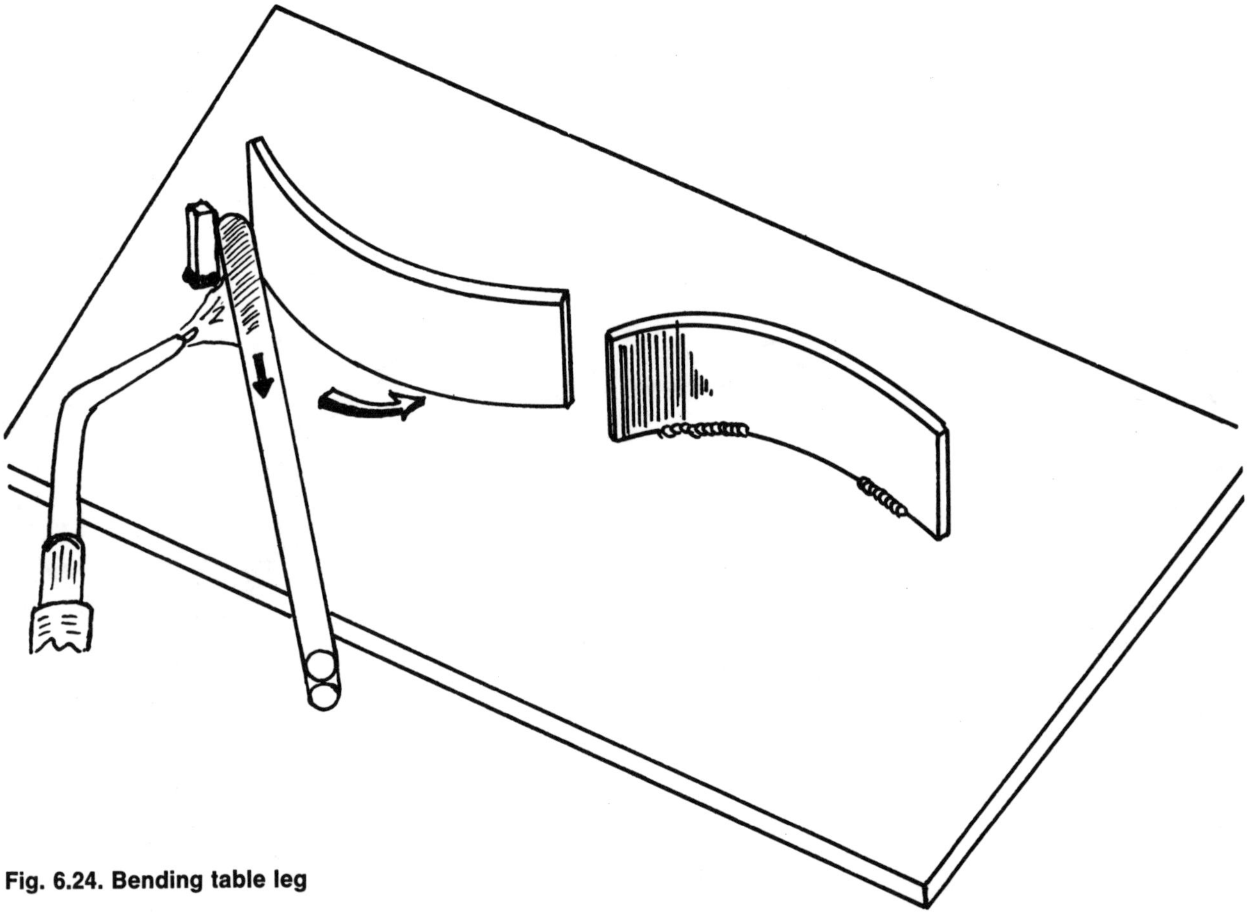

Fig. 6.24. Bending table leg

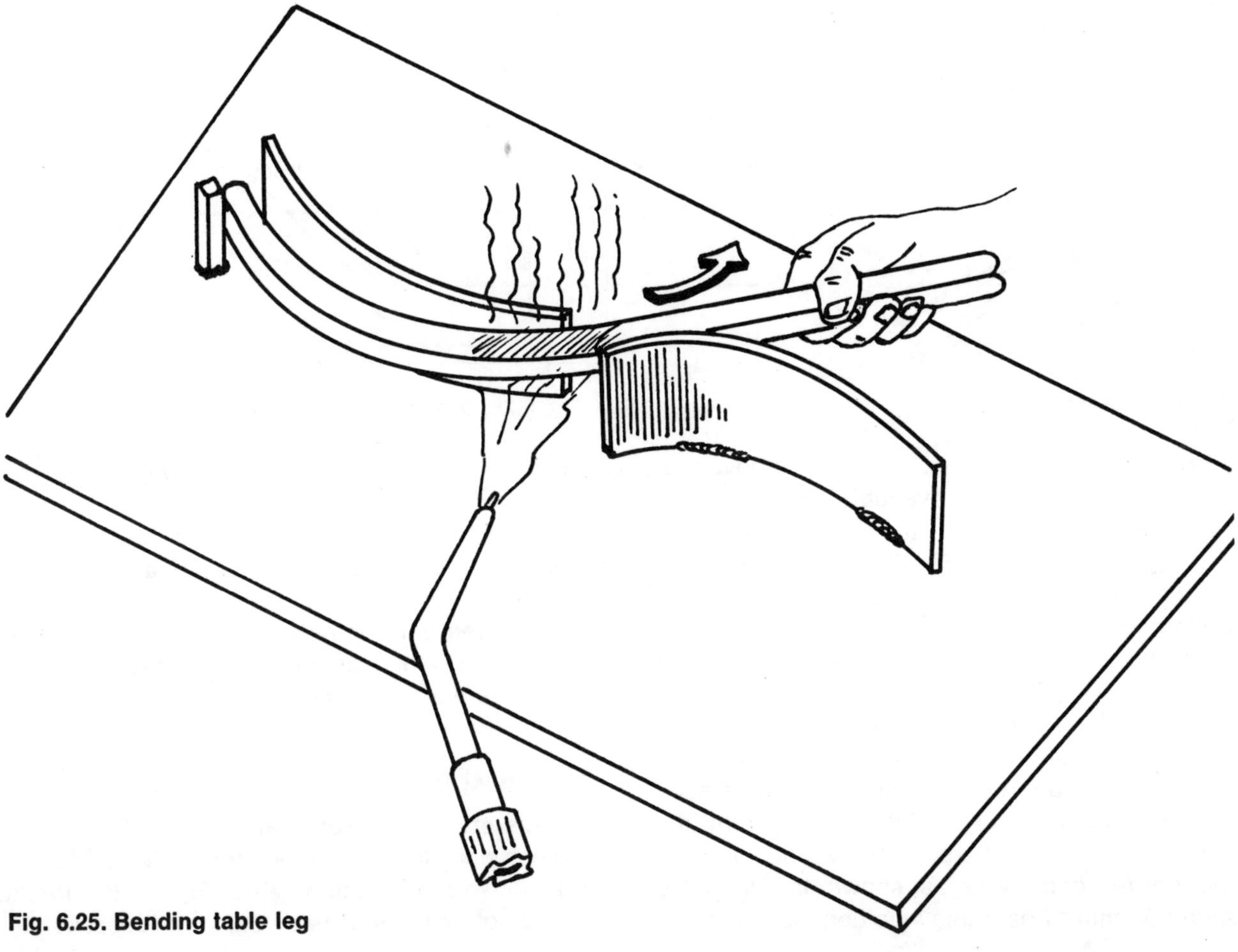

Fig. 6.25. Bending table leg

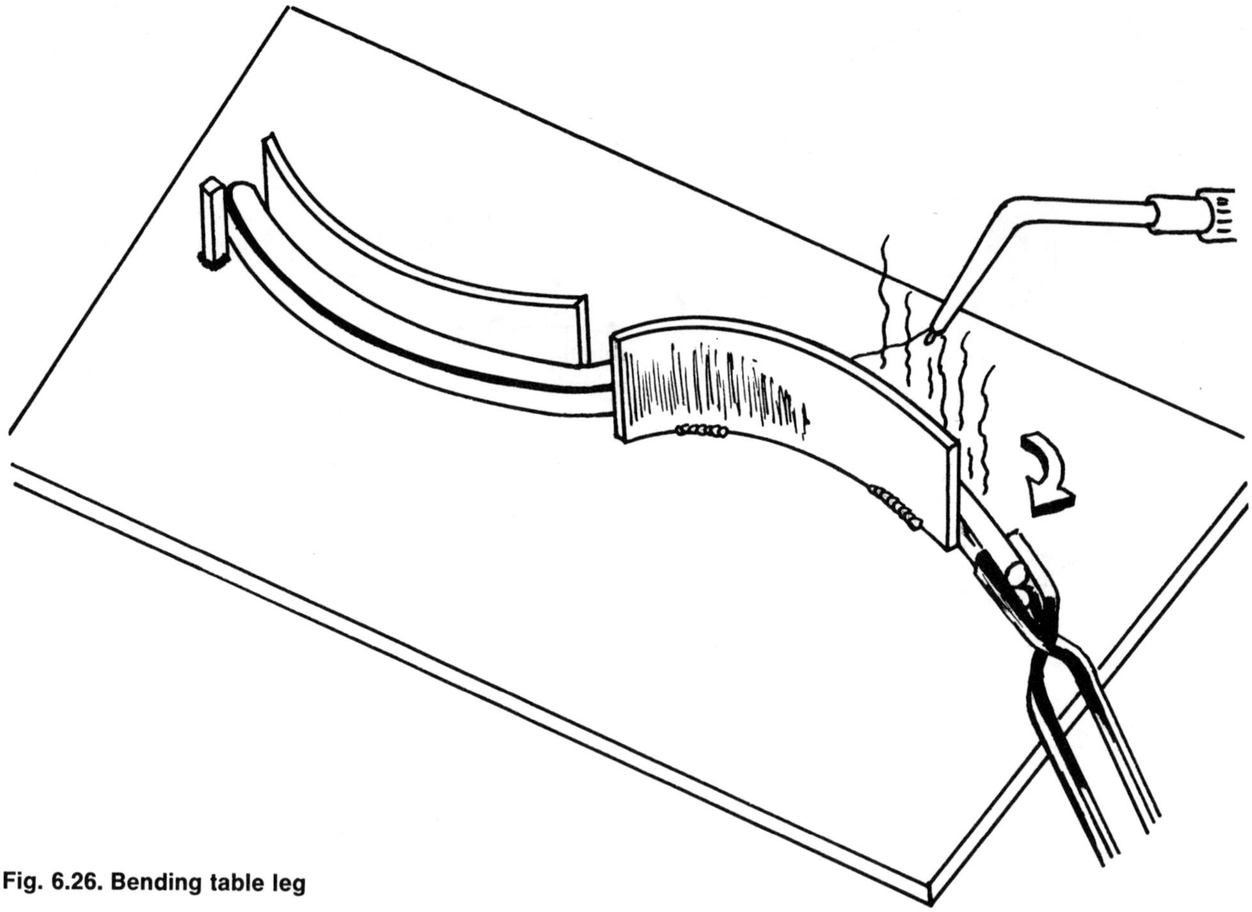

Fig. 6.26. Bending table leg

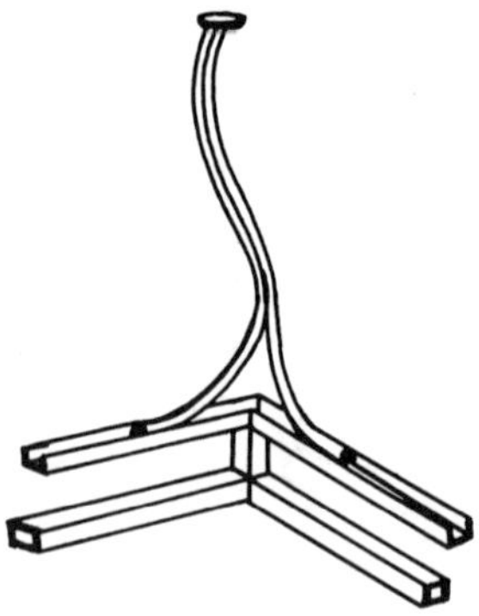

Fig. 6.27. Assembling leg

Fig. 6.28. Ornamentation for patio table

a short section of 12 in. diameter pipe about 75 mm wide that was available. It was cut in half and shaped, then tack welded onto a bench top as shown in Fig. 6.23. (p. 136.) A stop bar was added to hold the end of the leg section while it was being formed.

The legs were heated and bent by hand as in Fig. 6.24. (p. 137.), Fig. 6.25. (p. 137.), and Fig. 6.26. After forming all four legs they were checked for uniformity and any small discrepancies were altered by hand. The top of each leg was spread using heat so that it fit the corner of the table at a point about 150 mm from the corner each way as shown in Fig. 6.27. About 50 mm of each leg was bent so that it sat in the channel where the leg is welded to the frame.

After tack welding all four legs to the frame the table should be set on its legs and checked to see if it is level. Make sure that the bottom of each leg is directly below each corner. Tack weld a 1½ in. washer to each leg and add sufficient weld to strengthen the project.

ORNAMENTATION

At this stage the table can be used. However, it does not look very appealing without decoration or ornamentation. See Fig. 6.20. for a variety of designs or use your own ideas. Whatever design you choose it is necessary to make it fit both the length and the width. Draw the inside area of the spaces in the top to be decorated on a piece of paper or on a bench top. Sketch out your design in that area to see how many pieces are needed to fit it. Fig. 6.28. (p. 138.) shows a full-sized design that will fit the dimensions of this table. Five are needed for the length and three for the width. The main stem was made using ¼ in. round bar while the stems for the leaves used 3/16 in. These particular leaves were cut by hand. Many different shapes of leaves may be purchased.

Fig. 6.29., Fig. 6.30., Fig. 6.31., and Fig. 6.32. show a variety of alternate leg designs that may be used to change the design of this project. Remember that you may use angle iron for a frame and the angle may be up or down. See the section on assembling angle iron for information.

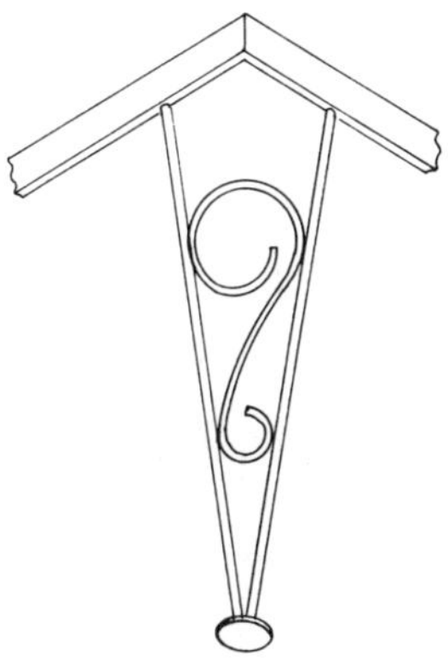

Fig. 6.29. Alternate leg design

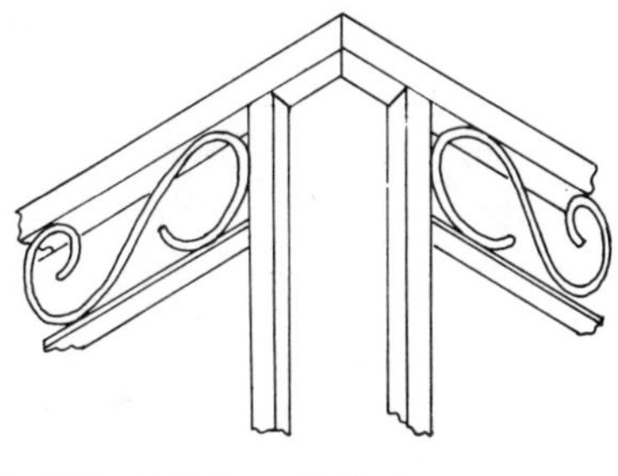

Fig. 6.30. Alternate leg design

Fig. 6.31. Alternate leg design

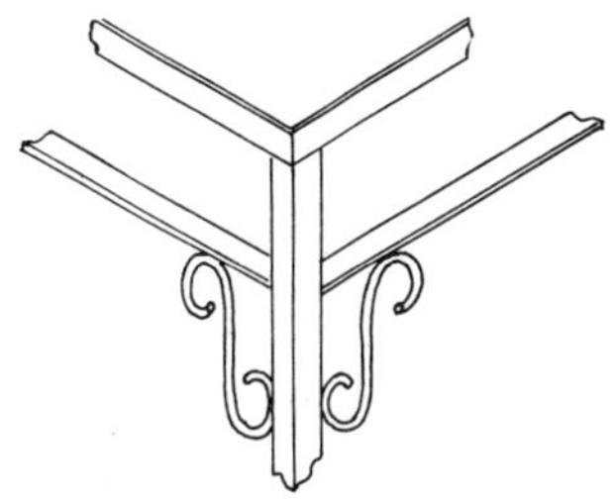

Fig. 6.32. Alternate leg design

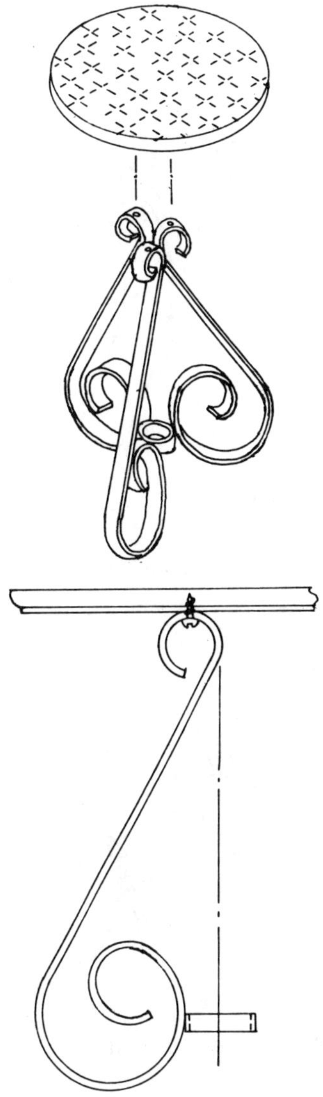

Fig. 6.33. Round patio table

ROUND PATIO TABLE

This table is simple and inexpensive to make. It is an excellent project for junior grade students. Materials required are 3 pieces of 3/16 in. x ¾ in. flat bar 890 mm long, 1 piece of 2 in. tubing or pipe ½ in. wide, and 1 sink cutout of arborite or formica-covered plywood about 380 mm in diameter. Choices of welding processes are 1. M.I.G. welding, 2. electric arc welding, 3. oxyacetylene welding (fusion), and 4. oxyacetylene welding (braze-welding).

PROCEDURE

Form "S" scrolls with the three pieces of flat bar, a 125 mm scroll on one end and a 100 mm scroll at the other. Place the three small scrolls together as in Fig. 6.34. and tack weld. Add the section of pipe or tubing as in Fig. 6.35. and tack weld to the scrolls. Weld the unit together, then add the top by drilling a hole in each top scroll. It may be necessary to level the table top on the scrolls. Place a level across the scrolls and bend the scrolls up or down slightly until all three are level.

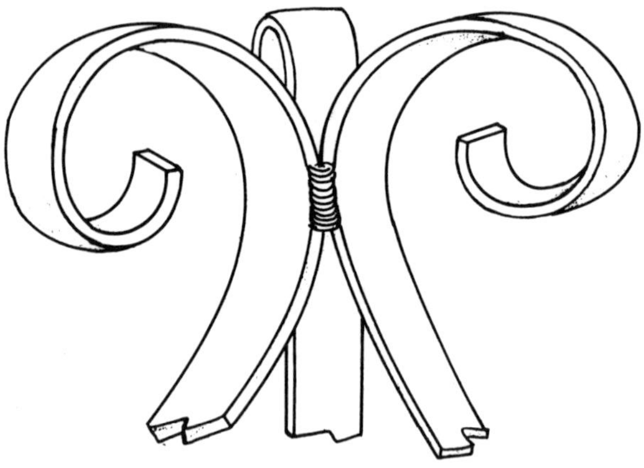

Fig. 6.34. Tack weld small scrolls

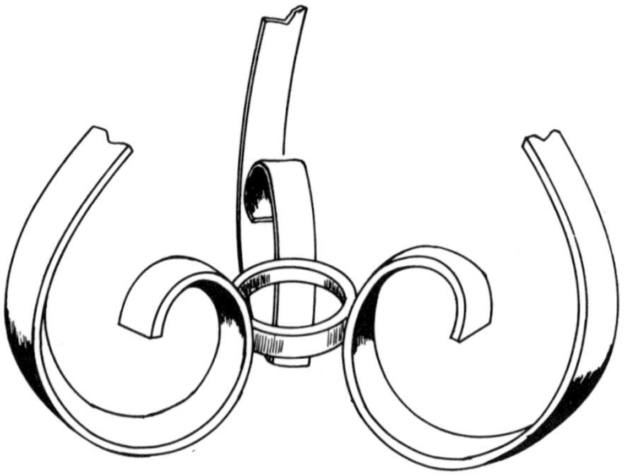

Fig. 6.35. Tack weld large scrolls

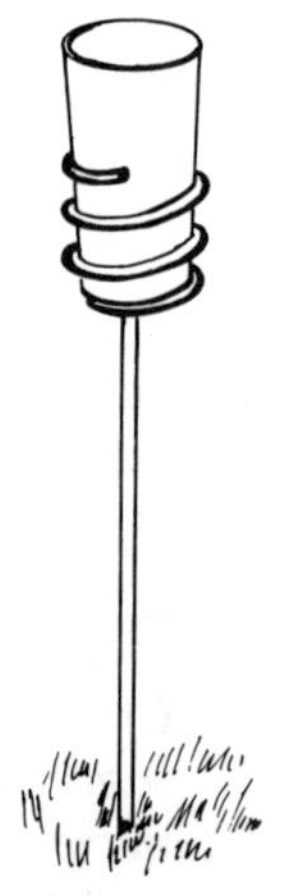

Fig. 6.36. Wire glass holder

GARDEN TABLE

Almost everyone has seen the small wire glass holders that are used in the garden. See Fig. 6.36. A more attractive unit such as shown in Fig. 6.37. can be made using the round formica top once again. Materials required are 860 mm of ½ in. round bar, 3 pieces of $\frac{3}{16}$ in. x ½ in. flat bar, 2 pieces of $\frac{5}{16}$ in. round bar 1200 mm long (this will fit a top 40 mm in diameter), and 1 piece of 14 ga. sheet steel 100 mm diameter.

PROCEDURE

Weld the centre of the 14 ga. plate to the end of the ½ in. bar. Grind the opposite end of the

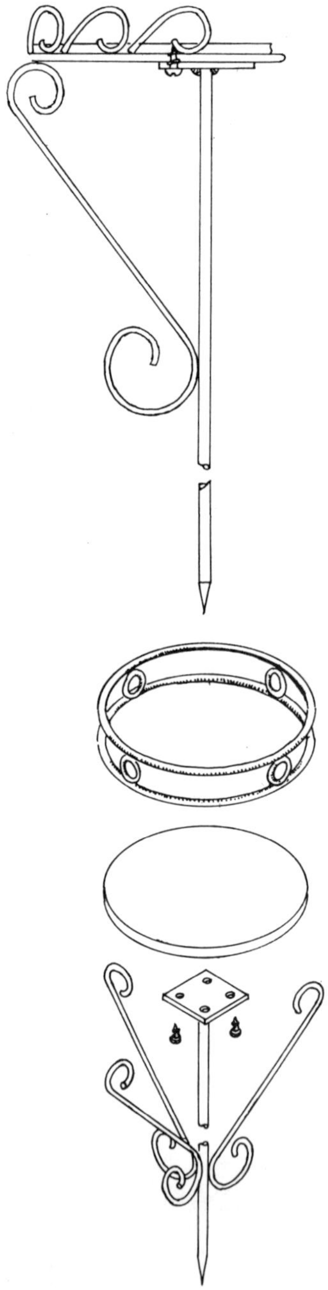

Fig. 6.37. Garden table

round bar to a point. This will make it easier to push into the ground. Make 2 rings using the 5/16 in. round bar. Scroll the 3/16 in. x 1/2 in. flat bar into "S" scrolls with a 100 mm scroll on one end and a 65 mm scroll on the other.

ASSEMBLY

Lay one ring of 5/16 in. round bar on a flat surface. Stand the assembled 1/2 in. round and 14 ga. plate on the flat surface in the centre of the ring, plate down. Place one scroll against the ring and the 1/2 in. round bar as shown in Fig. 6.37. with the smaller scroll against the ring and tack weld together. Add the other two scrolls 120° apart and weld.

Using either of the two designs shown in Fig. 6.37. or your own design add the other ring to provide a rim 50 mm-75 mm high. Before adding the top, paint the assembly in a bright colour such as green, yellow, or orange. Two or three of these tables in the garden look attractive as well as being functional.

The remaining illustrations in this unit have been included to let you exercise your judgment in deciding on the dimensions of a project as well as selecting the type of materials to use. The nesting tables must of course vary in dimensions to allow them to fit or stack together. The narrow table shown in Fig. 6.39. was originally made to fit along a wall between two beds in a bedroom. The height of the table, therefore, would have to correspond with the height of a bed from the floor. The other illustrations could be coffee or dining tables depending upon dimensions used. The types of tables that can be fabricated by students is almost endless and limited only by imagination.

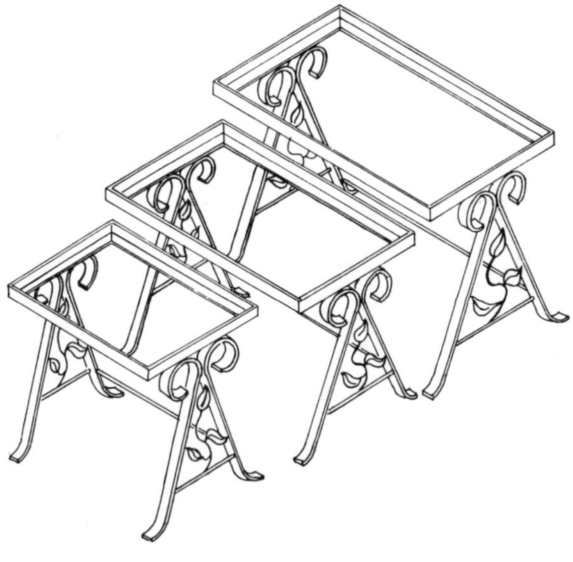

Fig. 6.38. Nesting tables

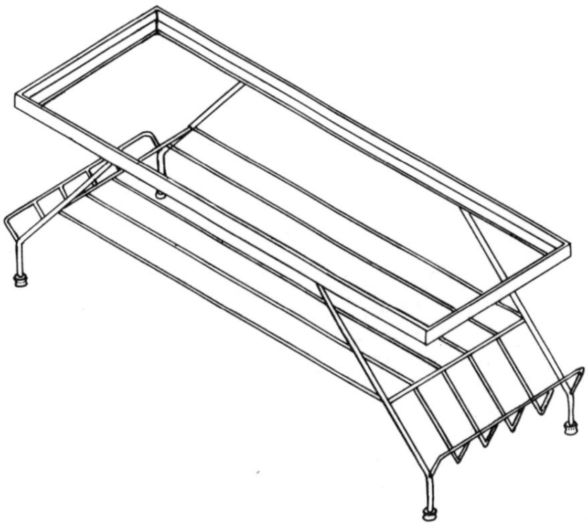

Fig. 6.40.

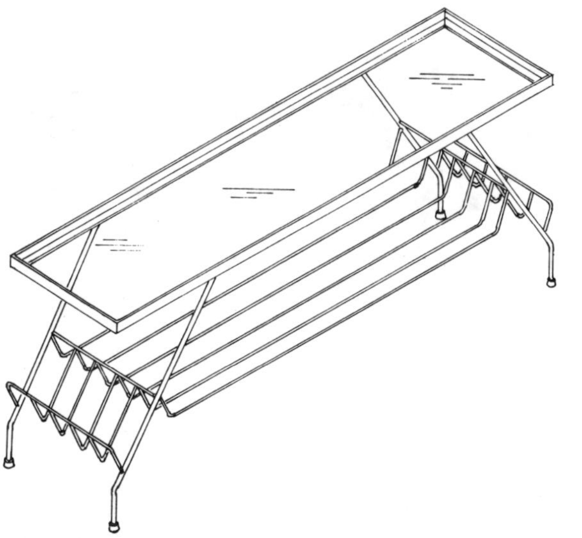

Fig. 6.39. Narrow table

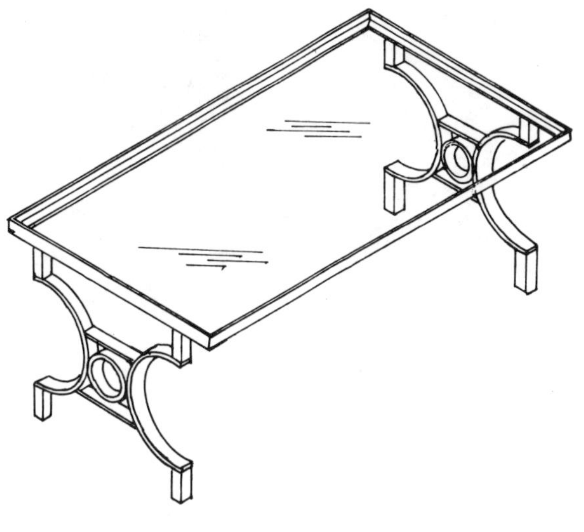

Fig. 6.41.

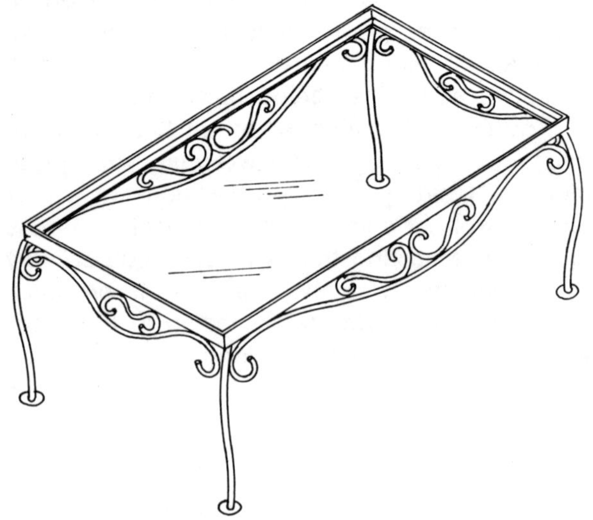

Fig. 6.42.

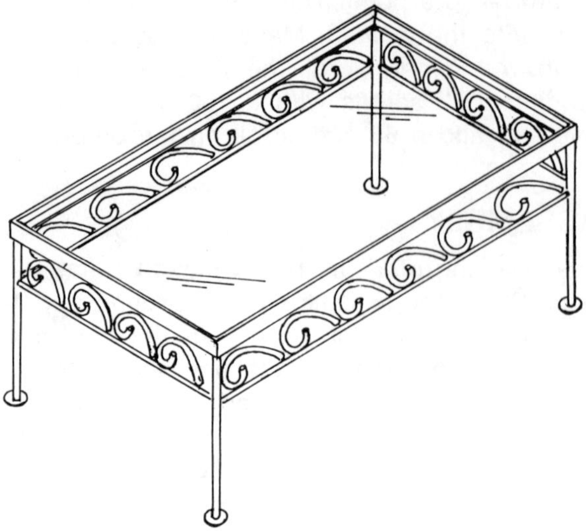

Fig. 6.43.

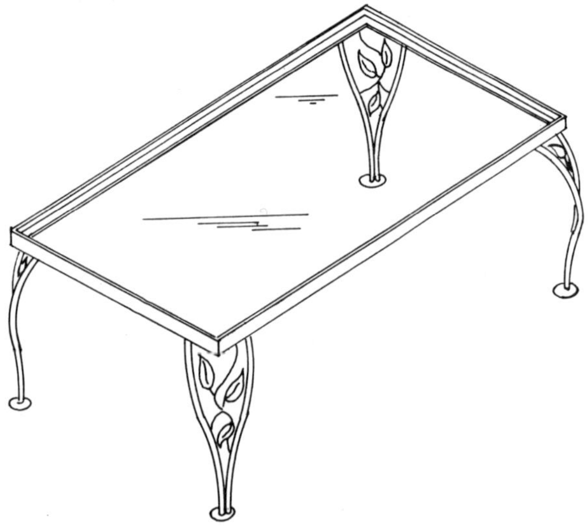

Fig. 6.44.

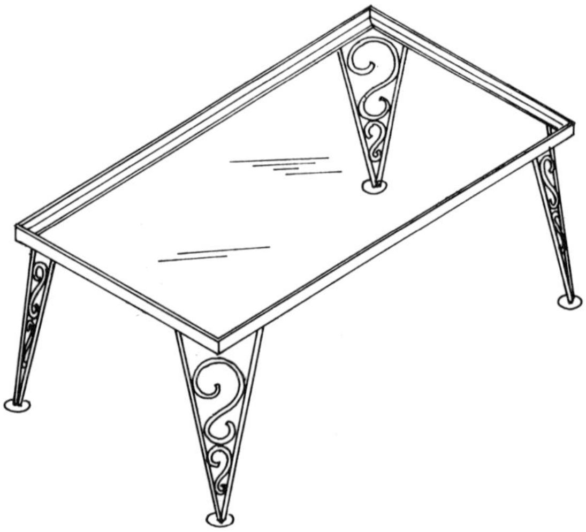

Fig. 6.45.

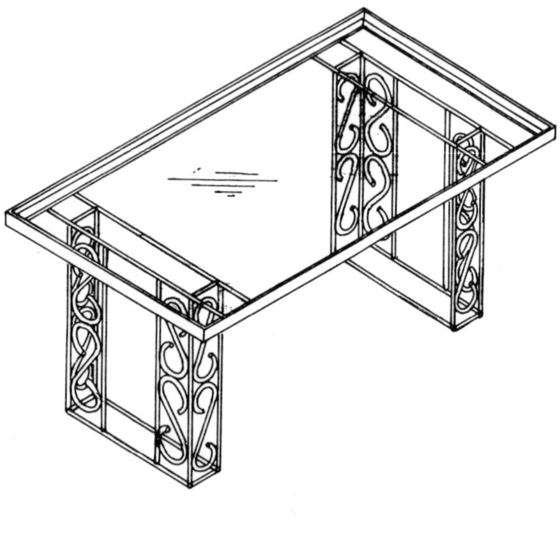

Fig. 6.46.

Fig. 6.47. Typical use of tables

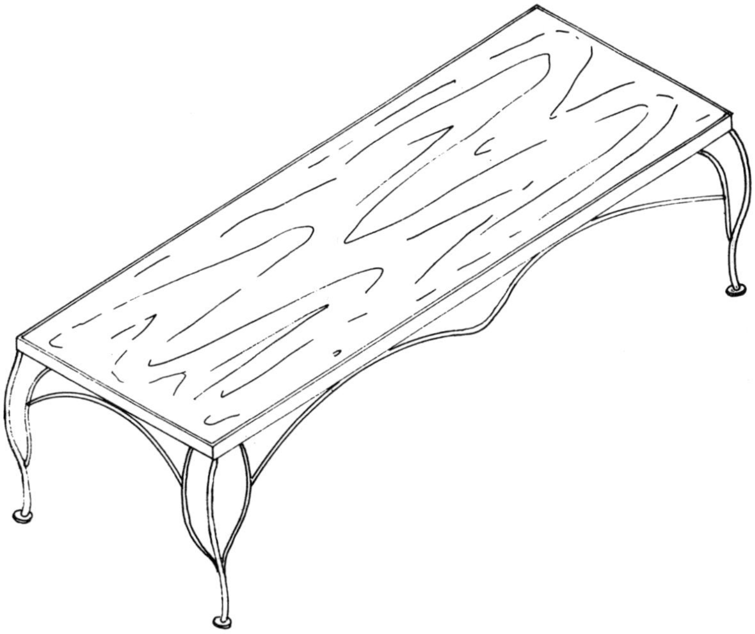

Fig. 6.48.

UNIT 7. CHAIRS AND SEATS

This unit contains information on the design and construction of seats and chairs. Some of the designs will be compatible with some projects shown in other units such as the unit on tables.

In most instances specific measurements and types of materials will not be given because you would have to make seat covers or pads to match. It is best to have the seat pad first then decide on the dimensions of your project to fit it. This will eliminate the added expense of having the seats custom made.

The basic seat or chair consists of the seat, legs, and back. It would seem simple to construct a project from only three basic parts, but consider the following. 1. Will the front legs be individual units? 2. Will the back and the back legs be made as one unit? 3. Will the front legs be made from one piece of metal? 4. Will there be arms on the chair? 5. Will the arms be separate from the other sections?

As can be seen from the few questions posed above the making of this type of project is not simple, but it can be done if you work at it systematically. Make a drawing, then proceed, taking one section at a time.

The first question you must answer before starting a project from this unit is "What type of chair do I want to make?" Fig. 7.2. (p. 144.) shows that the various types of chairs are made to different measurements. A good example is the comparison of a kitchen chair and a lounge chair. You will find it difficult to relax while sitting on a kitchen chair, but a lounge chair lets you lean back and stretch your legs. They both have basically the same parts. It is the variation in dimensions, heights, and angles that makes the difference.

CHAIR SEATS

Chair seats may be round, square, or rectangular in shape. They may be made using flat bar, angle iron, square bar, or tubing. If the seat pad to be added has wood in its construction it is only necessary to add small metal tabs to hold the seat in place. If a soft type of pad is to be used it will be necessary to fill in the seat with straps made from flat bar or expanded metal. See Fig. 7.8. for an example.

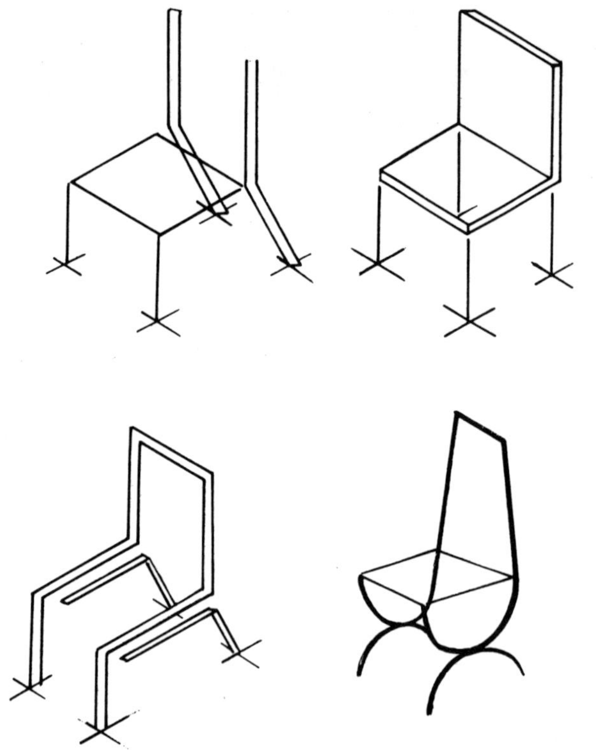

Fig. 7.1.

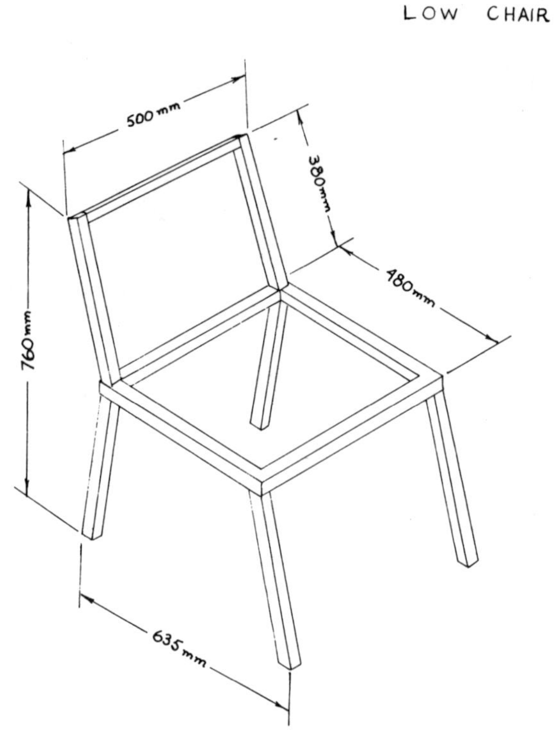

Fig. 7.2.b.

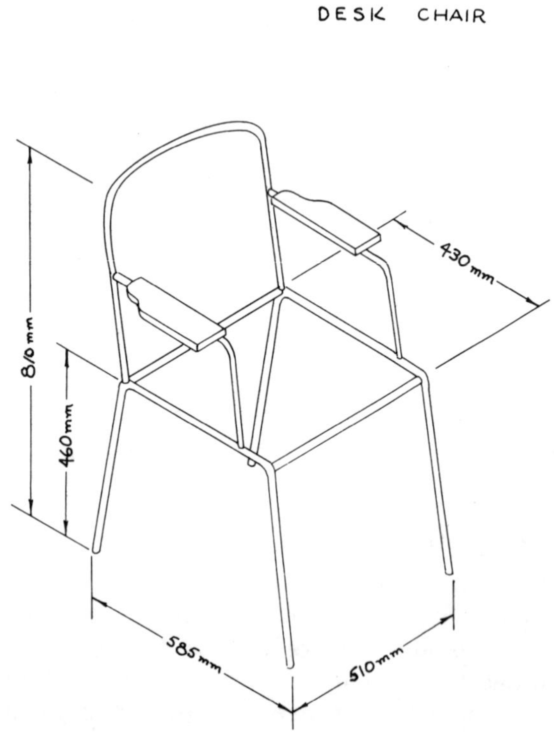

Fig. 7.2.a.

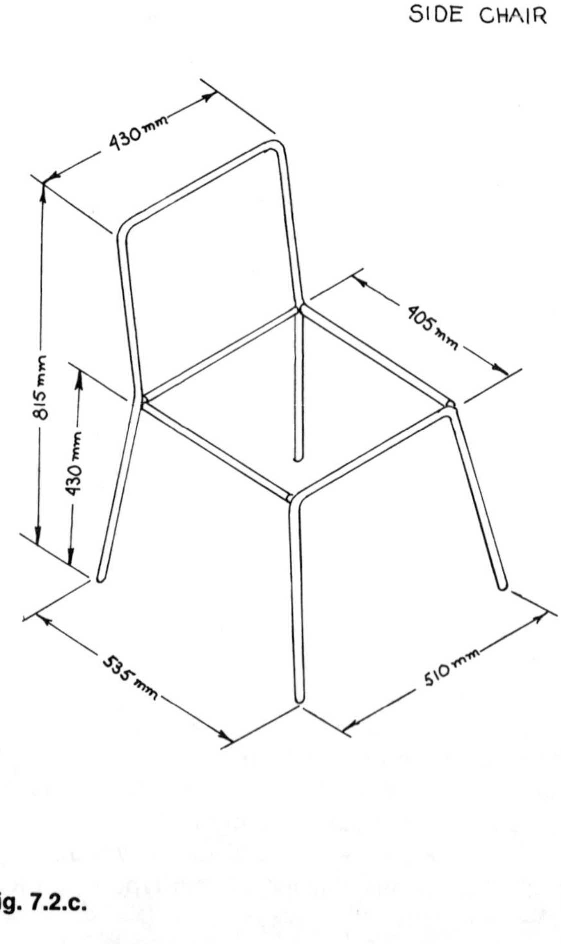

Fig. 7.2.c.

Fig. 7.3. Exploded view of chair

CHAIR BACKS

The back of the chair may be one unit including the back legs. It can be a separate unit joined to the seat. The seat and the back make one complete unit made with expanded metal. See Fig. 7.3. The back may also include a pad similar to the seat. See Fig. 7.4.

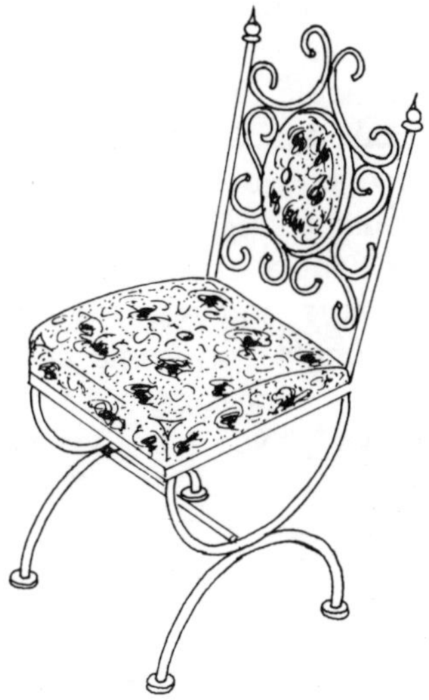

Fig. 7.4. Back pad may match seat

CHAIR LEGS

We all understand that the legs of a chair are simply extensions that raise the seat from the floor. The legs may be simply four pieces of metal added to the seat. The back legs may be an extension of the back. The front and back legs may be made in one section. The front legs and the arms may be one unit. The design variations are numerous. Check the many illustrations in this section and try to change the leg design mentally.

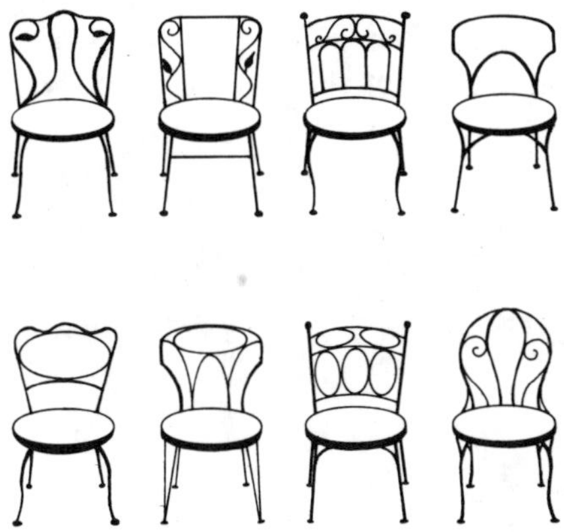

Fig. 7.5.

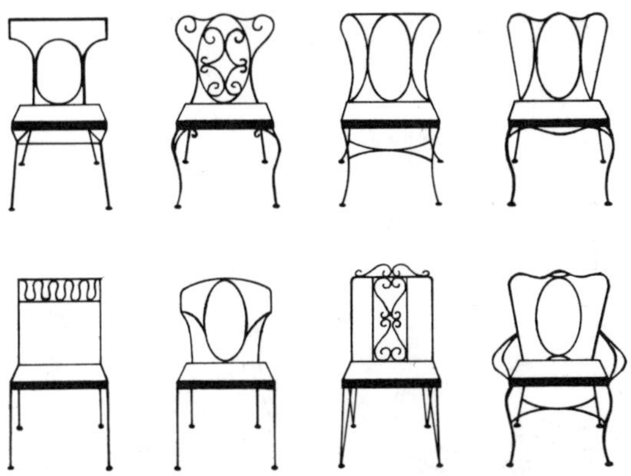

Fig. 7.6.

CHAIR ARMS

Depending upon the use to which the chair will be put you may wish to add arms to the design. Usually when a dining room set is made there are four chairs. Two of the chairs may have arms.

Almost all designs of lounge chairs include arms. Again a look at the illustrations will show that the arms are a dominant part of the design in some instances and almost an afterthought in others. However they all serve the same function.

Fig. 7.3. shows a chair with the parts separated around it. It has been chosen to illustrate how the various parts fit together. The seat and the back are combined in one unit made from expanded metal. In the case of a commercially made chair the expanded metal is made to a specific width with a safety border. If you decide to make a chair like this you will have to add a safety edge using ⅛ in. x ½ in. flat bar.

The back is also strengthened by extending the back legs up and using one length of metal, possibly ½ in. round bar. The front legs are individual units also using ½ in. round bar. The front and back legs are joined and strengthened using side rails made from ⅜ in. round bar. The seat part of the expanded metal is stiffened using the curved pieces made from ½ in. round bar. The expanded metal back is strengthened with two pieces of flat bar ⅛ in. x ¾ in. The

Fig. 7.7.

arms would be ½ in. round bar with the side extensions made from 5/16 in. or ⅜ in. round.

What we have done here is to take a specific chair and dissect the construction. It is possible to stay with this particular design and by changing the method of joining have an entirely different project to build each time. To figure out the dimensions for this project look at Fig. 7.2. and decide what type of chair it is.

HALL SEAT

This project may be used in a home as a hall seat, as a bedroom seat, or as a vanity seat in a bathroom. Materials required are about 3.4 m of ½ in. round bar, 2 m of ¾ in. angle iron, 1.8 m of ⅜ in. round bar, four 1 in. ball bearings, and 3.2 m of ⅛ in. x 1 in. flat bar.

PROCEDURE

Cut the ½ in. round bar to provide two pieces 840 mm long and two pieces 815 mm long. The two 840 mm pieces make the bottom of the seat. Roll or pull all of the pieces of ½ in. bar so that each forms a semi-circle 610 mm in diameter.

Using the ¾ in. angle iron make a rectangular frame 600 mm x 380 mm. Check the section on assembling angle iron. After welding the angle frame together it is necessary to bend it so that it curves 75 mm. See Fig. 7.8. If a Hossfeld bender with the proper dies is available the angle can be bent on it. We managed to do our bending by progressively pushing the frame between the bars of our layout table and pulling a little each time until we reached the centre of the frame. We then reversed the frame and repeated the operation.

Using the angle iron frame and rolled round bar, we laid out each side of the seat and tack welded the two halves together with the frame

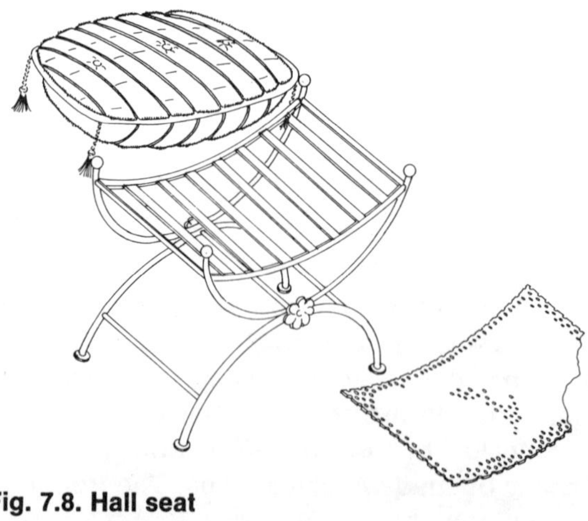

Fig. 7.8. Hall seat

in the middle as in Fig. 7.9. Cut three pieces of 3/8 in. round bar 380 mm long for the cross braces on the lower round bar. See Fig. 7.8. At this point you can assemble the seat. The angle iron frame is placed about 40 mm down from the top. One cross brace is placed in the middle and the two others about 115 mm up from the ground on each side. Make sure that the unit goes together squarely.

The ball bearings are best joined to the tips of the upper round bar by silver-brazing. We added two steel stampings (rosettes) for decoration.

The remainder of the 3/8 in. round bar is used to make the small braces that fit between the round bar arcs. See Fig. 7.9. Cut four pieces 165 mm each in length. Grind or draw both ends of each to a slight taper. This makes them blend in with the main legs. Heat and bend each in the middle to form a 90° angle. Weld to the main frame as shown in Fig. 7.9. Cut seven or nine slats using the 1/8 in. flat bar. Space them equally in the angle iron frame and weld. Use washers for the feet or cut 3/8 in. thick pieces from 1¼ in. round bar. As can be seen in Fig. 7.8. the angle iron frame may be filled with expanded or perforated metal to hold the seat cushion or pad.

Consider making this project using ½ in. square bar. How about making the top of each leg longer and forming a scroll? See Fig. 7.10. (p. 148.) The seat shown in Fig. 7.10. serves the same function as the previous project. It is longer (about 760 mm) and made with ½ in. square bar. Its width could be between 30 mm and 35 mm. Notice the alternate ideas for this project as shown in Fig. 7.10. Design your own hall seat from the basic shapes given. Fig. 7.11. (p. 149.) and Fig. 7.12. (p. 149.) illustrate a few designs that include arms. As you can see the variations are unlimited. If you do not wish to design your own chair you can use an opaque projector to project Fig. 7.13. (p. 149.) until the squares are 1 in. and have a full-size pattern for a chair.

A chair or seat may look like a simple project, but it does offer the most scope for individual design. If you start with the standard dimensions for the type of chair you want then either select a complete illustration or possibly parts from many, you should be able to produce excellent projects.

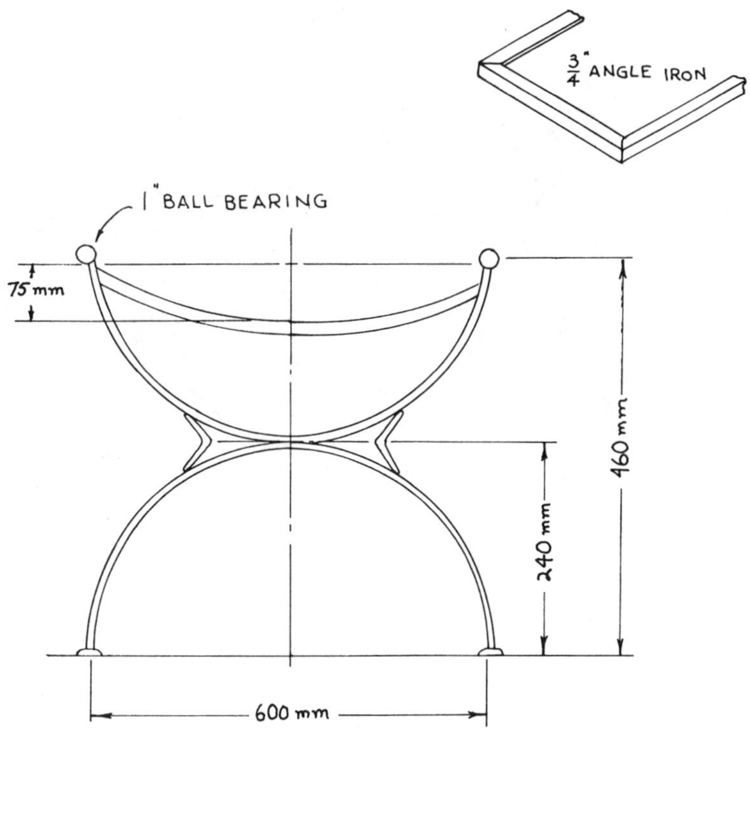

Fig. 7.9. Diagram of hall seat

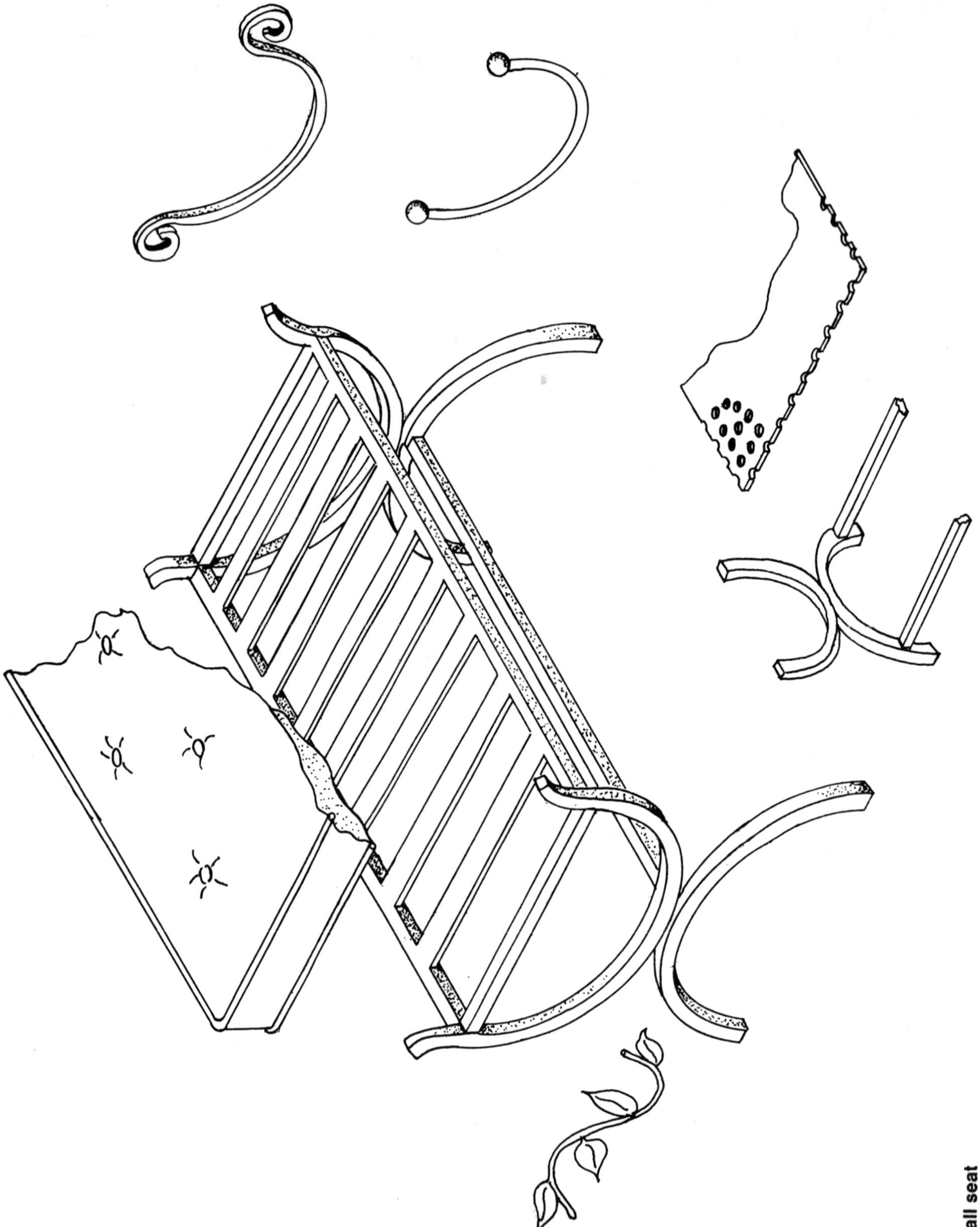

Fig. 7.10. Hall seat

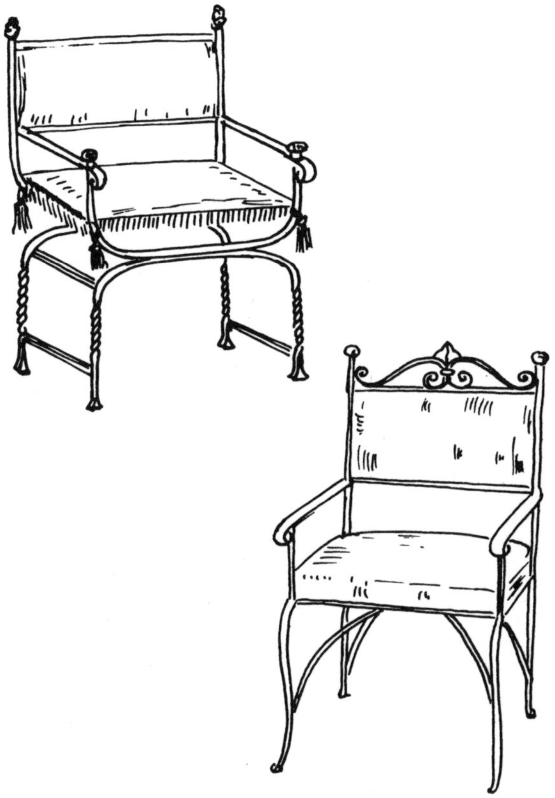

Fig. 7.11.

Fig. 7.12.

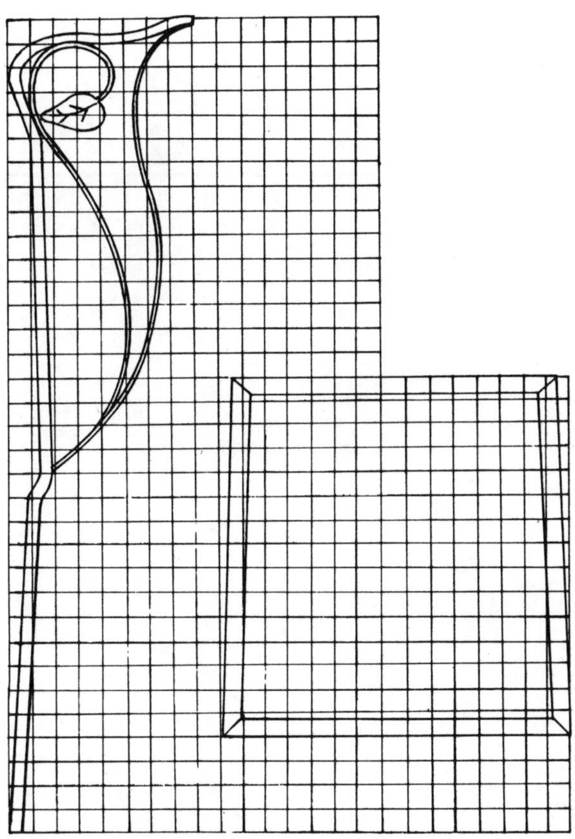

Fig. 7.13. Pattern for chair